HANDBOOK OF ENVIRONMENTALLY CONSCIOUS MANUFACTURING

HANDBOOK OF ENVIRONMENTALLY CONSCIOUS MANUFACTURING

edited by

Christian N. Madu, Ph.D.
Research Professor and Chair
Management Science Program
co-Editor, International Journal of Quality & Reliability
Management
Department of Management and Management Science
Lubin School of Business
Pace University
New York, New York, USA

KLUWER ACADEMIC PUBLISHERS
Boston / Dordrecht / London

Distributors for North, Central and South America:
Kluwer Academic Publishers
101 Philip Drive
Assinippi Park
Norwell, Massachusetts 02061 USA
Telephone (781) 871-6600
Fax (781) 681-9045
E-Mail <kluwer@wkap.com>

Distributors for all other countries:
Kluwer Academic Publishers Group
Distribution Centre
Post Office Box 322
3300 AH Dordrecht, THE NETHERLANDS
Telephone 31 78 6392 392
Fax 31 78 6546 474
E-Mail <services@wkap.nl>

 Electronic Services <http://www.wkap.nl>

Library of Congress Cataloging-in-Publication Data

Handbook of environmentally conscious manufacturing / edited by Christian N. Madu.
    p.cm.
  Includes bibliographical references.
  ISBN 0-7923-8449-0 (alk. paper)
   1. Industrial ecology. 2. Production engineering--Environmental aspects. I. Madu,
  Christian N. (Christian Ndubisi)

TS161 .H36 2000
658.5--dc21   00-048695

Printed on acid-free paper.

Printed in the United States of America

*This book is dedicated to my wife
Assumpta and our three boys -
Chichi, Chike and Chidi*

Contents

Preface

The Handbook of Environmentally Conscious Manufacturing is written as an introductory reference material on environmentally conscious manufacturing (ECM). The chapters as well as the authors were carefully selected. The authors of the chapters are respected figures in the field and have done extensive research and /or practice work in the field of ECM, it is important to organize this body of knowledge in a substantial form. Authors of the chapters have emphasized both theory and practice and m any of the chapters contain real life case studies.

The chapters are written with the reader in mind. The aim of the authors is to carefully organize and convey their thoughts to the reader. These chapters are readable and can appeal to a wide range of audience interested in environmental manufacturing issues.

The reader will find that the handbook is very comprehensive as it covers all the major topics in Environmentally Conscious Manufacturing. There are specific chapters to deal with sustainable manufacturing, recycling, eco-labelling, life cycle assessment, and IS0 14000 series of standards, as well as decision-making aspects of Environmentally Conscious Manufacturing. Decision-oriented topics on supply chain, decision models, quality initiatives, environmental costing and decision support systems are also covered. The influence of ECM on marketing imperative is also covered. The reader will find this book an excellent reference guide to ECM and perhaps, the most comprehensive ECM handbook in the market.

The Handbook could not have been possible without the devotion and dedication of the authors who have to follow strict guidelines and also abide by the publishers style of format. It took a lot of time commitment but the results are worthwhile. I am greatly indebted to the authors for producing an excellent masterpiece and I'm sure you will all be proud of the work. I will also like to acknowledge the support of my graduate assistant Lere Odusote, who spent innumerable hours to ensure the consistency of the chapters in following the publisher's style format. I am grateful for the excellent work you have done.

I thank all that have participated in one way or another to ensure that this project is completed. The aim of this book is not simply as an academic piece, but also, a way of creating awareness to the growing environmental problems that the world is facing. This book has addressed major strategies that can help manufacturers to conserve limited natural resources and protect the natural environment. It discusses in an unbiased way, how our quality of life and standard of living can be maintained without compromising the future of generations yet to come. It sends a strong message on reducing environmental burdens and making optimal use of limited resources. I am hopeful that after reading this book, the reader will be a different person and contribute in achieving the goals espoused in the book.

Christian N. Madu
Pace University
New York

Biographical sketch

Christian N. Madu is research professor and chair of the management science program at the Lubin School of Business, Pace University. He is the author/co-author of more than 100 research papers in several areas of operations research and management science that have appeared in journals such as *Decision Sciences, IIE Transactions, Journal of Operational Research Society, Applied Mathematics Letters, Mathematical and Computer Modeling, Journal of Environmental Planning and Management, Quality Management Journal, Long Range Planning, OMEGA, European Journal of Operational Research, Socio-economic Planning Sciences, Futures, Technological Forecasting and Social Change*, and several others. He is the author/co-author and editor of more than 10 books including the *Handbook of Total Quality Management*. Dr. Madu is also co-Editor of the *International Journal of Quality and Reliability Management*.

Chapter 1

Sustainable Manufacturing
Strategic Issues in Green Manufacturing

Christian N. Madu
Department of Management & Management Science
Lubin School of Business
Pace University, New York

1. INTRODUCTION

Earth's resources are limited. With the explosion in world population and the increasing rate of consumption, it will be increasingly difficult to sustain the quality of life on earth if serious efforts are not made now to conserve and effectively use the earth's limited resources. It is projected that the current world population of 5.6 billion people would rise to 8.3 billion people by the year 2025 [Furukawa 1996]. This is an increase of 48.21% from the current level. Yet, earth's resources such as fossil fuel, landfills, quality air and water are increasingly being depleted or polluted. So, why there is a growth in population, there is a decline in the necessary resources to sustain the increasing population. Since the mid-1980s, we have witnessed a rapid proliferation of new products with shorter life cycles. This has created tremendous wastes that have become problematic as more and more of the landfills are usurped. Increasingly, more and more environmental activist groups are forming and with consumer supports, are putting pressures on corporations to improve their environmental performance. These efforts are also being supported by the increase in the number of new legislatures to protect the natural environment. Thus, responsible manufacturing is needed to achieve sustainable economic development. Strikingly, studies have linked economic growth to environmental pollution [Madu 1999]. Thus, there is a vicious cycle between improved economic development and environmental pollution. This traditional belief on a link between environment pollution and economic growth often is a hindrance to efforts to achieve sustainable development. Sustainable manufacturing is therefore, a responsible manufacturing strategy that is cognizant of the need to protect the environment from environmental pollution and degradation by conserving the earth's limited resources and effectively planning for the optimal use of resources and safe disposal of wastes. In the past, manufacturers have been lukewarm about any strategy to develop sustainable manufacturing. They viewed such strategies as expensive and therefore, not economically viable. However, this mood is gradually changing as more and more big companies are developing environmentally conscious manufacturing strategies through their entire supply chain. Many have also seen that environmentally conscious manufacturing can become an effective competitive strategy. Thus, sustainable manufacturing can lead to an improved bottom-line and therefore, makes wise business sense. We shall in this chapter, trace the origins of sustainable development, which gave rise to sustainable manufacturing. Further, we shall identify different strategies to sustainable manufacturing and then present cases of successful implementation of sustainable manufacturing by multinational corporations such as Kodak and Xerox.

2.   SUSTAINABLE DEVELOPMENT

The origins of sustainable development can be traced to the United Nations publication in 1987 titled the Brundtland Report. This report is named after Mrs. Brundtland, Prime Minister of Norway who chaired the UN World Commission on Environment and Development. The report focused on the problems of environmental degradation and states that "the challenge faced by all is to achieve sustainable world economy where the needs of all the world's people are met without compromising the ability of future generations to meet their needs." This report received an international acclaim, as more and more people are concerned with the theme of the report on environmental degradation. Since its publication, the world community has convened several conferences on how to achieve sustainable development. In 1992, the UN organized the Earth Summit in Rio de Janeiro, Brazil with a focus on how to get the world community to cut down on the use of non-renewable resources in other to achieve sustainable development. This conference highlighted the disparate views between the industrialized and the developing countries on how sustainable development could be achieved with those from the Southern Hemisphere seeing dependence on the use of natural resources as a prerequisite to their economic growth. Several publications have emerged on sustainable development since the conference.

Duncan [1992] defined sustainable development as an "economic policy, which teaches that society can make the appropriate allocation of resources between environmental maintenance, consumption, and investment." However, such balance is difficult to achieve when a nation becomes completely dependent on the exploitation of natural resources to satisfy its social and economic needs [Madu 1996]. Furthermore, with the absence of a developed private sector, countries faced with harsh economic realities such as poverty and over population, are more likely to focus on exploitation of natural resources and deployment of inappropriate technologies for manufacturing. Such attempts may hinder the global efforts to achieve sustainable development. Following this debate, Singer [1992] argues that sustainable development is akin to a "New Economic Order" that may not encourage reasonable and realistic development from the Southern Hemisphere. Rather, it could be seen as an attempt to make the South financially dependent on the North. This he refers to as a Robin Hood effect, which may result in the transfer of funds from the poor in the rich countries to the rich in the poor countries. Clearly, achieving sustainable development is a goal for the entire world and marginal efforts by each country will be ineffective. Fukukawa [1992] pointed out "current global environmental problems may bring about a crisis that could never have been anticipated by our predecessors. Since the very inception of history, humankind has been pursuing technological development to protect itself from the threats and constraints of nature. However, economic activities triggered by these technological developments have grown large enough to destroy our vital ecosystem." This view is shared by many around the world and has been a motivating force in seeking for responsible manufacturing through sustainable manufacturing. While many companies in the industrialized countries have embarked on the road to sustainable development, it is important to achieve environmental conformance throughout the world. After all, non-compliance may affect the supply chain especially since some of the raw materials may be generated from the poorer nations. Getting these countries to participate in sustainable development will require understanding of their perspectives on economic and social development and how they could be assisted by the more affluent nations. The problems in developing countries are well explained by Mr. Kamal Nath, India's Minister of the Environment. He noted in the Rio de Janeiro conference that "Developed countries are mainly responsible for global environmental degradation and must take the necessary corrective steps by modifying consumption patterns and lifestyles; developing countries can participate in global action, but not at the cost of their development efforts... On climate change and greenhouse gases, India's stand is that global warming is not caused by emissions of the gases per se but by excessive emissions. The responsibility for cutting back on the emissions rest on countries whose per capita consumption is high. India's stand is that emission in developed countries be reduced to tally with the per capita emission levels of developing countries." This view obviously, is controversial in industrialized countries. However, what it points out is the link or the perception of a link between environmental pollution and economic growth. In fact, as Figure 1-1 shows, the emission levels of carbon tend to support such a link. This figure, tend to suggest a direct relationship between carbon emission and economic growth when the cases for OECD (Organization of Economic Co-operation and Development) are compared to the cases for non-OECD nations.

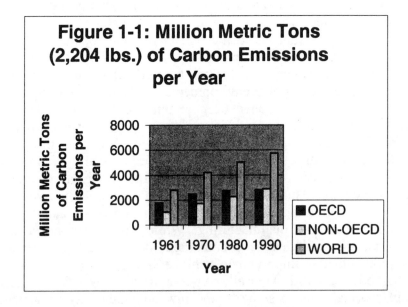

Figure 1-1: Million Metric Tons (2,204 lbs.) of Carbon Emissions per Year

In 1997, the UN conference on Climate Change was held in Kyoto, Japan. This conference further raised some doubts and disagreements between member nations, non-governmental organizations, labor unions, and environmental activists. A preamble placed on the Internet states as follows, "The threat of global warming has brought more than 140 governments together in intensive negotiations to try to limit the emission of carbon dioxide and other greenhouse gases that trap heat in the atmosphere. But history, geography, economics and politics are driving them apart. Island states fear the rising oceans that warming may cause. Oil producers fear what lessening the world's dependence on fossil fuels would mean to them. Big industrial nations worry that emission limits might slow their economies. Poor nations say they should not have to bear the same burden as the rich." Obviously, sustainable development is intertwined with politics and economics and these may impede the effort to achieve sustainable development. Strategies to achieve this goal must therefore, take into consideration these concerns. Clearly, sustainable development cannot be achieved without sustainable manufacturing. Sustainable manufacturing is one of the processes or strategies to achieve the goal of sustainable development. Sustainable manufacturing as a strategy will require the re-engineering of the organization to change design, process, work attitudes and perceptions. It will require the entire organization to be environmentally conscious and will require the support and participation of top management. More importantly, it will require investment into the future and retraining of the work force. Sustainable manufacturing is a capital venture that a company must undertake and this is a risk that some may not yet be ready for especially from the developing economies. Yet, from all indications, those corporations that have embarked on this bold step are reporting dramatic successes as we shall outline later.

While the Brundtland Report was instrumental in getting the world to focus on sustainable development, it was the formation of the Business Charter for Sustainable Development (BCSD) by a group of 50 business executives that provided the momentum for much of businesses involvement in sustainable manufacturing. BCSD was formed in 1990 in preparation of Business activities at UNCED. This group was headed by Stefan Schmidheiny and published a book titled "Changing Course." This book detailed with case studies, challenges facing business in a sustainable environment. In January 1995, BCSD merged with another influential group with strong business ties known as the World Industry Council for the Environment (WICE). WICE is an initiative of the International Chamber of Commerce (ICC) based in Paris while BCSD was based in Geneva. These two groups shared common goals and attracted executives from similar organizations although BCSD was an executive-based group. The result of this merge is the World Business Council for Sustainable Development (WBCSD). WBCSD is presently, a coalition of 125 international companies that share a commitment to environmental protection and to the principles of economic growth through sustainable development. Its membership is drawn from 30 countries and more than 20 major industrial sectors. The aims of WBCSD as listed in its web page are stated below as follows (http://www.wbcsd.ch/whatis.htm):

1. Business leadership - To be the leading business advocate on issues connected with the environment and sustainable development;
2. Policy development - To participate in policy development in order to create a framework that allows business to contribute effectively to sustainable development;
3. Best practice - To demonstrate progress in environmental and resource management in business and to share leading-edge practices among our members;
4. Global outreach - To contribute through our global network to a sustainable future for developing nations and nations in transition.

The participation and support of many executives and major industrial sectors in sustainable development issues gave the momentum to corporate focus on sustainable manufacturing or environmentally conscious manufacturing.

Schmidheiny in his 1992 article discusses the term ecoefficiency. He defines it as "companies that add the most value with the least use of resources and the least pollution." This definition clearly linked industrial production to achieving sustainable development and shows that ecoefficiency or sustainable development can be achieved only when limited natural resources are optimized and environmental waste and pollution are minimized. Thus, corporate responsibility for sustainable development is obvious and corporations and their executives by participating in WBCSD are leading the way to achieving sustainable manufacturing. Sustainable manufacturing is therefore, synonymous to ecoefficiency. We shall therefore, define sustainable manufacturing as a means for manufacturers to add the most value to their products and services by making the most efficient use of the earth's limited resources, generating the least pollution to the environment, and targeting for environmental clean production systems. Although we emphasize sustainable manufacturing, it should be apparent that the goal of environmentally clean production cannot be achieved if the service component of the manufacturing system is not environmentally conscious. The service sector must contribute by ensuring that its services are environmentally efficient. For example, can the purchasing and receive department conserve its use of paper for placing orders? Obviously, such a simple case can be achieved by using recycled papers and packaging, and by placing most orders through the computer in a paper-less environment. Thus, our focus is on both the manufacturing and the service sector working in harmony to achieve the goal of environmentally conscious manufacturing.

3. STRATEGIES FOR SUSTAINABLE MANUFACTURING

Several strategies have been developed to achieve sustainable manufacturing. We shall briefly discuss the different strategies. The aim of each of these strategies is to find a better way to make more efficient use of the limited earth's resources, minimize pollution and waste. Some of these strategies may appear in more details in subsequent chapters.

3.1 Inverse Manufacturing

This strategy is based on prolonging the life of a product and its constituent components. Umeda [1995] refers to this as a closed-loop product life cycle. Simply stated, the life of any product can be extended by disassembling the original product at the end of its original life into components that could be reused, recycled, maintained or up-graded. Focus is on limiting the amount of components that are disposed or discarded as wastes. When this is done, environmental costs are minimized [Yoshikama 1996]. Inverse manufacturing gets its name from the reverse approach to recovery of the components that make up a product. Due attention is given at the conception of the product to the ease of disassembly. This will make it possible to reclaim component parts for future use thereby prolonging the life of the product. There are many examples of Inverse Manufacturing. For example, older computers are frequently upgraded to give them more capabilities by retaining much of the computer unit and adding only the needed features, its life is further extended. Also, important precious metals present in some older computer units such as silver, platinum and gold can be extracted and reused in building newer models when it is no longer economical or feasible to upgrade the unit. These activities reduce waste through recovery, recycling and reuse of materials. In the paper industry also, the use of recycled paper rather than virgin pulp in new paper production prolongs the life of the original virgin pulp. Inverse manufacturing has obvious advantages in extending the life of the product, minimizing waste of materials and conserves the landfills. The goal however, should be to keep waste to a bare minimum.

3.2   Recycling

Recycling is one of the better-known strategies for sustainable manufacturing. In most communities, it is mandatory to participate in recycling programs. Many people identify with recycling of newspapers, packages, soda cans, bottles, and in fact, are required to separate them from other garbage for recycling purposes. Although there are arguments about the weaknesses of the current day recycling policies, however, the aim of recycling is to focus our attention to the finite resources available to mankind. The earth is composed of about 30% land and the rest is water. Our landfills are gradually filling up. If we continue to discard and dump wastes, the landfills will be filled up. We depend on the limited earth's resources for economic growth and if we are not able to thoroughly recycle and extend the lives of these resources, the future will be blink. Thus, a recycling policy that is efficient is needed. Such policy should be efficient and encourage more people and industries to participate in the program. Without recycling of paper for example, we will destroy many of the thick forests and thereby affect the eco-balance. This will have drastic consequence on survival on earth.

3.3   Re-manufacturing

This is the process of rebuilding a unit or machinery to restore its condition to "as good as new." This may involve reuse of existing components after overhaul, replacement of some component parts, and quality control to ensure that the remanufactured product will meet new product's tolerances and capabilities. The remanufactured product will normally come with a new product warranty. To make re-manufacturing effective, the following steps are normally taken:

1. Collection of used items - This could be achieved through a recycling program where used or expired original products are collected from the customer and reshipped to the manufacturer. Some examples of these are drum and toner cartridges for computer printers and photocopying machines, auto parts, etc.

2. These items on receipt are inspected based on their material condition and a determination can be made on the economic feasibility of re-manufacturing them.
3. Subsequently, the items are disassembled. If the full unit cannot be remanufactured, some components may be recovered for use in other components. Otherwise, the original item can be restored to a condition as good as new through repair and servicing. The recovery process must be efficient and focus on strategies that are conducive to the environment.

It is important that new products are designed for ease of disassemble and recovery of parts. This will make it more economical to conduct re-manufacturing activities since it will be easier to determine which parts need repair or replacement. This will also help in effective planning of the master production scheduling by minimizing the production planning time and parts inventory levels.

3.4   Reverse logistics

Reverse logistics requires that manufacturers take a "cradle-to-grave" approach of their products. This management of a product through its life cycle does not end with the transfer of ownership to the consumer and the expiration of warranty. Rather, the manufacturer is forever, responsible for the product. This is often referred to as "product stewardship." [Dillion and Baram 1991]. Roy and Whelan [1992] noted that this is a "systematic company efforts to reduce risks to health and the environment over all the significant segments of a product life cycle." Product stewardship is driven by public outcry about the degradation of the environment. This has led to new legislatures making manufacturers responsible for the residual effects of their product on the environment with no time limit. As a result, more and more companies are responding by developing environmentally responsible strategies. Some are also seeing that such strategies are good for business and may lead to competitive advantages. The concept of product stewardship as outlined by Roy and Whelan [1992] requires a focus on the following:
1. Recycling
2. Evaluation of equipment design and material selection
3. Environmental impact assessment of all manufacturing processes

4. Logistics analysis for the collection of products at the end of their lives
5. Safe disposal of hazardous wastes and unusable components
6. Communication with external organizations - consumer groups legislature, and the industry at large.

This focus is embodied in the reverse logistic strategy. It is a new way for manufacturers to view their products and develop a business model that could enable them profit from developing a product stewardship approach. Obviously, by using re-manufacturing strategy, the manufacturer can save significantly from the cost of labor and materials. Giuntini [1997] note that about 10 to 15 % of the gross domestic product could be affected by adopting reverse logistic as a business strategy. Furthermore, about 50 to 70 percent of the original value of an impaired material can be recovered from customers. In addition, the cost of sales (direct labor, direct material, and overhead) which currently, averages 65 percent to 75 percent of the total cost structure of a manufacturer, can be reduced by as much as 30 percent to 50 percent through reverse logistics. He identified the by-products of reverse logistics as follows:

- Industrial waste throughout the manufacturing supply chain, would be reduced by as much as 30 percent
- Industrial energy consumption would be noticeably reduced
- Traditionally underfunded environmental and product liability costs would be better controlled and understood.

He went further to offer the following 10 steps for a manufacturer to implement a reverse logistics business strategy:

1. Products must be designed for ease of renewal, high reliability, and high residual value.
2. Financial functions must be restructured to cope with different cash-flow requirements and significant changes in managerial accounting cues.
3. Marketing must reconfigure its pricing and distribution channels.
4. Product support services and physical asset condition monitoring management systems must be implemented to manage manufacturer-owned products at customer sites.
5. Customer order management systems must be implemented to recognize the need for the return of an impaired asset from a customer site.
6. Physical recovery management systems must be implemented to manage the return of impaired physical assets.

7. Material requirements planning management systems must be implemented to optimize the steps that will be taken upon the receipt of recovered impaired assets.
8. Renewal operational processes must be established to add value to impaired assets.
9. Re-entry operational processes must be established to utilize renewed assets.
10. Removal processes must be established to manage non-renewed assets.

3.5   Eco-labelling

The aim of eco-labelling is to make consumers aware of the health and environmental impacts of products they use. It is expected that consumers will make the right decision and choose products that will have less environmental and health risks. By appropriately labelling the product and providing adequate product information for consumers to make the choice between alternative products, it is hoped that manufacturers will move towards developing environmentally conscious production systems. Ecolabelling as a strategy is therefore, intended to identify the green products in each product category. It could be perceived as a marketing strategy that is partly driven by legislatures and partly driven by consumers concern for the degradation of the environment. Many of the eco-labelling schemes are based on the life cycle assessment (LCA) of a product and take the "cradle-to-grave" approach by evaluating the environmental impacts of the product from the extraction of the raw material to the end of the product's useful life. However, some of the popular eco-labelling schemes do not take this approach. The German "Blue Angel" mark, which is one of the best-known eco-labelling schemes, focuses on the environmental impacts of the product only at disposal and the Japanese EcoMark focuses on the contributions of the product to recycling efforts. [Using Eco-labelling, http://www.uia.org/uiademo/str/v0923.htm].

Eco-labelling is increasingly being used in many industries and consumers are paying attention, as opinion polls tend to suggest [Using eco labelling]. However, for eco-labelling to be effective, the public needs to be well informed and the labelling scheme must be credible. As has been suggested, it is important that all the major stakeholders (i.e., consumers, environmental interest groups, and producers) participate in developing the ecolabelling schemes. Also, information presented on the content of the product has to be valuable and understandable to consumers. There is a need for a standardized scheme in each product category to make it easier for comparative judgments. One of the major problems facing eco-labelling schemes is that it is voluntary and often, it is administered by third parties. Bach [1998] argued that mandatory eco-labelling schemes would be illegal within the context of the World Trade Organization and act as a barrier against international trade. He is of the opinion that regulatory measures will not reduce environmental degradation and further note that different countries have different environmental policies and standards as well as different economic policies and standards.

However, market forces and not government laws and legislatures drive eco-labelling. We operate in a global environment and without a standardized eco-labelling scheme, the entire supply chain will be affected. It is clear that many producers in industrialized countries source their raw materials and parts from different countries. If a standardized eco-labelling scheme is not developed, the entire supply chain will be affected and it will be difficult to implement an eco-labelling scheme that is based on a cradle to grave approach. Furthermore, the changes we have observed in the market economy since the 1980s as a result of the total quality movement and the subsequent development of the ISO 9000 series of product standards suggest that international standards on eco-labelling are not far from implementation. In fact, with the success of ISO 9000, the International Organization for Standards (ISO) has developed the ISO 14000 series of standards with a focus on guidelines and principles of environmental management systems. The technical committee (TC 207) charged with developing standards for global environmental management systems and tools, has environmental labelling as part of its focus. ISO 14020 deals with the general principles for all environmental labels and declarations [Madu 1998]. As expected, these standards will be widely adopted and when that happens, businesses will be expected to follow accordingly in order to compete in global markets. ISO already has classifications for ecolabelling schemes and the Type I ecolabels have the greatest impact on international trades. A third party to products that meet specified eco-labelling criteria grants certification. The issue is not to have each country develop its own plan for eco-labelling but for world bodies such as ISO to institute a standardized scheme that will be cognizant of the limitations poorer nations may face. Indeed, ISO has four standards dealing with eco-labelling. These are ISO 14020, ISO 14021, and ISO 14024 and ISO 14025. ISO 14020 has been adopted as a standard while ISO 14021 and ISO 14024 are at the final stages of being adopted as standards. ISO standards are voluntary, however, with its worldwide acceptance, it is expected that many companies and countries will accordingly work within the guidelines of these standards. Environmental protection should be a worldwide effort and with out such an effort, the whole idea will be marginalized. Finally, some have argued that ecolabels do not boost sales [Christensen 1998] but it is too early to verify this claim since the public has to be sufficiently aware. Also, sales should not be the single criterion for environmental protection. Otherwise, many companies may find alternative ways to invest in that may be more profitable. Due concern should be appropriated to the need of the consumer being aware of the content of the product and having the ability to make a purchasing decision based on that information.

3.5.1   ISO 14000

This is a series of international standards on environmental management. These standards are being put up by the International Organization for Standards (ISO) with the objective to meet the needs of business, industry, governments, non-governmental organizations and consumers in the field of the environment. These standards are voluntary, however, they continue to receive the great support of ISO member countries and corporations that do business in those countries. We shall not go into the details of these standards since ISO 14000 is a chapter in this book. We shall however, present a table that lists the ISO 14000 standards and other working documents at the time of writing. This is to help draw your attention to the work done by ISO on environmental management. However, the work of the ISO technical committee working on ISO 14000 family of standards is to address the following areas:

- Environmental management systems.
- Environmental auditing and other related environmental investigations.
- Environmental performance evaluations.
- Environmental labelling.
- Life cycle assessment.
- Environmental aspects in product standards.
- Terms and definitions.

Table 1 shows the listing of approved standards and drafts at their different stages of development.

Table 1: ISO 14000 family of standards and ongoing work

DESIGNATION	PUBLICATION	TITLE
ISO 14001	1996	Environmental management system - Specification with guidance for use
ISO 14004	1996	Environmental management system - General guidelines on principles, systems and supporting techniques
ISO 14010	1996	Guidelines for environmental auditing - General principles
ISO 14011	1996	Guidelines for environmental auditing - Audit procedures – Auditing of environmental management systems
ISO 14012	1996	Guidelines for environmental auditing - Qualification criteria for environmental auditors
ISO/WD 14015	To be determined	Environmental assessment of sites and entities
ISO 14020	1998	Environmental labels and Declarations – General principles
ISO/DIS 14021	1999	Environmental labels and declarations - Self declared environmental claims
ISO/FDIS 14024	1998	Environmental labels and declarations - Type I environmental labelling – Principles and procedures
ISO/WD/TR 14025	To be determined	Environmental labels and declarations - Type III environmental declarations – Guiding principles and procedures
ISO/DIS 14031	1999	Environmental management – Environmental performance evaluation - Guidelines
ISO/TR 14032	1999	Environmental management - Environmental performance evaluation - Case studies illustrating the use of ISO 14031
ISO 14040	1997	Environmental management - Life cycle assessment - Principles and framework
ISO 14041	1998	Environmental management - Life cycle assessment - Goal and scope definition and inventory analysis
ISO/CD 14042	1999	Environmental management - Life cycle

		assessment - Life cycle impact assessment
ISO/DIS 14043	1999	Environmental management - Life cycle assessment - Life cycle interpretation
ISO/TR 14048	1999	Environmental management - Life cycle assessment - Life cycle assessment data documentation format
ISO/TR 14049	1999	Environmental management - Life cycle assessment - Examples for the application of ISO 14041
ISO 14050	1998	Environmental management - Vocabulary
ISO/TR 14061	1998	Information to assist forestry organizations in the use of the Environmental Management Systems standards ISO 14001 and ISO 14004
ISO Guide 64	1997	Guide for the inclusion of environmental aspects in product standards

Source: Adopted from "ISO 14000 - Meet the whole family!" retrieved 3/11/1999 from http://www.tc207.org/home/index.html.

NOTE:

CD = Committee Draft;
DIS = Draft International Standard;
FDIS = Final Draft International Standard;
TR = Technical Report

3.6 Life cycle assessment

We shall adopt the definition provided by ISO for life cycle assessment (LCA). It is defined as "a technique for assessing the environmental aspects and potential impacts associated (with products and services)... LCA can assist in identifying opportunities to improve the environmental aspects of (products and services) at various points in their life cycles." This concept is often referred to as the "cradle to grave" approach. It requires that emphasis be placed on the environmental impacts of production or service activities from the product conception stage (i.e., raw material generation) to the end of the product's life (i.e., recovery, retirement or disposal) of the product). Thus, the manufacturer is responsible for the environmental impacts of the product through different stages in its life cycle. Life cycle assessment often involve three major activities [Affisco 1998]:

1. Inventory analysis - this deals with the identification and quantification of energy and resource use as well as environmental discharges to air, water and land.
2. Impact analysis - is a technical assessment of environmental risks and degradation.
3. Improvement analysis - identifies opportunities for environmental performance improvement.

Notice also that several of the ISO standards listed in Table 1 deal with Life Cycle Assessment. Already, ISO 14040 on principles and framework and ISO 14041 on goal and scope definition and inventory analysis have been adopted as standards.

3.7   Design for the environment

Consequent to the growing demand for improvement in environmental performance is the growing need to change the traditional approach to designing. This strategy calls for an efficient designing of products for environmental management. Products are to be designed with ease of disassemble and recovery of valuable parts. Such design strategies will conserve energy and resources while minimizing waste. In designing for the environment, tradeoffs are made between the different environmental improvements over the product life cycle. Three main design strategies are design for recyclability; design for remanufacture; and design for disposal.

1. *Design for recyclability* - this involves the ease with which a product can be disassembled and component parts recovered for future use. For example, with computer units, precious metals can be easily recovered for use in new computers. For chemical compounds, the focus is on separability of materials to avoid contamination and waste of energy in recovering these materials.
2. *Design for Remanufacture* - This recognizes the different stages of equipment or product wears. For example, certain parts of machinery (i.e., auto parts) could be recovered, remanufactured and restored to a state as good as new. Reusing them in newer products could further extend the lives of such parts. The challenge is how to design the original product for ease of recovery of those parts. We notice for example that newer computer systems are designed with the ease of upgrading them. Thus, new capabilities could be added to the system without having to dispose of the old unit.

3. *Design for disposal* - This recognizes the fact that many of the earth's landfills are filling up at an alarming rate. Further, many of the deposits are hazardous and unsafe. It is important to design the product with the ease of recycling and disposal. The final waste generated from the product should also be disposed safely.

4.   CASE STUDIES

The case studies presented here are some of the popular success stories from leading manufacturers to show that responsible design; production and packaging that are environmentally sensitive are profitable. Many of these companies have witnessed growth in sales and revenue and attribute these successes to their environmental management programs.

4.1   Kodak Single-Use Camera

The Kodak single-use camera (SUC) is perhaps, one of the most remarkable successes stories. Kodak first introduced this product in the U.S. in 1987 and it is now, the company's fastest growing product category. This product is now the company's centerpiece in its efforts in recycling, re-use, and product stewardship. Interestingly, the single-use camera was introduced as an inexpensive camera and not as an environmental product. It became known widely as a disposable camera and was even dubbed an environmental "ugly duckling." However, through innovation, commitment, and hard work, Kodak has transformed this product into an environmental success story. How did Kodak achieve this feat?

The product was designed for the environment and it is dubbed by some as the best example of closed-loop recycling. The recycling of Kodak single-use camera is a three-prong process that involves the active participation of photofinishers and a strategic partnership with other SUC manufacturers. In fact, Kodak credits photofinishers with most of the success achieved in recycling and reusing Kodak's SUCs. The new SUCs use 20 percent fewer parts from the design features to the actual film processing stage. Photofinishers return the camera after processing to Kodak and are reimbursed for each camera returned plus shipping cost. In the U.S., a 63 percent return rate has been achieved for recycling. This is equivalent to fifty million SUCs or enough SUCs to fill up 549 tractor-trailer loads.

The three-prong recycling process as detailed by Kodak in its website (http://www.kodak.com/US/en/corp/environment/performance/recycling/suc .shtml retrieved 3/2/99) is as follows:

1. Photofinishers ship the SUCs to three collection facilities around the world and Kodak maintains a recycling program in more than 20 countries. Through the strategic partnership with other SUC manufacturers such as Fuji, Konica, and others, they jointly accept each other's products even though Kodak cameras are in the majority. The products are sorted according to manufacturer and camera model. These cameras arrive at these facilities in recyclable cardboard.
2. Kodak cameras are shipped to a subcontractor facility for processing. The packaging is removed and any batteries in the camera are recovered. The camera is cleaned up and undergoes visual inspection. The process of re-manufacturing the SUC has begun and the old viewfinders and lenses are replaced. New batteries are also inserted. Those parts that could be reused are retained after rigorous quality control checks.
3. The SUCs are now shipped to one of Kodak's three SUC manufacturing plants. This is the final assembly where new packaging made from recycled materials (with 35% post-consumer content) is added. The camera is now ready for use.

There are some lessons that could be learned from the Kodak experience:

1. Kodak notes that by weight, 77 to 86 percent of Kodak SUCs can be recycled or re-used. Yet, these products maintain high quality, attract huge demand and are profitable. This suggests that responsible environmental programs can be competitive and help the firm achieve its profit motives. Recycling programs such as this can help the manufacturer to save significantly by cutting down the cost of material, labor, and to achieve faster response to the market. It is estimated that it takes about 30 days from the time of collection of an SUC to reclamation and re-introduction to the market.
2. Strategic partnership and working with vendors may be instrumental in effective recycling programs. Photofinishers have an incentive to participate and the cost of recycling can be shared through industry partnership as demonstrated in the case of Kodak's SUC.

3. Design for the environment is essential. Products for recycling, reuse, and re-manufacturing must be designed for ease of disassemble and recycle. For example, with the SUC, it is easy to reclaim the packaging and recover component parts. Such design cuts down on cost and therefore, makes designing for the environment attractive.

4. Conservation of resources is achieved through effective recycling programs. For example, the equivalent of 549 tractor-trailer loads of SUCs has been recycled. This is said to be equivalent to 3,333 miles of cameras laid end-to-end. Imagine the enormous pollution this will create if these cameras are 'disposables'. How much landfill space will be needed to contain them? Since the recycling and reuse program began in 1990, more than 200 million cameras have been recycled. Since 1990, there has been an exponential growth in the number of cameras recycled with the number increasing from 42.1 million in 1996 to 51.9 million in 1997. This is shown in the figure below from the data presented by Kodak.

5. The Kodak SUC is an example of a cradle-to-grave approach of a product. The manufacturer designs the product so that it has control over the life of the product. This is achieved by ensuring that the consumer ultimately, will return the camera to a photofinisher for processing and the photofinisher is given an incentive to participate in the environmental management program. The manufacturer takes the responsibility of disassembly and reuse and disposal of the product. This process ensures an effective environmental program that makes the manufacturer responsible for the life of the product.

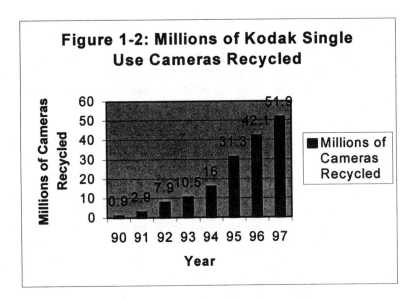

Figure 1-2 is adopted from
http://www.kodak.com/US/en/corp/environment/performance/recycling/suc.
shtml

The Kodak SUC is a success story that has helped to reduce environmental degradation and has achieved tremendous economic success. Next we look at the environmental program at Xerox.

4.2   Xerox

Xerox has a long history of developing sustainable products that dates back to 1967. Its strategy involves design foe environment and life cycle product valuation. In 1967, the company embarked on a metal recovery program from photoreceptor drums and continues today to reclaim metals for reuse or re-manufacturing purposes. Its design strategy today is known as "Waste-Free" design. How does this program work?

Machines are recovered from customers through trade-ins and lease options. Many of the components of the xerographic machines that can still perform at their original specifications are recovered for reuse and re-manufacturing. In 1997 alone, more than 30,000 tons of returned machines were used to remanufacture new equipment. Within the past five years, Xerox has more than doubled the number of machines it remanufactured. The remanufactured machines still meet Xerox's strict quality guidelines and are offered with the same

Xerox Total Satisfaction Guarantee. These machines are designed for ease of disassembly, and Xerox takes the responsibility of the product's life cycle. As a result of the company's environmental efforts, natural resources are conserved and new machines are designed with fewer replacement parts.

Xerox works with its customers to carry out the recycling program. Customers of copy cartridges are provided with prepaid return labels that enable them to reuse the packaging from the new cartridge to ship the used cartridges to Xerox. The reused cartridges are then remanufactured. In 1997, Xerox achieved a return rate of 65% for print and copy cartridges. This is now the industry benchmark. Xerox also maintains a Waste Toner Return Program. This program allows customers to return waste toners for re-manufacturing, reuse and recycle. This program is credited with the recovery of millions of pounds of toner which would have otherwise, been sent to landfill.

Xerox adopts a company-wide environmental program that tracks its product's life cycle and ensures environmental protection. Its recycling program works well because of the extensive network of people who participate in the delivery process to monitor the environmental and other potential impacts of the product on Xerox. A framework of its successful recycling program is shown below:

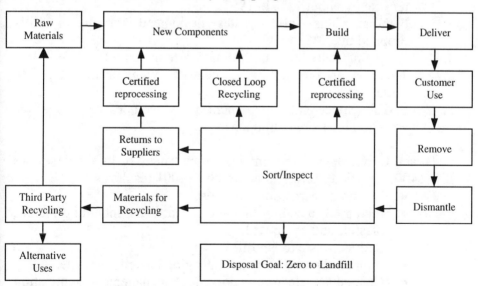

Figure 1-3: Xerox recycling program
(adapted from "Sustainable product development - The First 30 years",
http://www.xerox.com/ehs/1997/sustain.htm, page 8, retrieved 3/2/99)

Notice that like in the case of Kodak Single Use Camera, the Xerox recycling program adopts a closed-loop approach. This ensures that no wastes are incurred as the material is continuously recycled, reused and remanufactured. In the end, the goal of achieving zero disposals to the landfill can be achieved. However, this can only be feasible if the product is designed for the environment. In other words, recycling will become a profitable and an economical alternative to waste.

The approaches that Kodak and Xerox are taking are innovative in that both manufacturers develop effective collection systems for their expired products. Many of the existing recycling programs depend on the garbage industry and municipalities rather than on vendors and suppliers who have stake in the recycling process. The lack of a well-designed recycling program has created displeasure and dissatisfaction that often times denigrate the value of recycling programs.

Some of the achievements of the Xerox program are outlined below http://www.xerox.com/ehs/1997/iso.htm page 2:

1. There has been a dramatic reduction in the amount of hazardous waste generated and 1997 saw a reduction of about 20% from the 1996 level.
2. There is a reduction in the amount of solvents sent off-site for recycle by about 13% in 1997.
3. Non-hazardous solid wastes designated for the landfills dropped by 20% in 1997.
4. Recycling rate of non-hazardous waste increased to 85%.
5. There is an increase in monetary savings from waste-free initiatives.
6. SARA (Superfund Amendment and Reauthorization Acts) air emission levels remain unchanged.

To add to Xerox environmental program, many of its processes are ISO 14001 certified. Xerox noted its ISO 14001 results as
- Lowered energy consumption.
- Increased recycling of various materials, including metals, plastics and cardboard.
- Reduce waste to landfill.
- Improved procedures for managing hazardous waste.
- Created an electronic document management application now being developed as a product for customers implementing their own environmental management systems.

5. CONCLUSION

In this chapter, we have traced the origins of sustainable development to the United Nations publication in 1987 titled the Brundtland Report. This report provided further impetus for the world community to focus on environmental protection. Since this report was presented, several conferences have been organized by several world agencies such as the United Nations to focus the world attention to the need of pollution prevention and resource conservation. However, it was the formation of the Business Charter for Sustainable Development (BCSD) by a group of 50 business executives that provided the momentum for much of businesses involvement in sustainable manufacturing. As we mentioned earlier, the publication of a book "Changing Course" by BCSD, outlined the challenges facing business in a sustainable environment. The merging of BCSD with the World Industry Council for the Environment (WICE) further expanded the industry interests in sustainable manufacturing. These two groups shared common goals and attracted several executives and industries from around the world. The new group now known as the World Business Council for Sustainable Development (WBCSD) is presently, a coalition of 125 international companies that share a commitment to environmental protection and to the principles of economic growth through sustainable development. Its membership is drawn from more than 30 countries and from more than 20 major industrial sectors.

The interest of businesses and the world community at large has spurned a lot of interests on environmental protection since the 1990s. Subsequently, several strategies have been developed to deal with environmental protection issues. We have outlined some of these strategies in this chapter. Environmental protection is increasingly seen as a competitive weapon. Companies now pride themselves for the continuous efforts they are making to protect the environment and provide consumers with environmentally clean products. We continue to see this effort expanded to include not only the content of the product but also its tertiary value such as packaging and preparation of the order. Also, the entire supply chain is now involved with environmental protection efforts. Sustainable manufacturing is in vogue now and every progressive company should have a strategy to achieve its environmental protection goals. It is important to know that with many nations adopting ISO 14001, companies that fail to follow the lead may lack access into such markets and even when granted access, may not be able to compete in an environmentally conscious market.

We ended our discussion in this chapter by presenting two classic case studies from two of the world leading companies in sustainable manufacturing namely Kodak and Xerox. There are a host of other companies that have very good environmental strategies but the success stories of Kodak and Xerox may provide a motivation for more companies to re-evaluate their environmental programs and perhaps, benchmark the leaders.

REFERENCES

Giuntini, R., "Redesign it + Produce it + Rent it + Support it + Renew it + Reuse it = Reverse Logistics Reinventing the Manufacturer's Business Model" APICS—The Educational Society for Resource Management, 1997 (http://www.apics.org/SIGs/Articles/redesign.htm)

"Kodak Recycles Its 50 Millionth FUN SAVER" http://www.kodak.com/US/en/corp/environment/performance/recycling/suc.shtml, retrieved 3/2/99.

Furukawa, Y., "'Throwaway' mentality should be junked," *The Daily Yomiuri*, October 1, 1996, pp. 10.

Fukukawa, S., "Japan's policy of sustainable development," *Columbia Journal of World Business*, 27 (3 & 4), 96-105.

Umeda, Y., "Research topics of the Inverse Manufacturing Laboratory," http://www.inverse.t.u-tokyo.ac.jp, 1995.

Madu, C.N., "A decision support framework for environmental planning in developing countries," *Journal of Environmental Planning and Management*, Vol. 43 (2), 1999 (forthcoming).

Madu, C.N., Managing Green Technologies for Global Competitiveness, Westport, CT: Quorum Books, 1996.

Madu, C.N., "Introduction to ISO and ISO quality standards," in Handbook of Total Quality Management (ed., C.N. Madu), Boston, MA: Kluwer Academic Publishers, 1999, pp. 365-387.

Brundtland Commission, Our Common Future, Geneva, Switzerland: Report of UN World Commission on Environment and Development," April, 1987, p. 43.

Yoshikama, H., "Sustainable development in the 21st century," http://www.zeri.org/texter/ZERT_96_industries.html, 1996.

Singer, S.F., "Sustainable development vs. global environment -- Resolving the conflict," *Columbia Journal of World Business*, 27 (3 & 4), 155-162.

Schmidheiny, S., "The business logic of sustainable development," *Columbia Journal of World Business*, 27 (3 & 4), 18-24.

Roy, R., and Whelan, R.C., "Successful recycling through value-chain collaboration," *Long Range Planning*, 25 (4), 62-71, 1992.

Dillion, P., and Baram, M.S., "Forces shaping the development and use of product stewardship in the private sector," *Conference on the Greening of Industry*, The Netherlands, 1991.

Duncan, N.E., "The energy dimension of sustainable development," *Columbia Journal of World Business*, 27 (3 & 4), 164-173.

"Using Eco-labelling," http://www.uia.org/uiademo/str/v0923.html, retrieved 2/6/99.

Henriksen, A., and Bach, C.F., "Voluntary environmental labelling and the World Trade Organization," International Trade, Environment and Development, Institute of Economics, Copenhagen University, November 9, 1998.

Christensen, J.S., "Aspects of eco-labelling on less developed countries," International Trade, Environment and Development, Institute of Economics, Copenhagen University, December 7, 1998.

Affisco, J.F., "TQEM - methods for continuous environmental improvement," in Handbook of Total Quality Management (ed., C.N. Madu), Boston, MA: Kluwer Academic Publishers, 1999, pp. 388-408.

"Sustainable product development - The First 30 years," http://www.xerox.com/ehs/1997/sustain.htm, page 8, retrieved 3/2/

"What is the WBCSD," http://www.wbcsd.ch/whatis.htm, retrieved 3/9/99.

"Green business networks," http://www.epe.be/epe/sourcebook/1.14.html, retrieved 3/9/99.

Robinson, A., "Inverse manufacturing," March 10, 1998, http://mansci2.uwaterloo.ca/~msci723/inverse_mfg.htm, retrieved 3/2/

THE AUTHOR

Christian N Madu is Research Professor and Management Science Program Chair at Pace University. He is the author/co-author of more than 10 books including Handbook of Total Quality Management (Kluwer Academic Publishers, 1998), Statistics as easy as 123 with Excel for windows (Gordon & Breach Science Publishers, 2000), QFD in a Minute (Chi Publishers, 2000), Handbook of Environmentally Conscious Manufacturing (Kluwer Academic Publishers, forthcoming), Environmental Planning and Assessment (Imperial College Press, forthcoming), Managing Green Technologies for Global Competitiveness (Quorum Books, 1996). He is also the author/co-author of more than 100 research papers that have been published in all the leading journals in management and decision sciences. His work in maintenance modeling, statistical metamodeling, technology transfer and management, quality management have been widely cited.

Dr. Madu holds a Ph.D. in Management Science from the City University of New York, two MBAs and an MS in Industrial Engineering and Operations Research as well as BS in Industrial Engineering. He has also served as consultant for several organizations. He is currently, co-editor of the *International Journal of Quality and Reliability Management* and founder and former editor-in-chief of the *International Journal of Quality Science*.

Chapter 2

Concepts and Methodologies to Help Promote Industrial Ecology

J A Scott[1], I Christensen[2], K Krishnamohan[1] and A Gabric[1]
[1] *Center for Integrated Environmental Protection*
Griffith University, Queensland 4111, Australia
[2] *Brisbane City Council, Brisbane, Queensland, Australia*

1. INTRODUCTION

To help promote ecologically sustainable development (ESD), industrial ecology (IE) has emerged as a framework for proactive management of human impacts on the natural environment. IE focuses on borrowing natural ecological principles and applying them to the design and management of commercial systems, and the infrastructure required by them. IE principles can be relevant to the workings of both the private and public sectors, and have the potential to be applied to internal processes within a single enterprise and the interactions between a network of enterprises. As a concept, IE is finding favor at a time when traditional depollution approaches which rely on end-of-pipe solutions are increasingly being regarded as inefficient (Erkman, 1997), and has been defined as:

> "...An emerging framework for environmental management, seeking transformation of the industrial system in order to match its inputs and outputs to planetary and local carrying capacity. A central IE goal is to move from a linear to a closed-loop system in all realms of human production and consumption." (Lowe and Evans, 1995)

The philosophical underpinning of IE is the shift away from a reactive approach to environmental problems, in particular the 'poiluter pays principle' where organizations are penalized for a transgression of environmental legislation and governments (both local and state) are burdened with the heavy cost of regulation. IE does not accept the view that industry and environment are irrevocably opposed, but takes the position that pollutant waste generation results from process inefficiencies that reflect a loss of profitability (although this is not always perceived as such by orthodox corporate accounting practice).

The waste management problem has traditionally been approached through dilution of concentration (*e.g.* ocean outfalls), end-of-pipe treatment (*e.g.* stack-gas desulphurisation) and/or long-term sequestration in disposal sites (*e.g.* landfills). Traditional pollution control technologies do not eliminate waste, but rather shift the location of the waste. For instance, sewage disposal to coastal waters removes from urban populations the pathogenic threat, but can cause secondary, deleterious impacts on local fisheries through eutrophication and other changes in marine ecology.

The IE vision promotes a number of basic scenarios considered desirable, summarized as:

- industry recognizing its interdependence on the natural ecosystems which support it;
- material fluxes through local economies approaching a pseudo-closed cycle;
- widespread adoption of renewable energy sources;
- maintaining a balance between rate of consumption and rate of production of renewable materials;
- maintaining and improving the economic viability of industrial systems, while at the same time minimizing impacts on the biosphere.

IE principles and methods can be used by service as well as manufacturing operations, as they are intended to help organizations become more competitive by improving environmental performance through strategic planning. Their implementation should also help communities maintain a sustainable industrial/commercial base without compromising the quality of their environment. Proponents of IE claim that application should help government agencies design policies and regulations that improve environmental protection, whilst maintaining, and indeed building, business competitiveness. The IE principles can, therefore, be seen to compliment the aims and goals of triple bottom line accounting by simultaneously promoting economic, environmental and social benefit.

IE may sound as a consequence an ideal system for sustainable environmental and industrial management, but in practice it is on the international scale yet to be tested extensively. Those IE projects that have been initiated, are primarily to be found in the U.S.A, Canada and Europe, and are centred on the establishment of Eco-Industrial Parks, or recognition and support of existing inter-organizational synergies. In this chapter we outline the basic concepts behind IE, tools that can be used to support its implementation, examples from some of the success stories and finally through interpretation of a study of over 100 industries in Australia, an assessment of what is needed to increase IE awareness and application.

2. NATURAL ECOSYSTEMS AS A PARADIGM FOR INDUSTRIES

Before considering the IE paradigm it is useful to outline some of the basic properties of natural ecosystems, which in turn are used to model industrial ones. The ecosystem can be described as any unit that includes all the organisms (the community) in a defined area interacting with the environment so that energy flows lead to a trophic structure, biotic diversity and material cycles. The suitable size of the 'defined area' is always a matter of conjecture, with questions of scale key in the study of ecology.

All natural ecosystems use external sunlight and internal photosynthesis for primary energy generation, with abiotic elements (*e.g.* water and nutrients) also supplied externally. An autotroph creates biomass through the process of primary production and a consumer (heterotroph) uses the biomass of other organisms for their energy and material. Processes of respiration, death and decomposition release energy, much of which is recycled. Decomposers remineralize nutrients back into inorganic forms that can be re-used, but can also take up available nutrients, thus decreasing their availability for autotrophs. A prime example of the ability of living organisms to regenerate the basic materials (nutrients) needed to sustain life is the marine food web. Bacteria and other microbial decomposers utilize the waste generated by larger organisms (*e.g.* zooplankton) and in so doing regenerate dissolved nutrients for uptake by the primary producers (phytoplankton).

Not all ecosystems are closed with respect to materials and the import and export of material may be large or small compared with internal recycling. Understanding the mechanisms that promote the system property defined as the relative degree of recycling versus inputs of new nutrients is a key area of research in ecology. Typically most nutrients are recycled, often many times, before final export to another distinct ecosystem (*e.g.* the downstream flow of materials between river, estuarine and coastal ecosystems). Clearly when striving towards sustainable industrial activity, as with natural systems, optimizing this ratio is key.

Wastes generated by natural organisms are biodegradable and the ecosystem is, therefore, able to accept and consume the waste, and the system closes the loop. In the case of the industry, the waste produced is more often non-degradable and exceeds any natural emission rates. For example, estimated human-to-natural ratios for global emissions of heavy metals to the atmosphere range from 2 (Chromium) to 333 (Lead) (Galloway *et al.* 1982). This industrial sourced waste usually does not have a natural consumer, like bacteria, and hence accumulates and remains in the system without undergoing any beneficial change. By not putting waste back into the industrial system, the loop does not close and material flow remains linear, creating both a depletion of natural resources and a build up of wastes.

It should be noted however, that natural ecosystems do create waste energy as required by the Second Law of Thermodynamics. It has been estimated that only a small part (about 15%) of the energy contained in a resource organism (plant or animal) can be transformed into consumer biomass, the rest of the energy being dissipated as heat (Lindeman, 1942). Empirical evidence indicates that in nature, the number of trophic levels within an ecosystem seldom exceeds five or six, reflecting this energetic constraint (Pimm, 1982).

The main goal of IE is transformation of industrial systems into sustainable ones by adopting natural ecosystem principles, such as those above and others. To help achieve this, Tibbs (1992) listed certain characteristics of an ecosystem, which could be beneficially adopted by industry:

- in the natural system, there is little 'waste', with materials and energy being continuously recycled and transformed (*e.g.* the death and decay of other organisms provides nutrients for new growth);

- concentrated toxins are neither stored nor transported in bulk at the system level, but are synthesized by individual organisms and used only when needed;
- the natural system permits independent activity for each individual of a species, yet cooperatively meshes the activity patterns of all species (cooperation and competition exists amongst individuals, but they are balanced).

The IE perspective views each level of industrial activity as a living system embedded in, and interacting with, the larger natural ecosystem or biosphere. The aim of using an ecological metaphor for the design of industrial systems is to both to improve efficiency within the particular industry and to enable the planned reuse of otherwise wasted resources. That is, a major focus of IE is to mirror nature by identifying opportunities for reducing wastes in materials-intensive industrial systems by reuse of low-value/discarded by-products from one set of processes as raw materials for others (sometimes referred to as waste-mining and/or waste exchange).

There are, however, a number of significant unresolved issues in IE theory, which may impede widespread adoption. For example, an important characteristic of natural ecosystems (mostly neglected in IE theory) is the 'boom and bust' cycle that is commonly observed in predator-prey systems. This results in the size of natural populations varying between low and high extremes, which can also lead to species extinction. Whereas, it could be argued that an underlying reason for setting-up IE is to protect against or even out such variability and the risk of commercial extinction, but in doing so, runs contrary to a true ecological model.

Similarly, the notion of redefining Company A's waste as a marketable commodity, X, for Company B has implications which are also not often thoroughly examined by IE theorists. In particular, in its basic ecological model form, the IE approach can be at odds with pollution prevention or waste minimization programs. With an IE model in place, Company A will see little or no incentive to decrease the production of its waste. In some cases this could be argued as an advantage to Company A, as it may not now have to adopt expensive waste-treatment technology (Ayres and Ayres, 1996). That is, as in natural ecosystems, excessive practice of good housekeeping may not be necessary, as there exists an interdependent network of scavengers to deal with wastes.

In the commercial world, however, if a long-term binding agreement is not in place for Company A to remain 'wasteful', then at any time it may take up alternate technology and reduce and/or eliminate waste X. The consequence will be a collapse in its ecological interactions with Company B. Alternatively, Company A may over the course of time make market-driven product changes, a consequence of which might be that X no longer forms part of its waste stream, thereby again removing a relied upon commodity supply for Company B.

Equally, in a free-market economy Company B may at any time source X from another company which produces it more cheaply and/or conveniently. This can lead to a situation where Company A is left with a large oversupply of X, for which suddenly there is no longer a market. Commodity X then resumes its original classification as waste (*i.e.* pollution) for which Company A is liable, but not have the infrastructure in place to cope with.

It is probably fair to say, therefore, that IE assumes diverse companies will guarantee to collaborate in the future to an extent, which currently does not usually exist. At the present time in many regions of the world, it is not clear whether this level of co-operation can evolve naturally and how much government intervention (incentives) will be required to effect the change. Not least, as it is arguable that to sustain IE, companies are locked into inefficient modes and discouraged from free evolution. Whereas, in nature species and systems do continually change, leading to either adaptation or extinction of interdependent organisms. To be a true ecosystem mimic, in addition to understanding the necessary incentives and tools to promote itself, IE must also somehow address the problem of free form evolution.

3.   SOME INDUSTRIAL ECOLOGY TOOLS

With respect to helping to promote IE activity, there are a number of existing basic tools that practitioners can adapt and adopt. A common factor between them all, however, is that they require significant amounts of data on the operations of the industrial system, information which may not always be readily available. The following subsections summarize tools that are among those generally regarded most relevant to IE.

3.1   Cleaner Production (CP)

In many ways a precursor to IE, cleaner production (CP) is a well-developed field of environmental management that focuses on the change, design and redesign of industrial and/or commercial processes to improve both environmental and economic performance. Essentially, CP is about reappraising one-by-one the activities within an organization with a view to introducing smarter operation(s) that will lead to lower waste and hence a reduction on end-of-pipe solutions. There are numerous well documented CP examples, from the recycling of chrome previously discharged from tanning operations (Khan, 1999) through to savings of over 200 000 L/pa of otherwise spilt beer by improved filler design (Newland *et al.*, 1999). Dow chemicals by altering its manufacturing process for polystyrene foam completely eliminated use of cholorflourocarbons (CFCs) and volatile organic compounds (VOCs). CFCs and VOCs were used as blowing agents and have been 100% replaced by carbon dioxide. This redesign of the manufacturing process resulted in Dow preventing about 3.5 million pounds of CFCs from being released into the atmosphere each year (Anastas and Breen, 1997).

CP fits within the goals of IE, but implementation of cleaner technologies and cleaner practices can often be regarded as introspective approaches. That is, by focussing on internal individual company activities, this may deflect attention away from equally important wider (intra- and inter-organizational) relationships and opportunities.

3.2 Dematerialization

Dematerialization employs systematic reduction by design in the weight of materials or embedded energy in products in order to reduce both raw material/energy consumption and the magnitude of subsequent disposal problems (Ausubel, 1992). For example between 1972 and 1991, the average mass of a typical automobile decreased by more than 400 kg, with each kilogram of high-strength steel used in today's automobiles replacing 1.3 kg of carbon steel (Tibbs, 1992). The result is not only less material to handle, but better fuel efficiency. Similarly, our own work with a tinned fruit supplier showed that the average mass of a 450 ml can for processed foods in the 1960s was 85 g and contained about 1 g of tin. Improved technology has led to a reduction in the mass of the can to 65 g and it now contains only 0.5 g of tin, resulting in the twin benefits of economic savings and resource conservation. The trend towards dematerialization appears well established and is clearly environmentally favorable (Frosch, 1992). However, dematerialization should not be at the cost of quality. Poor quality will prompt early disposal of products resulting in a net increase in production to meet demand.

3.3   Design for Disassembly (DfD)

DfD facilitates recycling of a product's basic constituents by designing in such a way that the product once no longer required can be disassembled easily. For example, legislation in Germany mandates the ability to dismantle cars rapidly into homogenous component parts (Tibbs 1992) and indeed it was claimed that the Volkswagen Passat could be disassembled within 20 minutes for recycling (Ausubel, 1992). Similarly, the Z1 Roadster of the German Bavarian Motor Works (BMW) has doors, fenders and front, rear and side panels made of recyclable thermoplastic. The use of 'pop-in, pop-out' fasteners avoided the use of adhesives and screws. However, initially recycling was hindered as 20 different kinds of plastics were used to construct the vehicle, but BMW recognized the problem and subsequently reduced the types of plastics to 5 (Graedel *et al.*, 1993).

3.4   Industrial Metabolism (IM)

Similar to the idea of an environmental audit, although perhaps broader in scope, an industrial metabolism (IM) study traces material and energy flows from initial extraction, through industrial and consumer systems, to final fate. An IM study involves (a) mapping the material and energy flows, (b) analysing critical indices and (c) developing scenarios for performance improvement. IM can, therefore, highlight levels of inefficiency and waste, and to help a number of critical efficiency indices have been developed to assess an industrial system's efficiency. As such, IM can be considered to lead to incorporation of improvement strategies such as 'greening the supply chain' (Tang *et al.* 1999). To achieve this, IM uses indicators which include the virgin/recycled materials ratio, materials and energy/productivity ratios and resource input per unit of consumer output. It is as a consequence, a technique that requires substantial data, which may not always be available.

An IM study provides a framework for developing direct measures of community sustainability, but can also emphasize the difference between natural and industrial metabolic processes. In a natural system, materials tend to flow (and be recycled) in closed loops, but industrial systems are often very material dissipative, leading to concentrations too low to provide economic value (*i.e.* irrevocably lost to the system), but nevertheless high enough to pollute as they are not subject to 'natural' breakdown.

3.5 Life-Cycle Assessment (LCA)

LCA is a quantitative method to evaluate the environmental 'load' associated with a particular product, process and/or activity, It identifies and creates an inventory of all energy and material utilization, and any associated environmental releases (Koeleian, 1994). LCA has much in common with Industrial Metabolism (IM), although narrower in scope as it tends to be target specific. That is, LCA is often used to try and assess the relative merits of two (or more) competing options or products, such as the often-quoted example of using plastic or paper bags at the supermarket checkout. To achieve this, LCA attempts to combine various environmental performance indicators (Brennan, 1999) and sometimes also societal preference indicators (Lundie and Huppes, 1999) into a single evaluation mark in order that a comparative product by product ranking can be produced.

The technique is, therefore, data intensive and can be severely constrained by poor understanding of the local and regional environmental impacts of many substances. The main problems lie in identifying suitable indices that allow direct and objective universal comparisons. Are for example, the environmental impact indices for producing a ton of aluminum (*e.g.* energy consumed, green house gasses emitted, water consumed *etc.*) in temperate Canada, the same as those in sub-tropical Australia?

3.6 Design for Environment (DfE)

DfE designates a practice by which environmental considerations are integrated into product and process engineering design. That is, all stages in a product's manufacturing (including its raw materials) and end use (and subsequent recycling) would be considered in the design process. Some proponents see DfE as more qualitative than LCA (Graedel and Allenby, 1995), which requires a more quantitative approach, although DfE is still, as with LCA, hampered by a lack of sufficient data on the environmental impacts of materials. However, if correctly implemented DfE would make LCA largely redundant in the future (*e.g.* there would be no need to assess between paper and plastic bags, as only the DfE based option would be available).

3.7   Product life extension and the service economy

Finally, perhaps at odds with the current commercial trend for rapidly introducing new and changed production models, is the argument for lower demand for energy and materials through the design of durable and upgradable products with a long-life span. The question of 'how do manufacturing companies remain profitable?' is addressed by promoting that they refocus their mission to delivering customer service (emphasizing performance and customer satisfaction, rather than products). The result is that companies usually maintain ownership of their products as the means of guaranteeing that they provide this service. A classic example of this strategy is that operated by Xerox (see Section 4.4).

With product-life extension, designers seek to ensure that their products are optimized to:

*   be durable and difficult to damage;
*   be multi-functional;
*   have sub-components that are standardized and modular, as well as easy to repair and be upgradable;
*   have sub-components that can be reused in new systems (*i.e.* can be easily reconditioned and remanufactured).

4.   CASE STUDIES IN INDUSTRIAL ECOLOGY

There are in operation, only a limited number of what could be considered true IE sites and as a result they all have been extensively reported and analyzed in the literature. Below are summaries of three of the most often discussed locations, along with IE as applied to a single corporate identity, in this case Xerox.

4.1   Kalundborg, Denmark

Kalundborg is a small town 75 miles west of Copenhagen with a population of around 15000. It is probably the most often cited location of IE in practice, although the symbiotic cooperation between a network of industries is more the consequence of evolution over the last two decades, rather than systematic planning (Gertler, 1995; Ehrenfeld and Gertler, 1997; Richards, 1997). The resulting bilateral exchanges of waste materials were primarily motivated by economic benefits (Table 1), although as a result, tangible environmental benefits have been gained due to a reduction in pollution, particularly of the nearby Tissφ Lake.

The move towards industrial symbiosis was initiated by five main partners: a large coal fired power station (Asnaes Power); a pharmaceutical company (Novo Nordisk); a plasterboard manufacturer (Gyproc); a refinery (Statoil) and the municipality of Kalundborg (Richards, 1997). As an example, cooling water and treated effluents from Statoil are used by Asnaes, while waste steam and heat from Asnaes, is used as a source of energy by Statoil, Novo Nordisk, local fish farms and private residences. Indeed, a goal of the Kalundborg municipality is to provide energy to heat all residences from the waste steam of the power station by 2005. The overall success of Kalundborg is well demonstrated by the economic gains as it has been estimated that an investment of US$60 million over a 5 year period produced US$120 million of revenue and additional cost savings (Lowe and Evans, 1995).

A review of this Danish experience suggests a number of reasonably straightforward pre-conditions for successful industrial symbiosis to occur:

- the industries must be different (not competitors), but complement each other in their materials production and use;
- the partners must have well established process, such that wastes and other byproduct production is reasonably consistent and guaranteed (in quantity and quality) for the mid to long-term;
- management at the different industries must build a personal rapport;
- bilateral agreements must be voluntary and make commercial sense;
- close proximity between partners to enable effective transportation of materials;
- the local community should be small enough, that they all feel that they have a direct stake (environmental and economic) in successful outcomes.

Table 1. Reduction in the consumption of resources, generation of Emissions and wastes due to the industrial symbiosis at Kalundborg

Source of reduction	Reduction (T/year)
Resources	
Oil	19,000
Coal	30,000
Water	1,200,000

Emissions	
Carbon dioxide	130,000
Sulphur dioxide	25,000
Wastes	
Flyash	135,000
Sulfur	2,800
Gypsum	80,000
Nitrogen from biosludge	800
Phosphorous from biosludge	400

(Adapted from Grann, 1997)

4.2 Styria, Austria

In the early 1990's, Erich Schwarz of Karls-Franzens University, Graz in Austria, identified the existence of a large industrial recycling network. The network was not formally recognized by the managers in the industries involved and was only uncovered by Schwarz's research on tracing materials inputs and outputs in the region. By using this methodology, eventually a complex network of exchanges involving over 50 companies was identified. As with the Kalundborg experience, evolution not pre-planning predominated and in many ways the research exercise did little more than to formally recognize and label a 'naturally' developed process. Following on from the Styria example, it is highly likely that there exists similar waste exchange networks throughout the world, which are as yet unrecognized by the various participants.

From the experience of Styria, it would appear prudent, therefore, for far more systematic surveys of local industrial regions to be undertaken in order to help uncover such de-facto industrial ecosystems. Although it can be argued that formal identification could provide both positive and negative results. Recognition of the various interactions may lead to more stable and efficient relationships, but by the same token, also lead to a constraining of natural free form and flexible 'ecological' processes.

For systems similar to Styria to be promoted, then conditions for optimizing materials exchanges are required. Lowe *et al.* (1997) summarized the results of current IE research in this area as highlighting the need for:

▪ awareness and understanding by local companies of the industrial ecosystem concept - there is a clear role for education provision through short courses *etc.*;

- existence of an organizational structure that manages the whole network and assures participants of ongoing support if, for example, a particular by-product supplier leaves the system;
- clear support of government agencies for the IE concept (*e.g.* economic development and environmental protection, especially with regards to the previous point).

4.3   The Burnside Eco-Industrial Park, Canada

A similar exercise to that carried out in Austria, but more geographically focussed, was the study of the Burnside Industrial Park in Dartmouth (Nova Scotia, Canada). The project surveyed 278 existing businesses and collected information on the nature of raw materials used, wastes generated, types of facilities available at the park, attitude towards opportunities to prevent pollution and factors that motivate companies to change their attitude towards better environmental practices (Cote and Hall, 1995). The survey covered over 5100 full-time and 480 part-time employees representing, one-third of the Burnside workforce.

The project focussed on several key issues including:

- encouraging waste exchange through co-location of companies and by establishing a waste exchange database;
- attracting decomposer and scavenger companies for use of second-hand material;
- integrating buildings in the park with the surrounding environment by using passive solar heating and wetlands to filter runoff and sewage;
- providing participating companies with information on material and energy inputs and outputs;
- developing feedback between companies, park management and regulatory agencies.

The survey revealed that encouragingly:

- 90.4 % of the companies were willing to participate in cooperative waste reduction mechanisms;
- 95.4 % would like to be informed of opportunities to improve efficiency and minimize waste;
- 92.3% would support opportunities that make use of their waste for productive and environmentally acceptable activities.

But, in terms of actual practice, of the industries surveyed only:

- 33% had energy conservation measures;
- 22% had considered use of alternate sources of energy;
- 36% had considered using substitute materials to reduce waste;
- 45% had contemplated alternative processes to cut down on waste (although recycling was taking place in 75% of the industries).

Although the Burnside project is based around one well defined industrial park, the approach taken could be readily adapted to clusters of commercial zones, and importantly include (the often forgotten) 'mobile' companies trading in second-hand goods. As in any ecosystem, adaptable and flexible scavengers and decomposers are perhaps the single most vital component to ensure that the loop is closed.

4.4   The Xerox Asset Recycle Management (ARM) Program

IE concepts can be also applied to the functions of a single corporation, of which the Xerox Corporation's ARM initiative, is probably one of the most successful and studied examples. This initiative reflects the equipment design strategies of product-life extension and the primary focus on customer service not products. According to Xerox, ARM is intended to provide the necessary leadership, strategy, design principles, operational and technical support to maximize global recycling of parts and equipment, resulting in a major competitive, as well as environmental advantage for Xerox.

This mission has been built into the company's global organization with an ARM Vice President responsible for achieving 100% recyclability of all manufactured parts and assemblies. A key aim is remanufacturing to high quality standards and resale to new users in order to extend the life of equipment several-fold and reduce demand on virgin resources. The initiative, backed by an extensive customer service program, is designed to streamline the process by which returned machines are reconditioned, thereby increasing return on investment. As a result, the company has estimated that it has added hundreds of millions of dollars to its bottom line since ARM was formally started in 1991.

5. HOW IS INDUSTRIAL ECOLOGY VIEWED BY INDUSTRY

Research and activity based on the IE paradigm is to be primarily found in Europe and North America, but even there, has been only developed in a relatively limited number of locations. Nevertheless, the concept is attracting global attention particularly in developed countries, not least as it is seen as a vehicle to help bring together and advance existing methodologies and tools for enhancing industrial sustainability.

Australia is a prime example of a developed industrial society with a strong environmental consciousness and where IE could be usefully employed as a model for encouraging sustainable development. It is a large country with a relatively small population (around 19 million) and has a strong commitment to a progressive growth economy whilst protecting its diverse environment, and regards environmental sustainable industrial development as the desirable goal. The country has a strong history of production and processing of traditional commodity products (minerals and agriculture), as well as growing hi-tech and service (tourism) industries. The emphasis is very much on ensuring that the industry base, whilst being allowed to prosper, does not negatively impact on what are widely regarded Australia's greatest assets, its wide-range of unique on-shore and off-shore ecosystems.

A key factor that could benefit from IE would be better energy usage due to the country's predominant dependence on fossil fuels and its subsequent relatively high contribution to green house gas emissions. There is strong popular and legislative pressure to meet its obligations under the Kyoto Protocol. That is to limit its green house gas output by 2010 to 108% of 1990 emission rates (which could be considered a relatively modest target considering other developed countries have agreed to actual reductions of 5-20%).

However, IE essentially remains a concept residing in academic and government circles, and so far making little on no inroads into systematic planning of industrial activity, despite significant public support for implementation of environmental protection programs. As a consequence, the country serves as an ideal location to research how best to encourage a movement towards sustainable ecosensitive industrial activity through IE, by assessing current attitudes in order to determine what factors need to be in place to promote adoption.

Work was carried out, therefore, to see how traditional industrial sectors currently view their activities within the environmental management framework, with a specific a view as to identifying what could help promote greater adoption of industrial ecology principles. As a first stage to help achieve this, questionnaires aimed at gaining an insight into existing environmental management attitudes and procedures were distributed to 500 manufacturing companies throughout Australia (Krrishnamohan, 1999). These covered five industrial sectors (Table 2) generally regarded as major sources of energy consumption and waste generation, and associated pollution problems, within Australia (Herat, 1994).

Of the 500 questionnaires sent out, a final return of 132 (26.4%) was obtained (Table 2). These 132 companies had a total workforce of over 100000, of which 85% were in full-time employment. They were asked to self-classify their company depending on their production capacity with respect to their own industry sector. As a result, overall 10% considered themselves small-scale, 62% medium-scale and 28% large-scale.

Table 2. Number of responses to the questionnaire by industry sector

Industry sector	Responses	Relative % of total responses
Food products	57	43.2
Paper and allied products	10	7.5
Chemical and allied products	29	22.0
Petroleum and coal products	5	3.8
Fabricated metal products	31	23.5
Total	*132*	*100*

All the companies that responded had identified personnel responsible for environmental management (EM), but overall only 48% had a separate environmental management department (EMD). With respect to company size, 17 % of small, 49 % of medium and 54 % of large-scale companies had an EMD, figures in line with the expectation that with an increase in company size, a dedicated EMD becomes more supportable and indeed necessary. Among the industry sectors, the presence of an EMD was found to be highest in the chemical industry (69%), followed by petroleum and coal (60%), food processing (44%), paper manufacturing (40%) and lastly the metal fabrication at 36%.

The company attitude towards EM will be a significant factor in deciding the environmental policies of an industry and specifically, the possibility of encouraging application of IE. With respect to EM attitude, six predefined statements were supplied (Table 3) and the respondents asked to choose the single most relevant one describing their company.

Table 3. Main company attitude towards Environmental Management (EM)

Attitude towards EM	Relative response (%)
Our company considers EM as good business ethics	38
Our company considers efforts for EM as proactive and not reactive	25
Our company considers investment in EM as worthwhile and rewarding and therefore has invested in environmental management	12
Our company considers EM as a social responsibility	12
Our company addresses EM issues only when required by legislation	10
Our company doesn't need to address any environmental issues	3

The chemical industry was the only sector in which more respondents (43%) chose a 'proactive and not reactive' attitude over 'good business ethics'. This response reflects that the chemical industries probably run the most serious risk of major (high profile) environmental incidents, an issue best tackled by adopting proactive prevention measures, rather than reactive clean-up. (Company attitude was also analyzed (using cross-tabulations) with respect to company size, but there was no discernable linking pattern).

The actual reasons for EM adoption by the company were then sought by asking respondents to select from one or more of six predefined provided options. Table 4 presents the results by sector, but overall, the most common selection was 'legislative requirement' (61%), followed by 'environmental consciousness' (52%), 'social responsibility' (43%), 'corporate image' (42%), 'economic benefits' (41%), 'pressure from surroundings' (19%) and 'other reasons' (12%). This is in contrast to the previous answers in Table 3, where only 10% of respondents indicated 'legislative requirements' as underlying their company's attitude towards EM. Therefore, irrespective of a company's own individual corporate attitude, legislative requirements are likely to dominate decisions behind the actual adoption of environmental practices. The one clear exception was the petroleum industry, where corporate image clearly predominates, again most likely a refection of preempting the potential negative impact arising from high-profile incidents.

Table 4. Reasons for adoption of EM by industry sector

Reason	Industry sector				
	Food	*Paper*	*Chemical*	*Petroleum and Coal*	*Fabricated metal*
Legislative	79	70	52	40	53
Economic benefit	53	50	41	60	23
Corporate image	44	50	48	100	23
Environment awareness	49	30	69	80	43
Social responsibility	44	30	66	80	33
External pressure	32	10	21	20	3

5.1   Awareness of Industrial Ecology and its components

Respondents were also asked if they had come across the term Industrial Ecology before seeing the questionnaire (Table 5). Overall, around a third of the respondents were aware of IE before the questionnaire most aware. Not surprisingly, however, irrespective of sector, respondents whose company had a proactive attitude towards EM were most aware of IE (50%).

Table 5: Awareness of IE by industry sector

Industry sector	No. of industries	Awareness of IE (%)
Food	57	33
Paper	10	30
Chemical	29	38
Petroleum and coal	5	20
Fabricated metal	31	39

Whilst a third of the industries considered themselves aware of the principles of IE, further questions revealed that the majority of all companies were actively participating in one or more programs that can be viewed as components of the IE paradigm. Specifically, they were supplied with the following statements and asked if they participated in one or more of the activities:

- *cleaner production* - implementation of procedures and designs for reducing pollution at source, in contrast to an end-of-pipe control approach.

- *design for disassembly* - facilitate recycling of a product's basic constituents by designing in such a way that it can be easily disassembled for recycling.

- *design for environment* - consideration of all potential environmental implications of a product, to include energy and materials used in manufacture, packaging, transportation, end use, reuse/recycling and disposal.

- *dematerialization* - aim at a decline in weight of materials used in products in order to decrease raw material consumption.

- *waste exchange* - symbiotic relationship between industries in which by-products that cannot be recycled/reused by the company are traded with other companies which are able to use them as raw materials for their manufacturing process.

Table 6. Active participation in component programs of IE.

Reason	Industry sector				
	Food	*Paper*	*Chemical*	*Petroleum and Coal*	*Fabricated metal*
Cleaner production	88	100	79	100	90
Design for disassembly	52	38	40	0	44
Design for environment	54	63	62	100	53
Dematerialization	56	63	61	75	60
Waste exchange	51	50	39	80	58

Clearly from the results given in Table 6, the commitment to proactive actions to improve environmental and economic sustainability is there, not least as evidenced by the embracing of cleaner production. However, there was little evidence of the more holistic approach demanded by IE.

Awareness (or lack of) of IE by all industrial managers and employees clearly influences application of IE. According to one respondent "Industrial Ecology requires the commitment and education of all employees and directors", and to another "...the operators have to be trained in IE". Therefore, access to comprehensive education on the benefits of IE is arguably one, if not the, most influential factor dictating its application by companies.

Information is of paramount importance in the application of innovative concepts, not least for industrial and environmental management systems. As a consequence, the respondents who claimed prior awareness of IE were asked to indicate where they gained this knowledge by selecting from one or more of seven sources (Government Department of Environment (or equivalent), consultants, university (or equivalent), scientific journals, trade magazines, the internet and industrial associations). Fifty respondents answered this question and 17 (34%) identified trade magazines and industrial associations. While for 32%, scientific journals were also a source of information on IE. Other sources, such as the Department of Environment, consultants, universities and the internet accounted for only 24%, 20%, 18% and 4% respectively. Clearly, the current information situation is relatively *ad hoc* and lacking are mechanisms for well-structured dissemination of information on IE.

5.2 What could encourage the practice of IE?

To identify factors that encouraged practice of IE components in companies already practicing them, they were asked to identify one or more influential factors. The results are given in Table 7 and not surprisingly the main encouragement came from economic benefits, with long-term resource conservation receiving a low score. However, considering that waste exchanges are considered a popular mechanism for identifying and initiating potential inter-industry symbiosis, their apparent low availability (or maybe just lack of awareness of their existence) would be disappointing to advocates of such schemes.

Table 7. Factors considered as influencing the application of IE components

Factors	Influential (%)
Economic benefits of in-house waste reuse and recycle	63
Economic problems of waste disposal and emissions	52
Public image	46
Top management involvement in promoting novel EM practices	39
Availability of capital investment	36
Availability of technical expertise	27
Pressure from stakeholders (shareholders, financiers, public)	25
Resource (raw materials) conservation	17
Research outcomes	13
Local availability of facilities (*e.g.* waste exchange networks)	8

The same respondents were then asked to select from 11 statements, the benefits that they perceived from applying holistic IE (as opposed to individual components). The results (Table 8) suggest that anticipated benefits from practicing IE were again mainly directly related to minimization of waste through reuse/recycling, although resource conservation was now seen as much more relevant.

Interestingly, when it then came to identifying what factors are likely to encourage widespread industrial involvement in IE programs (Table 9), changes in regulatory requirements was the most popular choice, which was not evident by the reasons for carrying out IE component activities.

Table 8. Perceived benefits from the application of the IE concept

Benefits	Selected (%)
In-house waste minimization	61
External waste recycling and waste exchange	46
Resource (raw material) conservation	42
Energy conservation	40
Improved process efficiency	37
Enhanced public image	34
Competitive edge	31
Reduced cost of products, thus increased sales	30
Enhanced product performance	11
Enhanced business opportunities	11
Enhanced employment	8

Table 9. Factors considered as helping to encourage adoption of IE by industry

Reason	Industry sector				
	Food	*Paper*	*Chemical*	*Petroleum and Coal*	*Fabricated metal*
Regulatory requirement changes	96	100	96	80	96
Economic incentives (tax breaks)	86	100	78	100	68
Availability of technical expertise	79	67	0	0	68
Public image	70	0	62	0	54
Access to information	61	0	56	0	64
Waste exchanges	52	0	70	0	62
Research	48	0	40	0	54

Finally, to answer the question as to what are currently the perceived obstacles for greater implementation of IE. It is evident that from Table 10 that whilst lack of capital investment just topped the list, lack of adequate information is a real problem, which ties in with both lack of technical expertise and lack of trained manpower. Encouragingly, current government legislation and company policy are not so significant. However, it would be wrong to dismiss the potential (beneficial) impact of legislation on the implementation of IE (other than simply just discouraging waste dumping and end-of-pie solutions). Legislation may, for example, be written in such a way that it encourages waste reuse/exchange only as a last (but nevertheless acceptable) resort. That is, if there is a possibility of waste reduction by the modification of other aspects of the company's ecology, like a change in manufacturing process (DfE, CP *etc.*), then these options have to be investigated before resorting to waste exchange.

Table 10. Obstacles faced by industry in the application of the concepts of IE

Factor	Consider an obstacle (%)
Lack of capital investment	40
Lack of adequate information	39
Lack of access to technical expertise	34
Lack of trained personnel	30
Current legislation	27
Company policies	24

Where IE is applied, it is evident that economic factors have the greatest influence and provide the major benefits, but the major obstacle was not economic (lack of capital investment), but access to information and informed practitioners. This is substantiated by the fact that 65% of the respondents were not aware of IE, but after being supplied some information on the principles, almost 75% then wanted more information. There is, therefore, a pressing need for providing more structured information on IE to manufacturing industries.

6.   CONCLUSIONS

Industrial ecology, can succinctly summarized as:

"...a new and innovative strategy for sustainable
industry. It involves designing industrial systems so as to
minimize waste and maximize the cycling of materials
and energy" (Karamanos, 1995).

The rationale behind IE is to move away from the traditional pollution
prevention (risk reduction) focus of single industrial operations to the
establishing of long-term regional sustainability though creating an
integrated web of beneficial interchanges between two or more organizations
(Table 10).

Table 11. Pollution Prevention versus Industrial Ecology

Criterion	Pollution Prevention	Industrial Ecology
Primary goal	Prevent pollution and reduce risk	Promote sustainability by optimizing resource flow
Focus	Single company	Network of organizations
Core concept	Planning process	Systematic integrated networks
Role of recycling	On-site	On-site and intra-company
Role of Government	Technical assistance	Barriers removal
Mode of evaluation	Cost savings	Cost benefits

(Adapted from Oldenburg and Geiser, 1997)

However, despite the wealth of research and subsequent reports, very few
examples of demonstrable success exist. The over exposure of sites, such as
Kalundborg in Denmark is in danger of being self-defeating as skeptics can
point to a clear failure to carry forward and expand the concept forward
beyond acting as an academic research tool. Equally, the primary reasons for
the successes of Kalundborg are fairly transparent in there being a limited
number of existing (already commercially successful) and non-competing
large companies interacting with a relatively small community. As a
consequence, it was not difficult to get the whole community involved and

committed, as each member would have a direct (*e.g* employee of one of the companies) or indirect (shopholder, family connection to employee *etc.*) stake in the success of the project.

From our study in Australia, where the IE paradigm remains relatively unknown, the single most apparent barrier to stimulating application was lack of knowledge. That is simply, the majority of potential industrial exponents of IE were unaware of the overall concept. Whilst they may practice some of the components, in particular cleaner production, they could not place these actions within a holistic IE system. Clearly, despite the numerous reports available, far too few are reaching, or are accessible (intelligible or convincing) to the potential customers of the concept.

Even when informed of the thinking behind IE, environmental practitioners within industrial organizations raised a list of obstacles and (needed) incentives for application. The main obstacle was, as mentioned above, lack of IE information at the decision making managerial level. Several respondents also voiced the concern that it is difficult to get approval from government environment departments for new 'waste' orientated projects, as such projects are viewed 'suspiciously'.

From this study and other work carried out, to promote IE the following are desirable in order to encourage new, successful and commercially viable projects:

- Provide systematic education aimed at all levels of industrial management (this will be aided if systems that compartmentalize technologies into departments are replaced by a holistic approach wherein interdepartmental programs can be developed).
- Involvement of technical personnel rather than politicians in decision making at the government level for environmental clearance and approval of new projects (as is often the perception, politicians considered as unable to appreciate new concepts by lacking the technical knowledge required to understand a project).
- Clear tax benefits for research and development
- Maintain a high-profile waste exchange register (the service should not be '*gratis*', but at a price that ensures that quality, independent service is provided)
- Support for research to identify possible existing byproduct exchange symbiosis between clusters of companies.
- Once symbiotic IE based relationships are identified, governments may have to take responsibility of helping to negotiate contracts and assist in guaranteeing a consistent supply of wastes/by-products in terms of

quality and quantity (possibly through the set-up of a centrally run quality assurance facility (or network of)).

- Existing clusters or purpose designed eco-industrial parks, should be clearly identified to the community and managed by a corporate structure rather than a bureaucratic structure (seen to be more technical than political).
- Legislation put in place preventing the storage of wastes on company premises.
- Government to provide incentives to companies that can consume waste materials as raw materials (could be in the form of providing education, land for establishing the facility, tax benefits, financial and technical assistance *etc.*)
- Maintain a transparent account that is accessible to all stakeholders.
- Government should encourage 'risk' ventures in addition to projects that have already been proved to be commercially viable. Industries, universities and research institutions should be encouraged to come up with innovative solutions to problems of pollution, depletion of natural resources and energy conservation.

As in Kalundborg, the whole community needs to be involved and convinced from the outset, and crucially kept regularly informed during the operation of the project. Perhaps the most important message to get over is that IE, through creation of a network of cooperating industries, is not only an efficient economic exploitation of resources, but also an excellent urban partner (both economic and environmental).

There is a problem, if not a dichotomy relating as to how to get the message over. As most respondents to the Australian questionnaires alluded to a need for government incentives and directives, then at first sight, State and local Government would seem the obvious choice to take the vanguard in promoting the cause. However, this would then lead to suspicion of motives and a feeling of imposed bureaucratic dogma (backdoor excuse to regulate industrial activity and also penalize pollution yet harder). Hence the reason why many felt that it could only succeed if run by corporate sourced individuals, with a minimum of government intervention (other than presumably suitable tax breaks *etc.*).

What is always needed for countries with potential but no application, such as Australia, is demonstrable success. It would seem that the most likely chance of this success is to review existing operations within a defined region and 'reengineer' them into an IE model. In many cases, this may simply be a matter of renaming existing practice to fit the ecological concept and hence highlighting the benefits accrued. This activity would have to be run in parallel with an extensive education program and the whole project seen from the outset as run by the industrial/commercial concerns, with other players, in particular government, acting solely as facilitators.

A suitable model to follow may, therefore, be that of the Burnside Industrial Park in Canada. Although this project was based around one industrial park, the approach taken could be readily adapted to clusters of commercial zones, and importantly include (the often forgotten) 'mobile' companies trading in second-hand goods. As in any ecosystem, adaptable and flexible scavengers and decomposers are absolutely vital to ensure that the loop is closed.

Readily demonstrable success is most likely where sustainable commercial activity already occurs, the companies take a proactive operational role and the failure of one member will not necessarily have a catastrophic impact on the others within the defined industrial ecosystem.

REFERENCES

Anastas P.T. and Breen, J.J. (1997) Design for Environment and Green Chemistry: the Heart and Soul of Industrial Ecology. *Journal of Cleaner Production* 5(1-2), 97-102.

Ausubel, H.J. (1992) Industrial Ecology: Reflections on a Colloquium. *Proceedings of the National Academy of Sciences of the USA*, 89(3), 879-884.

Ayres, R.U and Ayres, L.W (1996) Industrial Ecology: Towards closing the materials cycle. Edward Elgar Publishing, London.

Brennan, D (1999) Application of Life Cycle Analysis in Developing Cleaner Process - Some Pointers from Case Studies in Desulphurisation.. In *Global Competitiveness through Cleaner Production*, Eds. J.A. Scott and R.J. Pagan, Australian Cleaner Production Association Inc., Brisbane, Australia, 461-468.

Cote, R. and Hall, J. (1995) Industrial parks as ecosystems, *Journal of Cleaner Production*, 3(1-2), 41-46.

Ehrenfeld, J. and N. Gertler. 1997. Industrial Ecology in Practice: The Evolution of Interdependence at Kalundborg. *Journal of Industrial Ecology*, 1(1) 67.

Erkman, S. (1997), Industrial Ecology: a Historical View, *Journal of Cleaner Production*, 5(1-2), 1-10.

Frosch, R.A. 1992. Industrial Ecology: A philosophical introduction, *Proceedings of the. National Academy of Sciences*, USA. 89, 800–803.

Galloway, J.N (1982) Trace metals in atmospheric deposition: a review and assessment. *Atmospheric Environment*, 16(7), 123-130.

Gertler, N (1995) Industrial ecosystems: Developing sustainable industrial structures. MSc Thesis, MIT, Cambridge, USA.

Graedel, T.E., Allenby, B.R. and Linhart, P.B. (1993) Implementing Industrial Ecology, *IEEE Technology and Society Magazine*, 12(1), 18-26

Graedel, T.E. and Allenby, B.A. (1995), *Industrial Ecology*, Prentice Hall, New Jersey.

Grann, H. (1997) *The Industrial Green Game*, National Academy Press, Washington DC.

Herat, S. (1994) Use of cement kilns in managing solid and hazardous wastes: implementation in Australia. PhD thesis. Griffith University, Australia.

Karamanos, P. (1995) Industrial Ecology - An Organizing Framework for Environmental Management. *Total Quality Environmental Management*, 3(1), 73-85.

Keoleian, G.A (1994) *Product life-cycle assessment to reduce health risks and environmental impacts.* Noyes Press, New Jersey.

Khan, A.U. (1999) Cleaner Production Experiences in Major Industry Sectors of Pakistan. In *Global Competitiveness through Cleaner Production*, Eds. J.A. Scott and R.J. Pagan. Australian Cleaner Production Association Inc., Brisbane, Australia, 131-142.

Krrishnamohan, K. (1999) Opportunities and Constraints in the Application of Industrial Ecology to Manufacturing Industries in Australia. MPhil Dissertation, School of Environmental Engineering, Griffith University, Australia.

Lindeman, R.L (1942) The trophic-dynamic aspect of ecology. *Ecology*, 23, 399-418.

Lowe, E.A. and Evans, L.K. (1995) Industrial Ecology and Industrial Ecosystems, *Journal of Cleaner Production*, 3(1-2), 47-53.

Lowe, E.A., Warren, J.L and Moran, S.R (1997) *Discovering Industrial Ecology: An executive briefing and sourcebook*, Batelle Press, Columbus, Ohio.

Lundie, S. and Huppes, G. (1999) Product Assessment based on a range of Societal Preferences. In *Global Competitiveness through Cleaner Production*, Eds. J.A. Scott and R.J. Pagan,. Australian Cleaner Production Association Inc., Brisbane, Australia, 441-451.

Newland, P., Wilczek, E. and Bilsborough, K. (1999) Cleaner Production - the South Australian Experience. In *Global Competitiveness through Cleaner Production*, Eds. J.A. Scott and R.J. Pagan,. Australian Cleaner Production Association Inc., Brisbane, Australia, 153-159.

Oldenburg K.U. and Geiser K. (1997) Pollution Prevention and ...or Industrial Ecology. *Journal of Cleaner Production*, 5(1-2), 103-108.

Pimm, S.L. (1982) *Food Webs*. Chapman and Hall, Publishers, London.

Richards, D.J. (1997) *The Industrial Green Game: Implications for Environmental Design and Management, National Academy of Engineering*, National Academy Press, Washington DC,

Tang, Y.-H., Tseng, R.-K., Chiu, S.-Y. and Le, C. (1999) Promotion of Industrial Waste Minimization through Corporate Synergy Systems in Taiwan. *Journal of. Cleaner Production*, 7(5), 351-358.

Tibbs, H. (1992) Industrial Ecology: An Environmental Agenda for Industry, *Whole Earth Review*, 77, 4-19.

THE AUTHORS

Ashley Scott

Ashley holds the Chair in Environmental Engineering at Griffith University and is the Director of Grifith's Centre for Integrated Environmental Protection, as well as a Director of Queensland's Consortium for Integrated Resource Management. He is active in the area of sustainable development in industry, tourism and government organisations, Vice-President of the Australian Cleaner Production Association Inc. and an invited member of several Environmental Protection Agency reference panels. He has over 100 publications, including editing *Global Competitiveness through Cleaner Production*.

Ian Christensen

Previously, with the State Government's Environment Policy Coordination Department, Ian is now head of Community Health and Safety with Brisbane City Council's Community and Economic Development Division. He coordinates the activities of Environmental Protection, Waste Management, Public Health, Emergency and Disaster Management, and Community Safety. Ian developed the framework from which Council's Environmental Management System was developed, a framework which has since been extended into a model Environmental Management system for enterprises licensed under the *Environmental Protection Act*.

Kanduri Krishnamohan

Krishna has a Masters in Environmental Toxicology from the University of Madras, India and has over ten years work experience in the fields of Environmental Toxicology, Environmental Management and Education. He is currently an Industrial Ecology Research Assistant at the Centre for Integrated Environmental Protection at Griffith University, and is working on a research project to develop materials flow analysis for the service industry using a large Metropolitan City Council as a model.

Albert Gabric

Al is a Senior Lecturer in the Faculty of Environmental Sciences at Griffith University where his research interests include environmental and ecological modelling. He has over 50 refereed journal publications and held visiting scientist positions at the European Union Joint Research Centre, Ispra (Italy) and Scripps Institute of Oceanography, La Jolla, California.

Chapter 3

ISO 14000 and Environmentally Conscious Manufacturing

John F. Affisco
*Department of Business Computer Information Systems
and Quantitative Methods, Frank G. Zarb School of
Business, Hofstra University*

This chapter examines the ISO 14000 environmental standard, the basic components of Environmentally Conscious Manufacturing, and gives some examples of how they may be integrated in support of an organization's environmental policy

1. ISO 14000 – AN INTERNATIONAL ENVIRONMENTAL STANDARD

ISO 14000 is an international environmental standard, completed in the third quarter of 1996, whose goal is to make organizations more attentive to the environment by helping them manage and evaluate environmental aspects of operations. This standard provides a construct for demonstrating commitment to environmental protection, sustainable development, and continual improvement of environmental management. The standard is universal and can be used by all countries and organizations. Special attention has been paid to make the standard accessible to small and medium-size organizations and enterprises because of their increasing importance.

ISO 14000 is composed of two groups of standards. The first group is concerned with organization processes while the second group deals with analyzing and characterizing the environmental attributes of products. The focus of this chapter is on the first of these groups, which consists of a series of five standards. The first two, ISO 14001 and ISO 14004, are concerned with establishing guidelines and principles for the management of environmental matters by organizations through the establishment and operation of an environmental management system (EMS). The remaining three standards are focused on auditing EMS. The specific titles of the five standards are:

ISO 14001 -	Environmental management systems-specifications with guidance for use.
ISO 14004 -	Environmental management systems-General guidelines on principles, systems and supporting techniques.
ISO 14010 -	Guidelines for environmental auditing – General principles of environmental auditing.
ISO 14011 -	Guidelines for environmental auditing – Audit procedures - part 1: Auditing of environmental management systems.
ISO 14012 -	Guidelines for environmental auditing – Qualification criteria for environmental auditors.

We turn our attention to the first two standards - ISO 14001 and ISO 14004. ISO 14001 is the core standard. It presents the specifications for the core elements of the EMS to be used for certification/registration. The general purpose of ISO 14004 is to provide assistance to organizations for implementing or improving an EMS based on the specifications set in ISO 14001. This standard outlines the elements of an EMS and provides practical advice on implementing or enhancing such a system. It also provides organizations with advice on how effectively to initiate, improve or sustain an EMS.

The design of an EMS is an ongoing, interactive process that consists of defining, documenting and continually improving on the required capabilities. Key principles for managers implementing an EMS include:

- Recognize that environmental management is among the highest corporate priorities.
- Establish and maintain communications with internal and external interested parties.
- Determine the legislative requirements and environmental aspects associated with the organization's activities, products and services.
- Develop management and employee commitment to the protection of the environment, with clear assignment of accountability and responsibility.

- Encourage environmental planning throughout the product or process life cycle.
- Establish a disciplined management process for achieving targeted performance levels on an ongoing basis.
- Evaluate environmental performance against appropriate policies, objectives and targets and seek improvement where appropriate.
- Establish a management process to review and audit the Environmental Management System and to identify opportunities for improvement of the system and resulting environmental performance.
- Encourage contractors and suppliers to establish an EMS.

Section 4 of this standard presents the principles and elements of an EMS. Figure 1 depicts the elements and flow of an EMS that conforms to the standard. Each block in the figure coincides with a general principle of EMS. Let us look at this congruence more closely.

The first principle is that organizations should ensure commitment to the EMS and define its environmental policy. The first step in developing or improving an EMS is obtaining a commitment from top management to improve the environmental performance of the organization in managing its activities, products and services. Next the current position of the organization with regard to the environment should be established through an initial environmental review. The process and results of the initial environmental review should be documented and opportunities for EMS development should be identified.

An environmental policy establishes an overall sense of direction and sets the global objective against which environmental performance of the organization will be measured. A good place to begin in developing environmental policy are the guiding principles put forth by a number of organizations such as the Rio Declaration on Environment and Development of the United Nations, the Business Charter for Sustainable Development of the International Chamber of Commerce, and the Responsible Care Program of the Chemical Manufacturers' Association, for example. An environmental policy should consider the following:

- the organization's mission, vision, core values and beliefs;
- requirements of and communication with interested parties;
- continual improvement;
- alignment with other organizational policies (e.g., Quality, Health & Safety); and
- specific local or regional conditions.

The second principle is that an organization should formulate a plan to fulfil its organizational policy. The planning process includes identification of environmental aspects; legal and other requirements; environmental objectives and targets; and environmental management programs.

Environmental aspects are elements of an organization's activities, products and services that are likely to interact with the environment. These aspects must be identified, and an assessment of the significant environmental impacts associated with them must be made. This should be an ongoing process that determines the past, current, and potential impact of the organization's operations on the environment.

The organization must develop procedures to identify and understand all legal and other requirements applicable to its activities, products and services.

The assessment of environmental aspect impacts and an understanding of regulatory requirements are inputs to the process of establishing environmental objectives and targets. An additional input is the relevant findings from environmental reviews. These objectives are the broad overall goals for environmental performance identified in the environmental policy. Environmental targets are specific, measurable environmental performance indicators. These targets may then be set to achieve the environmental objectives within a specified time frame. In addition to being consistent with the environmental policy, the objectives and targets must include a commitment to pollution prevention.

Environmental management programs should be established to achieve the environmental objectives and targets. An environmental management program identifies specific action steps, schedules, resources and responsibilities that are required to achieve the stated targets. Typical of such programs are the Pollution Prevention Pays program of the 3M Corporation, the packaging redevelopment program of McDonalds, and the many municipal waste recycling programs that are springing up in cities around the U.S.A.

The third principle states that for effective implementation an organization should develop the capabilities and support mechanisms necessary to achieve its environmental policy, objectives and targets. These include: providing the human, physical, and financial resources required by the EMS; top management developing structure and assigning responsibilities for the environmental effort; and identifying the knowledge and skills necessary to achieve environmental objectives and providing the appropriate training to all personnel within the organization. Further, the organization must establish procedures to communicate and report internally and externally on its environmental activities.

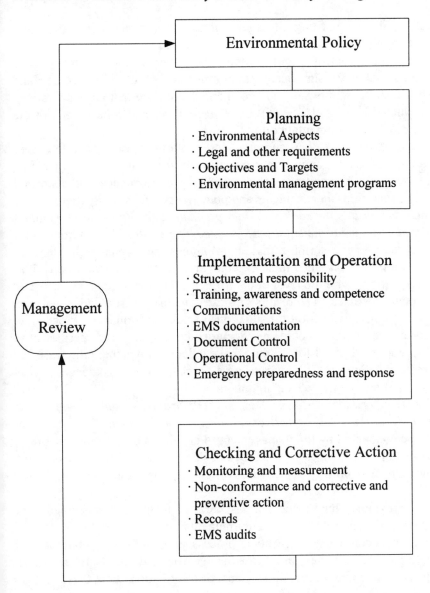

Figure 1. Environmental Management System

Documentation of the EMS is a critical factor in supporting the organization's environmental policy. Typically the key item of EMS documentation is an Environmental Management Manual. This document can serve as a permanent reference to the implementation and maintenance of the EMS. All EMS documents must be strictly controlled so that the information they provide may be used to assist in effectively operating the EMS. Good information management includes means of identification, collection, indexing, filing, storage, maintenance, retrieval, retention and disposition of pertinent EMS documents and records.

Operational control requires that the organization identify those operations that have significant environmental impacts. Procedures and controls must be established and documented to ensure that these operations are conducted at a level of environmental performance consistent with the organization's environmental policy.

Finally, emergency plans and procedures should be established to ensure that there will be an appropriate response to accidents. The procedures would take into account incidents arising as consequences of abnormal operating conditions and accidents and potential emergency situations.

The fourth principle states that an organization should measure, monitor and evaluate its environmental performance. To control the environmental plan a system for measuring actual environmental performance should be in place. Actual performance must be compared to environmental objectives and targets to measure the level of conformance of management systems and operational processes. The results of these comparisons should be analyzed and lead to corrective action for situations of non-conformance. Procedures for corrective action should be documented and revised periodically. Finally, audits of the EMS should be conducted on a periodic basis to determine whether the system conforms to planned arrangements and has been properly implemented and maintained.

The fifth principle states that an organization should review and continually improve its overall environmental performance. This review should be conducted by top management to ensure its continuing suitability, adequacy and effectiveness. As a result of this review changes may be made in environmental policy, objectives and targets, and other elements of the EMS.

As previously mentioned, the last three standards are concerned with environmental auditing. An environmental audit is defined as the systematic, documented verification process of objectively obtaining and evaluating evidence to determine whether specified environmental activities, events, conditions, management systems, or information about these matters conform with audit criteria, and communicating the results of this process to the client. ISO 14010 sets out the general principles for environmental auditing while ISO 14011 fleshes out them out by spelling out acceptable procedures for auditing EMS. ISO 14012 complements these standards by establishing qualification criteria for environmental auditors.

2. ENVIRONMENTALLY CONSCIOUS MANUFACTURING

2.1 From Waste Management to Pollution Prevention

Historically in the U.S.A. the problems of environmental pollution have been attacked by an approach that can be broadly defined as waste management. Waste management includes site remediation, pollution control, and waste handling/disposal. In remediation, techniques are developed for the reclamation of previously generated wastes from land, water, and air, conversion of these wastes into benign substances, and proper disposal. Pollution control uses 'end-of-pipe' measures to prevent discharge of waste. Examples of control techniques include cyclone separators, electrostatic precipitators, fabric filters, and particulate scrubbers. Although effective in preventing direct discharge of waste, these methods are often guilty of simply changing the state of the waste and delaying, not preventing, introduction of the waste into the environment. Pollution control is very costly and does not address the root cause of environmental degradation, which is the production of waste in the first place (Watkins & Granoff, 1992).

Recognizing this shortcoming of pollution control, forward thinking organizations have established the idea of pollution prevention as the most effective way to deal with the problems of the environment. With the Pollution Prevention Act of 1990, the U.S. Congress established pollution prevention as a "national objective" and the most important component of the environmental management hierarchy. Pollution prevention is the use of materials, processes, or practices that reduce or eliminate the creation of pollutants or wastes at the source. It includes practices that reduce the use of hazardous and non-hazardous materials, energy, water, or other resources as well as those that protect natural resources through conservation or more efficient use. Companies that adopt the pollution prevention approach typically find that they reduce both their operating costs and their potential liabilities, in addition to helping preserve the environment.

The Pollution Prevention Act of 1990 reinforces the Environmental Management Options Hierarchy as presented in Figure 2. The highest priorities are assigned to preventing pollution through source reduction and reuse, or closed loop recycling.

Preventing or recycling at the source eliminates the need for off-site recycling or treatment and disposal. Elimination of pollutants at or near the source is typically less expensive than collecting, treating, and disposing of wastes. It also presents much less risk to workers, the community and the environment

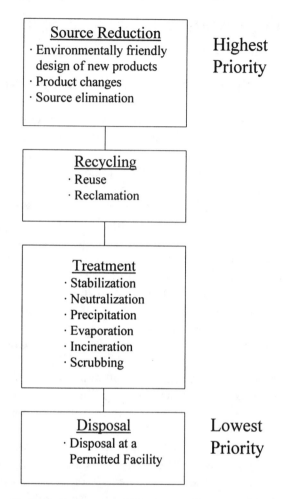

Figure 2. Environmental Management Options Hierarchy

2.2   Environmentally Conscious Manufacturing

Environmentally conscious manufacturing (ECM) is an approach developed by industry to meet contemporary environmental challenges. The approach is based on the premise that pollution prevention rather than waste management is the most environmentally friendly course of action. That is, the foundation of ECM is the acceptance of the environmental management options hierarchy as a blueprint for a manufacturing strategy that is wholly consistent with sustainable development. ECM may be defined as the transformation of materials into environmentally conscious products through a value-added process that simultaneously enhances economic well being and sustains environmental quality. An environmentally conscious product is one that, at the ultimate end of its useful life, passes through disassembly and other reclamation processes to reutilize non-hazardous materials. ECM strategy deliberately attempts to reduce the ecological impacts of industrial activity without sacrificing quality, cost, reliability, performance, or energy utilization efficiency (Darnall *et al*, 1994). Another salient feature of ECM is that it adopts the viewpoint of lifetime stewardship. That is, the manufacturer has 'cradle-to-grave' responsibility for the products it produces. Therefore, environmental factors must be considered in all the stages of a products life cycle from design through manufacture to ultimate disposal.

Figure 3 presents the structure of environmentally conscious manufacturing. The goal of ECM is waste minimization through pollution prevention. ECM seeks to minimize waste from production, mainly materials and energy. The ECM framework is defined by two major categories: Design and Analysis, and Management.

2.2.1   Design and Analysis

The first category, design and analysis, incorporates the ideas of pollution prevention into the initial process and product design. Decisions made during the design and analysis phase have a profound impact on the entire life-cycle of the product, from product proposal, through manufacturing, distribution, installation, servicing, reuse, recycling, to its ultimate disposal. The components of design and analysis are Design for Environment(DFE), Life-Cycle Assessment(LCA), and Process Modification(PM).

2.2.1.1   Design For Environment.

DFE integrates environmental considerations and constraints into existing process and product design practices. For example, some criteria for environmentally conscious products that incorporate DFE are:
- Use renewable natural resource materials.
- Use recycled material.
- Use fewer toxic solvents or replace solvents with an alternative material.
- Reuse scrap and excess material.
- Use water-based inks instead of solvent-based ones.
- Produce combined or condensed products that reduce packaging requirements.
- Produce fewer integrated units (i.e. more replaceable component parts).
- Minimize product filler and packaging.
- Produce more durable products.
- Produce goods and packaging reusable by the consumer.
- Manufacture recyclable final products (US EPA, 1992) .

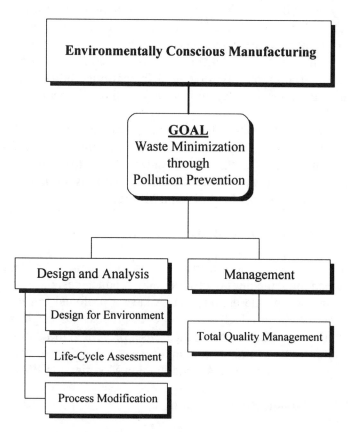

Figure 3. Environmentally Conscious Manufacturing Framework

To be most effective, DFE must be implemented up front in the conceptualization and design stages, when new ideas and design changes are easier and less costly to accommodate.   DFE is most successfully implemented when an organization adopts a concurrent product design philosophy.   Concurrent design embraces the notion of designing for the product life cycle as a source of competitive advantage. Concurrent design incorporates simultaneous product and process design and emphasizes that downstream product criteria should be integrated into the design process. This is accomplished, in large part, by integrated design teams that include all people involved in the product realization process.   Therefore, unlike the traditional linear approach, this design process seeks to facilitate input from numerous functional areas across the company.

In a concurrent design system, the design team can focus greater attention on DFE (Oakley, 1993). DFE strives to design environmental improvements into each stage of the product life cycle. Four stages, Raw Material Extraction and Processing, Manufacture, Consumer Use and Maintenance, and Product Retirement, are associated with a product and its packaging. By systematically considering each of these stages in the design process, environmental compatibility can be designed into the product and its packaging.

The environmental impacts of extracting and processing raw materials should be considered during this stage. Simply using renewable resources or substituting recycled materials in the product for virgin material can measurably reduce the environmental cost of this stage.

By considering the manufacturing process during the design stage, processes and materials that may be harmful to the environment can be avoided. Typical approaches include design for waste minimization, design for energy efficiency, and design for reduced use of toxic materials. For example, products may be designed to help reduce defects and rework thus minimizing wastes.

Design-based initiatives that consider environmental performance in the use of products are responsible for reducing their environmental impact. Examples include federally mandated fuel efficiency and exhaust emission standards that have been imposed on automobile manufacturers, and energy efficiency standards for computers, refrigerators, and air conditioners.

Design for product retirement involves consideration of the following four criteria:

1. Design for Reusability and Recyclability (DFRR)- focuses on reducing a product's impact on the environment when it reaches the end of its useful life. DFRR concentrates on the separability of product components. The more complete separation, the purer the component, hence, the greater its value. A corollary of this is the easier the separation, the more cheaply it can be achieved.

2. Design for Remanufacture - takes into account the fact that parts in a product wear out at different rates, making some discarded products very valuable for remanufacture. Thus, the challenge here is to design products that will allow us to easily recapture this value through design upgrades and manufacture.

3. Design for Disassembly (DFD) – is the process of designing products so they can be easily broken down into like materials for recycling, reuse, or remanufacturing. DFD's intent is to design products for maximum recyclability and efficient disassembly, offering an easier remanufacturing process. Fasteners are a core issue in designing for disassembly. Simple design changes such as leaving a slot on the side of

a product for a screwdriver can greatly speed the disassembly process, while not adding to cost or affecting performance (Darnall *et al*, 1994).
4. Design for Disposal - focuses on ways to ensure that the product can be disposed of safely. If, for example, the product has to be constructed of non-recyclable thermoset plastic, it should be compatible with incineration and other disposal practices.

2.2.1.2   Life-Cycle Assessment.

Life-cycle assessment (LCA) is the process of looking at all the aspects of a product design from the preparation of its input materials to the end of its use. An LCA of the product design evaluates the types and quantities of product inputs such as energy, raw materials, and water, and of product outputs such as atmospheric emissions, solid and waterborne wastes, and end product. The process can be used as an objective tool to identify and evaluate opportunities to reduce the environmental impacts associated with a specific product, process, or activity. The four basic components of LCA are inventory analysis, impact analysis, life-cycle costing, and improvement analysis.

Inventory analysis is a technical, data-based process of quantifying energy and raw material requirements, atmospheric emissions, waterborne emissions, solid wastes, and other releases for the entire life-cycle of a product, package, process, material, or activity (US EPA, 1993). The data are aggregated in uniform units so that they may be totaled for all components and life-cycle processes. The analysis is performed by identifying the process waste for each stage of the product life-cycle, then measuring or estimating its quantity.

Impact analysis builds on the information gathered by the inventory analysis. Impact analysis is the assessment of the consequences wastes have on the environment. It includes both quantitative and qualitative assessments. The analysis should address ecological and human health impacts, resource depletion, and possibly social welfare. Other effects, such as habitat modification and heat and noise pollution are also part of the impact analysis. The key concept in the impact analysis is that of stressors. The stressor concept links the inventory and impact analysis by associating resource consumption and releases documented in the inventory with potential impacts. Thus, a stressor is set of conditions that may lead to an impact. For example, a typical inventory will quantify the amount of sulfur dioxide released per product unit, which then may produce acid rain and which in turn might affect the acidification in a lake. The resulting acidification might change the species composition to eventually create a loss of biodiversity (US EPA, 1993).

Life-cycle costing (LCC) is a methodology in which all costs are identified for a product throughout its lifetime, from raw materials acquisition to disposal. This method of cradle-to-grave product accounting attaches a monetary figure to every effect of a product: landfill costs, potential legal penalties, degradation of air quality and so on. LCC should be performed before the product is manufactured so that changes in its design can be made easily. LCC requires establishing procedures to keep track of the expenses associated with the harmful by-products of manufacturing at each step in the value chain. As environmental costs are identified, managers can more effectively evaluate the full spectrum of choices and the costs and benefits of business actions that minimize waste and environmental liability (Darnall *et al*, 1994).

Improvement analysis is the evaluation and implementation of opportunities that bring about environmental improvements. Improvement analysis systematically documents periodic reviews of a facility's operations, ensuring waste minimization and pollution prevention. Improvement analysis is essentially the practice of environmental audit. This step in LCA ensures that any potential waste reduction strategies are optimized and that improvement programs do not interfere with human health or the environment. Improvement analysis identifies problems, measures progress over time, and offers a means for continuous improvement. This process is typically the driving force behind design and process changes to minimize waste (Darnall *et al*, 1994).

Information technology has proven to be a useful improvement analysis tool. Decision support systems are particularly useful in tracking materials through the product life-cycle. For example, such systems have been developed which track new vendor products that use environmentally friendly raw materials. Expert systems have been developed to evaluate disassembly procedures, and to identify waste minimization opportunities for industry.

2.2.1.3 Process Modification.

An ECM process modification is any change to a current process for the purpose of reducing scrap or by-products that are the result of poor quality operations. ECM process modifications typically fall into two categories: input substitution and reuse/recycle improvement.

Typical input substitutions include the use of a less hazardous or toxic solvent for cleaning or as a coating, the purchase of raw materials that are free of trace quantities of hazardous or toxic impurities, and limiting the variety of plastics used in a manufacturing process to make recycling easier.

Typical reuse/recycle improvements include the redesign of product packaging, and the segregation of waste streams to avoid cross-contaminating non-hazardous materials with hazardous ones. For some products, packaging costs may be as high as 70 percent of the product cost.

Many companies now limit packaging to what is necessary for safe transport. Many more use recycled materials in their packaging. Other companies are reusing polystyrene and foam used as packaging materials by suppliers. Printed circuit board manufacturers have segregated various rinse water streams to allow for their reuse and the easier reclamation of toxic materials.

2.2.2   Management

It is now clear to management that environmental responsibilities have become a strategic issue. That is, in the global economy of the new millennium, successful organizations will recognize that environmental responsibilities must have a place in strategy development along side manufacturing, marketing, and financial issues. ECM suggests that the way to achieve this goal is through the application Total Quality Management (TQM) to the environment. TQM is an evolving system of practices, tools and training methods for managing companies to provide customer satisfaction in a rapidly changing world. The application of TQM to environmental management has been coined Total Quality Environmental Management (TQEM).

The emphasis in TQM is on continuous improvement of business processed. This improvement manifests itself in a number of areas: eliminating product defects, enhancing attractiveness of product design, speeding service delivery and reducing costs, among others. In TQEM the emphasis is on improvement of environmental quality through pollution prevention. The idea is that processes and products that generate pollution and waste result in unnecessary costs and undesirable re-work. Therefore, prevention of environmental degradation through continuous improvement of products and processes is a foundation of TQEM. Therefore, TQEM can form an essential component of ECM, where constancy of purpose for improvement of process, product, and service is its guiding principle (For a more complete discussion of TQEM, see Affisco, 1998).

3.   ISO14000 AND ECM – SOME COMMON FEATURES

In reviewing the ISO 14001 standard, one can see a distinct strategic focus. In fact, all the major elements of strategic planning – policy, external and internal analyses, objectives and goals, programs, performance measurement and control, and management review – are present in the EMS model. The focus here really is on the *quality of environmental management.* An optimal audit result would show an effectively functioning EMS in pursuit of an environmental policy based on pollution prevention.

On the other hand, ECM is essentially programmatic in nature. That is, its focus is the development and implementation of programs aimed at waste reduction through pollution prevention. As we have seen, such programs may include DFE, LCA, and process modification. From a reading of the ISO 14001 standard, it is apparent that all of these programmatic options have been considered in its development. The clearest indication of this is the inclusion of ECM in the planning and implementation phases of an EMS. Indeed at least two of the key principles for managers implementing an EMS indicate the place of ECM in an EMS. As listed earlier in this work they are: (1) Encourage environmental planning throughout the product or process life cycle; and (2) Establish a management process to review and audit the EMS and to identify opportunities for improvement of the system and resulting environmental performance.

In the following sections we take a more specific look at the relationship between ISO 14000 and EM.

3.1.1 Environmental Aspects and LCA

To illustrate the integration of ECM into an EMS, we first look at the planning phase of an EMS. An important component of this phase is the identification of environmental aspects. Environmental aspects are elements of an organization's activities, products, and services that can interact with the environment. What makes an environmental aspect is whether the activity, product or service creates a potential for an environmental impact under reasonably foreseeable conditions. Although there is a connotation that environmental impacts are usually negative, environmental aspects can also be positive. For example, activities such as paper recycling or water conservation may result in positive environmental impacts. Essentially, in identifying environmental aspects an organization is trying to answer two questions, "What is our environmental situation?" and "How does what we do affect the environment?"(Cascio, 1996)

Environmental aspects can be classified into two broad categories. The first is environmental aspects directly regulated by applicable laws and regulations. This category includes regulations on process and plant operations, regulations on products and services, and regulations regarding packaging and transportation. Examples of regulations on operations include facility air emissions governed by local, state, and national air pollution control regulations and emission limits, regulated wastewater discharges, and hazardous waste management regulatory requirements, including requirements that govern reporting and remediation of soil and groundwater contamination. Environmental attribute product regulations are appearing more regularly across the world. In addition to the typical gas mileage regulations for automobiles produced in the U.S., there are regulations on the amount of phosphates in laundry detergent, energy use by computers, and on product disposal. Finally, regulatory requirements for the packaging and transportation of hazardous materials exist to prevent the release of hazardous materials during transportation, and in the event of transportation accidents, to minimize damage to human health and the environment.

The second class of aspects is unregulated environmental aspects. Unregulated environmental aspects fall into two general categories. There are those that are regulated up to specified limits, and those that are not regulated at all. In the first class is the limited regulation of nitrous oxides. Typically a facility is allowed to release nitrous oxides up to a prescribed limit. As long as the nitrous oxides emissions do not exceed the regulatory limits, the facility is in full compliance. But nitrous oxides emissions released while in full compliance nevertheless represent an environmental aspect of your operating processes. Nitrous oxides releases legally below regulatory limits may provide an opportunity for further environmental performance improvements. In the second class, examples of totally unregulated environmental aspects include water and energy consumption, carbon dioxide releases, and wastes such as many forms of non-hazardous solid wastes, including packaging and office paper waste.

After all environmental aspects have been identified, their impact on the environment must be assessed. ISO 14001 defines an environmental impact as "any change to the environment, whether adverse or beneficial, wholly or partially resulting from an organization's activities, products or services." Further, the significance of each of the aspects and the corresponding impacts must be determined. In this process of identifying environmental aspects and their impacts, the strong presence of the components of LCA is evident. Essentially the identification of environmental aspects is an implementation of the first three components of LCA – inventory analysis, impact analysis, and life-cycle costing.

We illustrate the process of identifying environmental aspects for the case of a manufacturer of metal storage systems. The manufacturing process for these products includes operations typical of a metal working facility. Some of these are stamping, forming, cleaning, drilling, welding, painting, staging, final assembly, packaging, and shipping. There are environmental aspects associated with each of the processes in the manufacturing flow. For example, we focus on the painting process.

In the painting process, cabinets and drawer parts are painted in a robotic paint facility. Prior to painting all metal surfaces are cleaned using an appropriate solvent. The paint facility uses both oil-based and water-based paints. For oil-based paints, polyurethane thinner is used for paint thinning and equipment cleaning. A thinner containing isopropyl alcohol and xylene is used with water-based paint. Parts move through the paint operation on a conveyor system. The paint operation concludes with some time in an oven to dry and harden the paint. Cabinets and parts may be painted to order in approximately twenty different colors.

Table 1. Environmental Aspects for Painting Process

Origin	Environmental Aspect	Environmental Impacts
Process Inputs		
Solvents	Storage – Tank evaporation; spills; system leaks; damaged containers.	Contamination of air quality. Contamination of soil, water.
Paints	Storage – Tank evaporation; spills; System leaks; damaged containers.	Contamination of air quality. Contamination of soil, water.
Cabinets	Storage and/or transport damage.	Contamination of soil, water. Solid waste disposal.
Parts	Storage and/or transport damage.	Contamination of soil, water. Solid waste disposal.
Process		
Cleaning	Solvent Emissions – evaporative losses. Cleaning solvent excess. Product changeover solvent wastes.	Contamination of air quality. Degradation of work environment – Impact on human beings. Contamination of soil, water. Liquid waste disposal.
Painting	Excess paint or overspray. Paint evaporation.	Contamination of air quality. Degradation of work environment – Impact on human beings.
Drying	Excess paint evaporation. Excess solvent evaporation.	Contamination of air quality. Contamination of air quality.
Outputs		
Painted Cabinets	Excess paint. Damage in moving to staging area.	Solid waste disposal. Solid waste disposal.
Painted Drawer Parts	Excess paint Damage in moving to staging area	Solid waste disposal. Solid waste disposal.

An initial attempt at identifying the environmental aspects associated with the painting process is made with the help of the matrix presented in Table 1.

The first column of the matrix takes a simple systems view of the painting process by breaking it into its component parts – input, process, and output. For each phase, the corresponding materials, activities, and products are listed. In the second column of the matrix, the environmental aspects associated with each item in the first column are identified. In the third column, potential environmental impacts for each of the aspects are given. To complete the analysis the relative significance of the impacts should be evaluated. To arrive at this measure the severity of the impact and its frequency of occurrence should be considered. This would be accomplished by collecting data using the methods of life-cycle inventory analysis. A final determination of the environmental impact can then be based on the data obtained as a result of the inventory analysis. In fact, the International Organization for Standardization has recognized the importance of LCA to an EMS and is working on a series of supporting guideline standards for LCA.

An interesting approach to classifying the environmental impact of materials has been developed by Polaroid Corporation. Polaroid's Toxic Use and Waste Reduction program was voluntarily developed to reduce toxin use and waste sources as a means of preventing pollution. A critical element of this program was to assign environmental impact categories to the chemicals that Polaroid uses in its production lines. After evaluating each material based on toxic characteristics, physical attributes, and chemical properties, Polaroid assigns it to one of the following categories:

1. Known human carcinogens, teratogens, and toxic reproductive agents; highly acutely toxic; or a great environmental threat;
2. Known animal carcinogens, teratogens, and toxic reproductive agents; chronic toxicity; or an environmental threat;
3. Suspected animal carcinogens, moderately-toxic chemicals, or corrosive materials;
4. Chemicals that cannot be classified in 1,2, or 3.
5. Other materials such as plastic, paper, and cardboard.

Based on this classification, Polaroid targets chemicals for either reduction (1 and 2) or recycling (3,4, and 5). As new chemical information becomes available, Polaroid evaluates and reclassifies its chemicals as appropriate. New chemicals are assessed before they can be introduced into production lines. Polaroid uses incentive plans to encourage the reduction of Category 1 and 2 materials and the recycling of category 3 and 4 materials in production lines. In addition, reduction and recycling goals are factored into each program manager's performance evaluations.

Although various regulations have established chemical lists, none meet the requirements of Polaroid. By comprehensively addressing and grouping chemicals specifically for its activities, Polaroid ensures that all its materials are included. Managers have a reliable and comprehensive source for identifying which chemicals should be eliminated and which can be managed through recycling (BMP, 1999).

3.1.2 Environmental Programs , DFE and Process Modification

According to ISO 14000, the planning phase of an EMS requires the development of environmental management programs. Environmental management programs should address schedules, resources and responsibilities for achieving the organizations environmental objectives and targets. An environmental management program identifies specific actions in order of their priority to the organization. These actions may deal with individual processes, projects, products, services, sites or facilities within a site. Environmental management programs should be dynamic and revised regularly to reflect changes in organizational objectives and targets.

DFE and process modification are major building blocks of the majority of environmental management programs. Some examples from a number of organizations and industries follow

3.1.2.1 Polaroid Corporation.

Polaroid typically develops dozens of new products each year. Polaroid has used a formalized structured process for product development called Product Delivery Process (PDP) since the late 1980s. The process focuses on seven steps: idea exploration; concept; feasibility; product development; design pilot; manufacturing pilot; and commercialization. In 1992 Polaroid modified PDP by integrating the process with DFE elements and manufacturability efforts.

Specific DFE element changes to PDP were included in the concept and feasibility steps. Additions to the concept step included assessing environmental issues; examining environmental impact by the development program; identifying potential chemical, hardware, and packaging issues; and assuring that the product and its production comply with Polaroid's environmental objectives. For example, Polaroid eliminated ozone-depleting substances by removing the Teflon coatings on the friction points in its Captiva camera. Through its efforts to reduce the amount of silver needed for film processing, Polaroid developed a medical imaging product that was completely silver free. Additions to the feasibility step included specifically looking for opportunities to eliminate environmental problems; identifying substitutes for targeted chemicals; and examining the product for maximum usage of post-consumer waste. For example, Polaroid was able to use approximately 63% of post-consumer waste content in its corrugated product packaging (BMP, 1999).

3.1.2.2   Northrop Grumman Corporation.

In the late 1980s, Northrop Grumman was spending nearly $7 million to dispose of its 7,500 tons of hazardous wastes. In 1990, an environmental corporate goal was set to achieve a 90% reduction in hazardous waste disposal by1996. As a result of environmental efforts the goal was achieved in many cases and surpassed in others. Several approaches used to achieve this pollution prevention objective included changing processes or equipment; substituting chemicals; eliminating hazards; and reducing toxicity and emissions. Key process modifications included:

— On demand, multi-component paint dispensing machines to reduce the amount of paint purchased for touch-ups as well as excess waste generation;
— Specialized HEPA filter banks on paint booths to reduce chromate emissions;
— Replacement of photographic chemistry with electronic imaging;
— Replacement of air agitation with eductor-assisted fluid agitation;
— Use of solvent distillation to aid in solvent recovery;
— Replacement of large, wasteful, open-topped solvent tanks with enclosed spray gun cleaning machines;
— Creation of specialized tables to limit the amount of alodine wasted during specific touch-ups
— Replacement of chromate processing chemicals with non-chromated equivalents.
— Building on this initial success, Northrop Grumman set a new objective of eliminating 50% of the remaining hazardous waste by the year 2001.

The search for improved technology and environmentally improved processes continues. Additionally, aggressive steps have been taken to reduce non-hazardous solid waste at all Northrop Grumman facilities (BMP, 1999).

3.1.2.3   Computer and Electronics Industry.

Over the past ten years the computer and electronics industry has taken a number of actions toward improving its environmental performance. These include the following DFE and process improvement activities:

— In-line chemical reprocessing systems that reduce chemical usage;
— Closed loop cleaning processes without emissions;

- Systems that alarm higher than normal chemical usage or waste generation based on statistical process analysis;
- Use of water-based, biodegradable materials instead of solvents such as methylene chloride in the production of printed wire boards;
- Use of an aqueous solution instead of ozone depleting chemicals in the board cleaning process;
- Use of a "no clean" process which eliminates the need for the cleaning step in electronic board assembly;
- Elimination of metal barcodes or other metals in computer packaging to aid in plastics recycling;
- Identification and use of recyclable materials in computer packaging;
- Use of recyclable field replacement materials;
- Use of a design which makes disassembly easier, so that parts and components can be recycled;
- Use of low-energy power supplies whenever practical (Woodside & Prusak, 1994).

3.1.2.4   General Motors and Steel Dynamics.

A final example is the changes that have occurred in the manufacture of steel due to the emergence of the mini-mill. Today mini-mills using electronic arc furnace technology manufacture new steel from recycled scrap steel. Statistics indicate that 98% of the steel manufactured by mini-mills is made from recycled steel whereas the same figure for integrated mills is 28.8%. Historically there has been open loop recycling of steel. That is, a recycled car could end up as a can, or vice versa. This creates some quality problems for the mini-mills. One solution is to create a closed loop recycling system, where cars end up being recycled into cars, etc. To achieve such a result, General Motors and Steel Dynamics have entered into a scrap steel recycling agreement. According to the agreement Steel Dynamics will recycle General Motors scrap and sell it back to General Motors as flat hot rolled steel at significant savings. This will improve the quality of the flat hot rolled steel based on existing steel scrap from General Motors fabricating plants where all scrap is from prime steel and sorted, and therefore, known and controlled (Cole, 1999).

4.   CONCLUSION

In this chapter we have examined the ISO 14000 environmental standard and the evolving practice of ECM. Programs for continuous improvement of environmental quality are at the heart of the ISO 14000 standard. The practices of ECM as part of a certified EMS offer great promise to achieve the marked improvement of environmental quality. As of July 1999, some 10,700 firms were certified to the IS 14001 standard. The largest percentage of these was in Asia and Europe. As with ISO 9000 certification, the U.S.A. has lagged behind many of its global neighbors. Mindful of the increasing importance placed on sustainable development by the peoples of the world, the rate of new certifications to ISO 14001 for U.S. companies has sharply increased in the first two quarters of 1999 to the point where 480 firms are certified (ISO 14000, 1999). Undoubtedly, ECM has made a signification contribution to this certification effort. Continuing research into new environmental technologies and better analytical modeling, along with an increased acceptance of environmental management as a strategic organizational concern, will go a long way to the increased acceptance of ECM. This will ultimately result in more organizations with an EMS in place, which, in turn, will be eligible for ISO 14000 certification.

REFERENCES

Affisco, John F. (1998). TQEM-methods for continuous environmental improvement, In C.N. Madu (Ed.), *Handbook of Total Quality Management*, The Netherlands: Kluwer Academic Publishers.

Best Manufacturing Practices (BMP) (1999). Retrieved 8/8/99 from http://www.bmpcoe.org.

Cascio, J. (Ed.) (1996). *The ISO 14000 Handbook*, Farirfax, VA: CEEM Information Services with ASQC Quality Press.

Cole, S. (1999). *The Competitive Advantage of Steel Mini-mills: A Successful Implementation of the Total Quality Philosophy*, Unpublished Master's Thesis, Kennedy-Western University.

Darnall, N.M., Nehman, G.I., Priest, J.W., & Sarkis, J. (1994). "A Review of Environmentally Conscious Manufacturing Theory and Practices." *International Journal of Environmentally Conscious Design and Manufacturing*, 3(2), 49-57.

ISO 14000 (1999). Retrieved 8/8/99 from http://www.iso14000.com.

Oakley, B.T, (1993). "Total Quality Product Design-How to Integrate Environmental Criteria into the Product Realization Process." *Total Quality Environmental Management*, 309-321.

US Environmental Protection Agency (US EPA) (1992). *Facility Pollution Prevention Guide*, EPA/600/R-92/088.

US Environmental Protection Agency (US EPA) (1993). *Life-Cycle Assessment: Inventory Guidelines and Principles*, EPA/600/R-92/245.

Watkins, R.D & Granoff, B. (1992). "Introduction to Environmentally Conscious Manufacturing." *International Journal of Environmentally Conscious Design and Manufacturing*, 1(1), 5-11.

Woodside, G. & Prusak, J.J. (1994). "Design for the Environment: A Challenge for the Computer Industry." *International Journal of Environmentally Conscious Design and Manufacturing*, 3(2), 3-5.

AUTHOR

Dr. John Affisco is Professor of Business Computer Information Systems and Quantitative Methods at the Frank G. Zarb School of Business and Director of Instructional Technology Integration for the Zarb School. He received a Ph.D. in Business from the City University of New York Graduate School. His work has appeared in a variety of academic and professional journals including *The International Journal of Production Research, European Journal of Operational Research, Simulation and Gaming, and the International Journal of Quality Science*, among others. Dr. Affisco has also contributed chapters to *The Handbook of Total Quality Management* and *Management of New Technologies for Global Competitiveness*.

Chapter 4

Communicating Product Recovery Activities
Processes, Objectives and Performance Measures

Geraldo Ferrer and D. Clay Whybark
The Kenan-Flagler Business School
The University of North Carolina at Chapel Hill

This chapter aggregates and clarifies various expressions in the product recovery literature. In some instances, practitioners and researchers have used these expressions with conflicting meanings. We intend to facilitate communication by establishing a common usage for the product recovery vocabulary.

1. INTRODUCTION

The production of durable goods represents an important contribution to the GNP of all developed countries. It employs large amounts of human resources, raw materials and energy. Most of the resources consumed are not renewable. The minerals that provide most of the raw materials and energy in the production of durable goods have been continually depleted since the beginning of the Industrial Revolution. Several scientists have shown evidence that even renewable resources have been consumed beyond Nature's ability to renew them. Examples include the exhaustion of fisheries and the increased demand for fertilisers and pesticides just to keep agricultural production at constant levels. While resources are depleted, many durable products are discarded in landfills at the end of their useful lives. Landfill space has been decreasing, and the price charged by those landfills still in operation has increased rapidly.

Hence, the production of durable goods is associated with the development of two major problems: the exhaustion of landfill space and the depletion of non-renewable resources. (Purser et al., 1995) and (Hart, 1995) have proposed an ecocentric rather than an anthropocentric approach to management. This would require a change in priorities for which most firm are not yet ready. The first significant step in that direction was the development of recycling programs in many countries in the world. (Biddle, 1993) reports the success that many firms have had with recycling their used products. However, recycling faces numerous technological barriers. That raises questions about remanufacturing, reuse and other product recovery methods. In the last few years, there has been an increased interest in product recovery activities, as can be witnessed by the number of articles, conferences and reports from small and large corporations engaged in these activities. *Figure 1* shows the material flow in an ideal recovery process using today's technology.

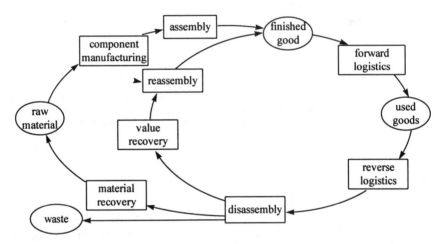

Figure 1. Recovery cycle of a durable good

One of the difficulties in communicating these developments is the variety of expressions that have been used to describe different practices. To some extent, this variety is natural given the wealth of vocabulary in the English language. Nonetheless, a certain lack of precision in the choice of words may lead to unnecessary misunderstandings. Communication could be improved by establishing a common definition for the main expressions in the jargon. (Thierry et al., 1995), (McConocha and Speh, 1991) and (Ferrer, 1997b) have provided some definitions that describe different recovery processes. This chapter addresses this issue by attempting to clarify and codify them. Moreover, it identifies practical issues that are specific to certain product recovery options, as well as research opportunities that still have to be addressed towards the development of a sustainable industrial ecology.

Section 2 discusses processes associated with the recovery of raw materials, and section 3 discusses value recovery methods. Section 4 describes the operating processes usually appearing in a product recovery operation. The output of these processes is only as good as the firm's ability to market the recovered good. Section 5 discusses marketing processes associated with recovered products and materials. It highlights the many opportunities that still have to be addressed to efficiently market recovered goods. Section 6 defines certain design objectives that have been mentioned in the literature, sometimes without the necessary clarity. In addition, it includes performance measures that evaluate the recovery fitness of a given product design. Appropriate nomenclature is proposed.

2. MATERIAL RECOVERY PROCESSES

Material Recovery is the process of recovering the material (or intrinsic) value in the product with the purposing of **recycling**. It is a destructive process with complete loss of the original form for which the product was designed.

Recycling is the term that should be used to describe a material recovery process. Recycling processes recover the original material used to make finished products. Once in the recycling stream, components lose their functional form as the process recovers the material used to make them. It is not uncommon to confuse product collection with recycling. For example, municipalities establish "curbside recycling" to reduce municipal solid wastes. However, collection processes merely make products available to recycling or other recovery options. It does not ensure that the collected used good will be recycled as intended.

Recycling is more efficient when the goods recycled are made from fewer materials. The process requires that components made of different materials be easily separated. Products made of complex composites are harder to recycle. If separation is too costly, it is unlikely that the recycling process will be economically interesting. Hence, the most common recycling processes involve ferrous, non-ferrous and precious metals. Other commonly recycled materials are glass and paper. Often a material is recycled as a matter of public policy, to reduce the amount of waste sent to landfills. This is the case for paper, plastic, tires and fabrics. The economics of recycling these materials are often not very attractive. Most of the benefit comes from the waste reduction.

Of all complex durable products, the automobile is the most recycled worldwide, with the recovery of nearly all metals. (Das et al., 1995) produced a very comprehensive report regarding the energy impacts of automobile recycling in the United States, where the recycling rate approaches 75%. In Europe, Auto Recycling Nederland BV is committed to recycling 86% of every end-of-life vehicle (ELV) in Holland. Such aggregate percentages are usually followed by a breakdown indicating the process used for each component. There are three material recovery processes: like-for-like recycling, cascade recycling and material reuse.

Eventually the material is reused in the absence of any industrial transformation. An example of **material reuse** is when car seat foam or newspaper is employed as sound damper. A **like-for-like recycling** process generates secondary material of the same quality as that obtained from primary sources. Like-for-like recycling is relatively easy to achieve when recycling pure metals (e.g. aluminium cans), but more difficult with recycling glass, metal alloys, paper or plastics. Some laboratories are working on fundamental research to develop like-for-like recycling for additional kinds of plastics (e.g. polypropylene) and rubber from tires. One of the many challenges resides in the large variety of plastics in the market, which are not always compatible from a recycling perspective. Fundamental research in like-for-like recycling is the key to improve the sustainability of material recovery processes.

A **cascade recycling** process obtains useful, but degraded, material from the recycled component. For example, the high quality plastic used to make automobile bumpers can be recycled to make less visible plastic components in the car, such as the radiator grill or air conduits. Likewise, the high quality rubber used in tires can be cascade recycled to make rubber mats. PVC from plastic bottles can be recycled as shoe soles.

Other things being equal, it is clear that like-for like recycling is preferable to cascade recycling. However, the technology required to implement like-for-like recycling processes is not always developed, so that choice is not always available. Consequently, it is important to maintain considerable research efforts in recycling processes that minimise material degradation.

Incineration is at the bottom of the recovery hierarchy. Sometimes, the cost of recovering the materials with the technology currently available is too high. If the product is made of organic material, incineration might be the single alternative to dumping. It does not allow material recovery (except, possibly for ash) but provides some **energy recovery**. For example, tires and some plastics can be incinerated in thermoelectric plants or in cement kilns as an additional source of energy. *Figure 2* illustrates the material flow in tire incineration for energy recovery, including the by-products. If the incineration occurs in a cement kiln, the by-products are entirely incorporated in the cement. Hence, no additional pollution control is required. If the incineration takes place in a thermoelectric plant, the process requires specific pollution control equipment designed to separate the toxins generated in tire incineration.

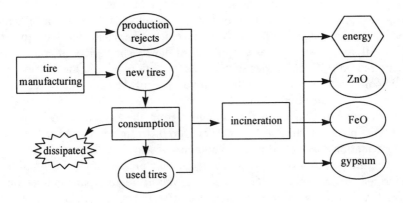

Figure 2. Material flow in the tire incineration process

3.   VALUE RECOVERY PROCESSES

Value Recovery is the process of recovering components, assemblies or whole modules. (Ayres et al., 1997) describes it as **asset recovery**. Both the material value and the value-added in the production of individual components are saved. There are several ways to recover the value -added in the product, each with a different denomination. For example, **remanufacturing** is applied to whole products (possibly with some technology upgrade); renovation is applied to structures; repair is applied to both structures and products. Other processes intend to recover some components within the original product, ignoring the remaining components in the original equipment (**retrofitting, rebuilding** or **reconditioning**). Finally, there are processes intended to recover a few components to reuse in a different finished product (**refilling, cannibalisation** or **overhauling**).

3.1   Recovery of components to reuse in a different product

Refilling (containers) is probably the most common recovery process for consumer goods. It is the process of recovering a container after the product itself is consumed. In the past, when any container was considered a valuable asset, it was reused a number of times until it was eventually damaged and discarded. For instance, oak wine-casks were much more valuable than their contents. Jars, bottles and barrels were reused a number of times to carry wine, beer, olive oil, milk or water. In recent times, the list of products delivered in reusable containers has greatly expanded. It includes soft drinks and pressurised gases (oxygen, helium, acetylene, butane, etc.) among other fluids. More recently, other containers have joined the refill stream, such as laser printer cartridges. The variety of containers being reused (refilled) varies with the consumption pattern, the distribution (and collection) network, and the relative value of the container in the local economy. A major limitation in this process is the complexity of the reverse logistics required to carry the container from the consumer back to the bottling plant or filling site. This problem increases when market fragmentation does not allow for economies of scale.

Cannibalisation is the process of recovering pre-determined parts from a damaged or obsolete machine as part of the effort to repair another one. Usually, it occurs in the maintenance of similar equipment, when the damaged machine contains valuable parts in good condition (sometimes requiring a minor repair), and the spare part is difficult to obtain. Successful cannibalisation implies that it is technically viable to retrieve the valuable parts without damaging them. (Gupta and Taleb, 1994) developed a method for determining disassembly processes for cannibalisation activities.

Overhaul is the process of repair of a mechanical module due to a major breakdown or excessive wear. The typical examples are the recovery of truck engines or airplane turbines. The overhaul process reviews the entire module. Some components are machined to new specifications and others are substituted.

Eventually the module is reinstalled in the same piece of equipment, either because of the owner's preference or due to regulation. This is always the case with airplane turbines: aviation authorities require that turbines and other critical components be overhauled and re-installed in the same airplane. In other markets, similar modules are managed collectively. When the overhaul of a given module is due, it is removed from the whole and replaced by another one from stock, so that the equipment becomes available again as early as possible. The removed module is then overhauled and kept in stock as spare. (Sherbrooke, 1968) wrote the seminal piece that analysed this process.

3.2   Recovery of some components within the product

Repair is the process of eliminating the defect in a product as part of the maintenance process. In particular, **reconditioning** is the basic work of eliminating defects in machine tools and other industrial equipment. The repair of a few parts may suffice to return equipment to its operational state, by replacing worn components. The process has limited objectives, in the sense that there is no intervention in systems that have not failed. A machine tool may be reconditioned as part of a preventive maintenance program. In this case, belts, hoses and fluids may be replaced, hydraulic systems are checked, the panel wiring is revised and reorganised, and some major assemblies are inspected.

Rebuilding is the process of restoring industrial equipment to its original specifications. It includes the examination of major mechanical sub-systems, but many used parts are untouched. Only those clearly outside specifications are affected. It is a recovery product **without identity loss**; the machine obtained from the recovery process is the same machine originally recovered.

Upgrading or retrofitting a machine requires the installation of new components or sub-systems that enhance performance, use new technology or add new features. (Toensmeier, 1992) reminds us that this choice is adopted when the machine is solid, and when the process significantly extends its expected life. Systems with older technology are replaced, to improve the equipment's performance, to abide to new safety regulation, to execute new tasks or to reduce its impact on the environment. It is also a process without identity loss. (Sprow, 1992) and (Stauffer, 1990) describe recovery processes appropriate for machine tools. Although there is no practical difference between the two terms, retrofitting is preferred when the process recovers machine tools and other mechanical equipment. Upgrading is the preferred term when the process recovers computers and other electronic systems.

3.3   Thorough recovery of the final product

Remanufacturing is the process of disassembly and recovery at the module level and, eventually, at the component level. It requires the repair or replacement of worn out or obsolete components and modules. Parts subject to degradation affecting the performance or the expected life of the whole are replaced. (Venta and Wolsky, 1978) were the first to address the economic impact of the automobile engine remanufacturing industry, and (Lund, 1984) was the first to describe the activity. It differs from other recovery processes in its completeness: a remanufactured machine will match customer expectation of new machines.

There are three types of remanufacturing activities, each with different operational challenges. This has led to a wide variety of research. Sometimes, remanufacturing takes place **without identity loss**. In this case, a current machine is built on yesterday's base, receiving all of the enhancements, expected life and warranty of a new machine. The physical structure (the chassis or frame) is inspected for soundness. The whole product is refurbished and critical modules are overhauled, upgraded or substituted. If there were defects in the original design, they are eliminated. This is the case for customised re-manufacturing of machine tools, airplanes, computer mainframes, large medical equipment and other capital goods. Because of its uniqueness, this product recovery is characterised as a project. *Figure 3* illustrates the material flow in remanufacturing computer mainframes without loss of the original identity.

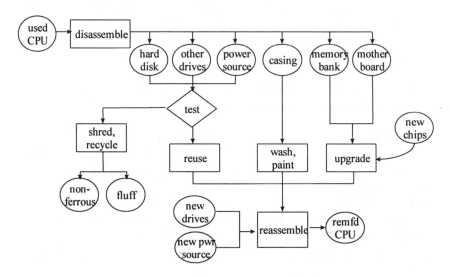

Figure 3. Material flow in computer remanufacturing

One may also have a **repetitive remanufacturing process without loss of identity**. In this case, there is the additional challenge of scheduling the sequence of dependent processes and of identifying the location of inventory buffers. (Guide and Spencer, 1995), (Guide, 1996) and (Guide et al., 1997) address these problems. There is a fine line between repetitive remanufacturing without loss of identity and product overhaul. Again, the critical difference is that remanufacturing is a complete process. The final output has a like-new appearance and is covered by a long warranty comparable to that of a new product.

For some products the remanufacturing process causes the **loss of original product identity**. It involves the disassembly of the used goods into pre-determined components. These are repaired to stock, ready to be reassembled into a remanufactured product. This is the case when remanufacturing automobile components, photocopiers, ready-to-use cameras and personal computers. Once the product is disassembled and the parts are recovered, the process concludes with a manufacturing operation not too different from original manufacturing. Disassembled parts are inventoried, just like purchased parts and made available for final assembly.

Remanufacturing with loss of original product identity encompasses some unique challenges in inventory management, and disassembly sequence development. Research is still necessary to address production planning issues specific to this process. Some of the open questions relate to the commonality of parts in products of different generations, the uncertainty in the supply of used products and their relationship with production planning.

Refurbishment or **renovation** requires a thorough revision of the equipment, with substantial elimination of small defects and, sometimes, aesthetic improvements. It is periodically executed on public transportation equipment (airplanes and buses), both to incorporate new safety requirements and to satisfy aesthetic requirements. (Panisset, 1988) introduced an MRP II approach useful in the refurbishment industry. In that study, it is clear that refurbishment can be characterised as a project that could be managed using PERT/CPM methods.

A refurbishment decision does not imply that the equipment is worn out. Usually it occurs when the most visible parts show significant signs of wear. In order to simplify the maintenance schedule in a large fleet, it is usually done in combination with a major intervention, such as the engine overhaul. (Jardine and Hassounah, 1990) developed a method for scheduling bus fleet maintenance programs. One such approach can accommodate the refurbishment events, in addition to engine overhaul.

Retreading is a very particular example of remanufacturing. It is the recovery process of adding a new tread to a used tire. It includes a thorough inspection of the casing structure, repair of any damages, removal (buffing) of residual rubber from the used tread, addition of a new tread, vulcanisation and final inspection. Tire retreading is a significant waste reduction activity, facilitating the recovery of approximately 90% of the weight of the used tire. The material flow appears in *Figure 4*. Given the fatigue accumulated in the casing, the life expectancy of a retreaded tire is smaller than that of a new tire.

Retreading is one of the most successful remanufacturing processes, once we recognise its market penetration and its impact on corporate waste reduction programs. The retreaded tire is expected to satisfy the same safety requirements as a new tire. In a comprehensive study of retreading and other tire recovery processes, (Ferrer, 1997a) shows that the newer generation of tires for commercial vehicles could be used for up to 600,000 miles if retreaded two or three times.

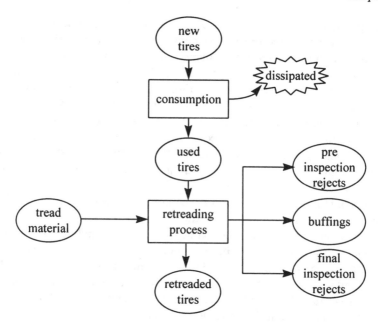

Figure 4. Material flow in tire retreading

4. OPERATING PROCESSES

Several researchers have worked on the optimisation of product recovery processes. (Fleischmann et al., 1997) produced a comprehensive literature review for quantitative models representing product recovery operations. (Johnson and Wang, 1995), (Gupta and Taleb, 1994), (Penev and de Ron, 1996) and (Ferrer, 1997b) are some of the authors that have helped clarifying the meaning of some of these processes.

Reverse logistics is the process of collecting used products from the consumer and transporting them to the plant where the recovery operation will take place. A reverse logistics operation is considerably different from forward logistics. It must establish convenient collection points to receive the used goods from the final customer. It requires packaging and storage systems that will ensure that most of the value still remaining in the used good is not lost due to careless handling. It often requires the development of a transportation mode that is compatible with existing forward logistic system.

Disassembly is the decomposition process involving the systematic removal of desirable constituents and separation of toxic or poisonous substances from the original assembly. It must ensure that there is no impairment of the useful parts, so that they can be reused.

Dismantling is the decomposition process in which the removal of some desirable components from the original assembly may result in the destruction of the rest of the assembly. It is a less precise destructive technique, usually carrying the advantage of being less costly. Dismantling is usually associated with cannibalisation processes when the valuable component is relatively inaccessible without the destruction of outer parts that have little value. The process is also useful in some recycling processes to separate parts made of different materials.

An even more destructive decomposition process is **shredding**. Shredding is the process by which the used good suffers complete loss of its original form. It is reduced into minute pieces of random size in a machine called the shredder. The resulting fragments are separated according to their material using a variety of techniques (e.g. centrifugation, magnetic separation, vibrating screening) before being further refined in a recycling process.

In the United States, about 94% of the end -of-life vehicles (ELV) are sent to dismantlers and shredders. Large shredding machines can process 300 scrapped cars at a time, to produce scrap steel. Considering the technology available today, an ideal automobile recycling process works as follows: a dismantler removes the fuel and water pumps, radiator, alternator, starter, engine and transmission to undergo specific reclamation processes. These are the more valuable components, which are sold to remanufacturers who will recover them in a batch-like process for the replacement market. In addition, most fluids, some plastic parts, the catalytic converter and the battery follow a dedicated recycling stream. The rest of the vehicle - the steel body and everything else that was not removed - goes to a shredder that will recover the steel and the non-ferrous metals, mostly copper, aluminum and zinc

The shredding process can also take place in other recycling operations. For example, the first step in recycling carpets takes place in the shredder. The ragged carpet strips are then fed to a hammer-mill to further crush and shred the carpet into minute pieces. The resulting material can then be separated. The minute nylon fibres pass through a vibrating screen leaving behind most of the dust and latex material once used to make the carpet backing. An air blow completes the separation to obtain colourless nylon fibre fluff, which can be used in the manufacture of other nylon products, including new carpets.

5.   MARKETING PROCESSES

In order to be successful, a recovery program must be accompanied by a well-designed marketing plan that will tackle the difficulties inherent to selling recovered products or recycled materials. (McConocha and Speh, 1991) and (Vandermerwe and Oliff, 1991) have discussed some of the challenges related to marketing remanufactured products. They also introduce two new expressions:

Reconsumption is the recovery process from a marketing viewpoint. The product is *reconsumed* when it is used following a recovery process such as recycling or remanufacturing.

Remarketing is the development of marketing and sales conditions that are suitable for recycled materials or recovered products. In addition to the usual marketing functions, it requires:

a) Identifying the market niche most suitable for recovered goods, and the development of its channel;

b) Managing customer expectations regarding the true performance of recovered goods;

c) Establishing a market channel that facilitates the collection of the used goods for material or value recovery, and

d) Defining a price for recovered goods that is compatible with the price of new goods of equivalent performance.

All areas in remarketing need considerable development. Two additional issues need to be addressed. One is the design of product lines and the relationship between successive product generations. Another important remarketing topic is the impact of planned obsolescence in the corporate environmental responsibility.

6.   DESIGN PROCESSES

Usually, a new product is designed to satisfy a certain function, constrained by performance specifications, safety and health regulations, manufacturing and assembly limitations, expected demand, cost and competition. A product is **designed for the environment** (**DFE**) if, in addition to these constraints, the designer has minimised the product's impact in the environment. The designer must ensure that production and utilisation generate minimal emissions, the design avoids harmful or poisonous chemicals and when the used good reaches its end-of-life, it can be appropriately disposed of, through remanufacturing or recycling.

Design for the environment closely relates to **green design**, which has been the object of an excellent study by (OTA, 1992). OTA describes green design as having two general goals: waste prevention (service life extension, reduction in weight, toxicity and energy use) and better material management (ease of product and energy recovery). Other earlier studies about DFE and green design include (EPA, 1993), (Owen, 1993) and (Horvath et al., 1995).

One of the design tools used to develop environmentally friendly products is **life-cycle assessment** (LCA). It was defined by (Heijungs et al., 1996) as "the process of evaluating the effects that a product has on the environment over the entire period of its life cycle". Life- cycle assessment is extremely complex since it evaluates the impact that a product has on the environment during its production, consumption and disposal. Every analysis starts with the definition of the boundaries of the study. Comprehensive LCAs should consider several alternatives for raw material, including their geographical sources, given the different environmental impacts that they might have. It requires a number of additional assumptions about consumption patterns and the product's final disposal, which makes it difficult to replicate any individual study.

Design for the environment is one of many possible **DFX**, where "X" is a design purpose. For example, a product is **designed for recycling (DFR)** when it can be economically disposed of with minimal generation of solid waste through product disassembly and material recycling. This requires the utilisation of easily recycled materials. Moreover, if an assembly is made of incompatible materials, their separation should be technically and economically feasible. (OTA, 1992) and (EPA, 1993) suggest several criteria compatible with DFR as intermediate targets to design for the environment. (Kriwet et al., 1995) suggest specific criteria for DFR, including material selection and separation. (Dillon, 1994) and (Fisher, 1991) offer a few examples of how DFR has influenced product design.

One requirement of successful product recovery is that it can be easily disassembled for the recovery of modules and components. A product is **designed for disassembly (DFD)** when the components that retain some value at the time of disposal can be easily separated while retaining their functionality. This should not compromise the component's integrity or its expected life. Generally, design for disassembly is achieved by adopting a modular design. The modules are assembled using connectors that can be quickly separated requiring at most simple tools. (Ferrer, 1996) discusses some of the pitfalls in product design that impair successful disassembly. (Kroll et al., 1996) proposes a method to evaluate whether a product was designed for easy disassembly. (Penev and de Ron, 1994) describe a disassembly line for refrigerators which addresses some of the challenges created by a design that is not concerned with product disposal. This highlights the need to evaluate how much a product design complies with DFD and DFR, and the development of respective performance measures .

6.1   Measures of design efficiency for recovery processes

(Ferrer, 1996) introduced economic definitions of some design measures, including analytical expressions that can be used to evaluate most assemblies.

Recyclability is a profitability measure of the recycling process.   It identifies how profitable recycling is, considering the market value of the raw material and the cost of shredding and refining the good. In its simplest form, it is given by the expression

$$Recyclability = \frac{raw\ material\ value}{shredding\ costs} - 1$$

Consider an assembly made of two parts. If the raw material value equals \$1, and requires a shredding operation costing 50¢ and generates fluff that costs 30¢ to landfill, the recyclability measure equals 0.25. The positive value implies that the process is profitable.

Disassemblability is a comparative measure of the profitability of the disassembly process. The disassemblability of an assembly compares the profitability of the disassembly process with the recycling alternative. Disassembly is a source of income, from the value of the components that can be retrieved, and a source of expense, from its operating cost and the loss of the original assembly. It is an important measure in remanufacturing operations, because it helps identify how thorough the recovery should be. If the disassemblability of a good is negative (positive), it means that recycling it is preferable (not preferable) to disassembly.

$$Disassemblability = \frac{value\ \text{-}\ added\ in\ parts}{disassembly\ costs} - 1$$

Consider the same assembly made of two parts. If the parts had to be purchased separately, they would cost \$1.50 each.   Since the total raw material value is \$1, the value that is added in order to make the two parts equals \$2. Let the disassembly operation cost 95¢, generating fluff that costs 30¢ to landfill. The disassemblability measure equals 0.6, a positive value, implying that the process is more profitable than recycling.

Reusability is a measure of endurance that compares the value of the used assembly with the alternative opportunity to undertake a disassembly-based recovery process. Hence, the reusability of a product is positive whenever the value of reusing it, subject to a simple renovation and minor adjustments, is greater than the value of its subassemblies, less the disassembly and renovation costs.

$$Reusability = \frac{value - added\ in\ assembly}{recovery\ costs} - 1$$

Let's consider again the same assembly that is made of two parts. If the whole assembly had to be produced from new parts, it would cost $5.50. Since the total raw material value is $1, and the value-added in the two parts equals $2, the value that is added at the assembly stage equals $2.50. Let the recovery operation, including washing, painting, and lubricating, cost $3. The reusability measure equals -0.17, a negative value, implying that the good should be disassembled to recover individual parts, rather than reused.

Remanufacturability is an intuitive indication of how easy it is to remanufacture a given product. Currently, there is no rigorous economic, environmental or physical expression to measure the remanufacturability of a product. The development of this performance measure would help evaluate the progress made in product design enabling remanufacturing the used product at the end of its lifetime.

(Graedel and Allenby, 1995) introduced physical efficiency measures of product recovery that relate the amount of materials recovered with its original input. **Manufacturing efficiency** relates production output with its waste. It is calculated as

$$Manufacturing\ efficiency = \frac{production}{production + production\ waste}$$

Where production could be initial manufacturing or remanufacturing. Lower production wastes increase manufacturing efficiency until it eventually reaches the maximum value of 1. Another useful measure defined by (Graedel and Allenby, 1995) is recovery efficiency. It would probably be clearer to call it **reverse logistics efficiency**, since it relates the amount of used products collected with the amount of products wasted. It is calculated as

$$Reverse\ logistics\ efficiency = \frac{returned\ goods}{returned\ goods + wasted\ goods}$$

Finally, (Graedel and Allenby, 1995) define **recycling efficiency** as the fraction of total material weight recovered by the recovery process. It is calculated as

$$Recycling\ efficiency = \frac{wgt.\ of\ recycled\ materials + reused\ parts}{total\ weight\ of\ inputs}$$

Each of these efficiency measures ranges from 0 to 1. For example, in the United States, automobile dismantling has a recycling efficiency of 75%, meaning that if an ELV weighs 1,200 kg, a shredder is expected to recycle or reuse 900 kg of its original weight.

From a systems perspective, it may be useful to measure the **product recovery efficiency** in *Figure 1* as the product of its manufacturing, reverse logistics and recycling efficiency:

$$Product\ Recovery\ efficiency = Manuf.\ eff. \times Rev.\ Log.\ eff. \times Rec.\ eff.$$

7.   CONCLUSION

The path to sustainability is a long one. It requires substantial effort from government, consumers and industry. Product recovery is an alternative that firms should adopt as the first step toward sustainability, to recover the value in discarded goods that the firm once produced and sold. The ease of product recovery implementation depends on the original product design, on the technology available, and on the selected process. Practitioners and academics have taken the challenge of addressing these and other questions in product recovery, as witnessed by some of the references in this chapter.

The progress on these fronts will be faster, however, if researchers in different fields and industries can report their findings clearly. This chapter contributes to the development of product recovery research by clarifying the difference between similar processes as well as identifying their similarities. In addition, it points to some of the research topics that are relevant to specific processes and those that need further investigation. Hopefully it will facilitate communication, not just within the field, but with interested individuals that want to familiarise themselves with the new developments in product recovery.

REFERENCES

Ayres, R. U., et al. (1997). "Eco-Efficiency, Asset Recovery and Remanufacturing." *European Management Journal* 15: 5, 557-574.

Biddle, D. (1993). "Recycling for profit: The new green business frontier." *Harvard Business Review* November-December 1993, 145-156.

Das, S., et al. (1995). "Automobile Recycling in the United States: Energy Impacts and Waste Generation." *Resources, Conservation and Recycling* 14: 265-284.

Dillon, P. S. (1994). "Salvageability by Design." *IEEE Spectrum* August 1994, 18-21.

EPA. (1993). "Life Cycle Guidance Manual: Environmental Requirements and the Product System." U.S. Environmental Protection Agency, Office of Research and Development, Report EPA600/R-92/226, January 1993,

Ferrer, G. (1996). "On the Economics of Remanufacturing a Widget." INSEAD-CMER, Working paper 96/64/TM, Fontainebleau, France.

Ferrer, G. (1997a). "The Economics of Tire Remanufacturing." *Resources, Conservation and Recycling* 19: 221-255.

Ferrer, G. (1997b). *Managing the Recovery of Value from Durable Products.* Ph.D. dissertation, Technology Management Area, INSEAD, Fontainebleau, France.

Fisher, A. (1991). "Reincarnation in the design studio." In *Financial Times,* London, UK, 3 Apr 91, pp. 29.

Fleischmann, M., et al. (1997). "Quantitative models for reverse logistics: A review." *European Journal of Operational Research* 103: 1-17.

Graedel, T. E.; and Allenby, B. R. (1995). *Industrial Ecology,* Upper Saddle River, NJ: Prentice-Hall.

Guide, V. D. R., Jr. (1996). "Scheduling Using Drum-Buffer-Rope in a Remanufacturing Environment." *International Journal of Production Research* 34: 4, 1081-1091.

Guide, V. D. R., Jr., et al. (1997). "Scheduling Policies for Remanufacturing." *International Journal of Production Economics* 48: 187-204.

Guide, V. D. R., Jr.; and Spencer, M. S. (1995). "Rough-cut Capacity Planning for Remanufacturing Firms." *Production Planning & Control*

Gupta, S. M.; and Taleb, K. N. (1994). "Scheduling Disassembly." *International Journal of Production Research* 32: 8, 1857-1866.

Hart, S. L. (1995). "A Natural-Resource-Based View of the Firm." *Academy of Management Review* 20: 4, 986-1014.

Heijungs, R., et al. (1996). *Life Cycle Assessment: What It Is and How to Do It,* Paris, France: United Nations Environment Programme.

Horvath, A., et al. (1995). "Performance Measurement for Environmentally Conscious Manufacturing." In *Manufacturing Science and Engineering ASME 1995,* pp. 855-860.

Jardine, A. K. S.; and Hassounah, M. I. (1990). "An Optimal Vehicle-fleet Inspection Schedule." *Journal of the Operational Research Society* 41: 9, 791-799.

Johnson, M. R.; and Wang, M. H. (1995). "Planning product disassembly for material recovery opportunities." *International Journal of Production Research* 33: 11, 3119-3142.

Kriwet, A., et al. (1995). "Systematic Integration of Design-for-Recycling into Product Design." *International Journal of Production Economics* 38: 15-22.

Kroll, E., et al. (1996). "A methodology to evaluate ease of disassembly for product recycling." *IIE Transactions* 28: 837-845.

Lund, R. T. (1984). "Remanufacturing." In *Technology Review,* February/March 1984, pp. 19-29.

McConocha, D. M.; and Speh, T. W. (1991). "Remarketing: Commercialization of remanufacturing technology." *The Journal of Business and Industrial Marketing*6: 1-2, 23-37.

OTA. (1992). "Green Products by Design, Choices for a Cleaner Environment." Office of Technology Assessment, Congress of the United States,

Owen, J. V. (1993). "Environmentally conscious manufacturing." In *Manufacturing Engineering*, No. 4, Oct 93, pp. 44-55.

Panisset, B. D. (1988). "MRP II for Repair/Refurbish Industries." *Production and Inventory Management Journal* Fourth Quarter, 1988, 12-15.

Penev, K. D.; and de Ron, A. J. (1994). "Development of disassembly line for refrigerators." In *Industrial Engineering*, November 94, pp. 50-53.

Penev, K. D.; and de Ron, A. J. (1996). "Determination of a disasembly strategy." *International Journal of Production Research* 34: 2, 495-506.

Purser, R. E., et al. (1995). "Limits to Anthropocentrism: Toward and Ecocentric Organisation Paradigm?" *Academy of Management Review* 20: 4, 1053-1090.

Sherbrooke, C. C. (1968). "METRIC: A Multi-Echelon Technique for Recoverable Item Control." *Operations Research* 16: 1, 122-141.

Sprow, E. (1992). "The Mechanics of Remanufacturing." In*Manufacturing Engineering*, No. 3, March 92, pp. 38-45.

Stauffer, R. N. (1990). "Making the right moves in machine make-over." In*Manufacturing Engineering*, No. 3, March 90, pp. 49-53.

Thierry, M., et al. (1995). "Strategic Issues in Product Recovery Management." *California Management Review* 37: 2 - Winter 1995, 114-135.

Toensmeier, P. A. (1992). "Remanufacture does more than save on investment." In *Modern Plastics*, April 92, pp. 77-79.

Vandermerwe, S.; and Oliff, M. D. (1991). "Corporate challenges for an age of reconsumption." *The Columbia Journal of World Business* Fall 1991, 6-25.

Venta, E. R.; and Wolsky, A. M. (1978). "Energy and Labor Cost for Gasoline Engine Remanufacturing." Argonne National Laboratory, September 1978, Argonne, IL.

AUTHORS

Geraldo Ferrer is Assistant Professor of Operations Management at the Kenan-Flagler Business School in the University of North Carolina at Chapel Hill. His research interests include Product Stewardship, Remanufacturing, Reverse Logistics and Remarketing. He earned his doctoral degree at INSEAD, where he completed a dissertation entitled "Managing the Recovery of Value from Durable Products". While at INSEAD, part of his work was developed in the context of the Centre for the Management of Environmental Resources. Geraldo holds an MBA from Dartmouth College and a Mechanical Engineering degree from the Military Institute of Engineering, Rio de Janeiro, Brazil. He is married with Ivanisa and has two children, Ivana and Alex.

D. Clay Whybark is Macon Patton Distinguished Professor of Management and Director of the Global Manufacturing Research Center at the Kenan Flagler School of Business, University of North Carolina at Chapel Hill. His research interests include manufacturing practices, remanufacturing and supply chain management. He has recently co-authored "Why ERP?" a novel on the implementation of SAP R/3 and is revising the 4th edition of "Manufacturing Planning and Control Systems." He has published more than 250 articles and business cases. He currently serves as President of the International Society for Inventory Research.

Chapter 5

Green Design and Quality Initiatives

Noellette Conway-Schempf and Lester Lave
Graduate School of Industrial Administration
Carnegie Mellon University
Pittsburgh, PA 15213

1. INTRODUCTION

Producing quality products - products that satisfy customer expectations, becomes more complex daily. Consumers increasingly expect high tech products that are low in price, durable, and attractive. Increasingly, consumers are also adding environmental criteria to their list of product expectations. Corporations are expected to comply with environmental regulations, carry out business with an environmentally conscious ethic, and manufacture and market products with reduced environmental impacts.

Companies that focus on their customers and their expectations are generally corporations that make the most forward-looking and aggressive business decisions. These are the companies most likely to have proactive environmental management approaches and active green design or environmental product design efforts underway. Here we define green design as the reduction of the environmental impacts of products, processes, and manufacturing activities through changes in product, process, and policy design (Conway-Schempf and Lave, 1998). Companies known for their customer focus and quality initiatives such as Xerox, AT&T, Hewlett Packard, and IBM are also corporate leaders in the field of green design. In many cases, the environmental management and green design efforts have grown out of corporate quality initiatives as the firm has recognized the new requirements of potential customers.

2.   THE LINK BETWEEN QUALITY MANAGEMENT, ENVIRONMENTAL MANAGEMENT AND GREEN DESIGN

Total Quality Management (TQM) principles were first stressed by W. Edwards Deming and Joseph Duran in the early 1950s. At that time, US companies were uninterested in the concepts, however Deming and Duran were successful in bringing their ideas to Japan. These new management principles were the basis for the revolution in quality and competitiveness of Japanese products post World War II. Today, TQM has gained acceptance in the US and Europe in addition to Asia. The basis of the TQM management philosophy is a strong customer focus – the provision of products and services that exceed expectations through a series of continuous improvements. Deming considered defective products or non-revenue products to be waste.

A natural extension of TQM principles is the management of environmental issues. In fact, most of the classic "tools" of TQM (benchmarking, goal setting, continuous improvement, etc.) are equally applicable to environmental management efforts (see Table 1). For example, in the 1980's many chemical companies began to scrutinize their process yields. Increasing yields had generally been considered to be an exercise in optimizing the use of raw materials. However, the realization grew that increasing yields increased manufacturing capacity without additional capital investment, and higher yields resulted in reduced waste disposal costs for the company. A quality initiative – more effective transfer of raw materials to products, also became an environmental initiative. The benefits of higher yields had not been recognized because the analyses regarding improved yields were omitting many financial and environmental impacts.

A number of industry guidelines and standards such as Europe's Ecomanagement and Auditing system (EMAS), and the ISO 14000 series of environmental management standards are providing companies with specific guidelines on how to set up environmental management programs which link well to quality initiatives. In fact, the ISO14000 series is modeled after the more familiar ISO 9000 family of quality management standards (www.iso.ch). In the case of the ISO14000 series, these environmental management system standards are linked to the development and use of green design tools such as life cycle assessment (ISO 140040), performance metrics (ISO 14031), full cost accounting, and product redesign. These, in addition to a standard for the elements of an environmental management system (ISO 14001). Greener product and process design are means by which the quality environmental management standards are linked to product design strategies, corporate benchmarking, and internal information system improvements.

Potential incentives/advantages of implementing total quality environmental management (TQEM) or other environmental management systems include:

1. Potential cost savings: As an example, several of the companies involved in demonstrating TQEM projects for the President's Commission on Environmental Quality showed cost savings after implementation of an environmental management framework (President's Council on Environmental Quality, 1993), of course, detailed cost analyses are necessary to track these benefits.
2. Technological innovation: Process changes resulting from implementation of environmental management systems may result in technological advances for the firm.
3. Increased public acceptance of corporate environmental management ethics.
4. Better relations with regulators, and in some instances the potential for alternative regulatory oversight mechanisms. For example, some US state regulatory agencies are exploring ways to integrate ISO14000 certification with regulatory requirements (BATE, 1999).
5. Better recognition for individuals and teams.
6. ISO 14000 certification.
7. Safer working conditions.
8. Reduced potential liability exposures.

In this chapter we describe a series of recently developed tools which link to both quality management and quality environmental management efforts at industrial sites. The discussion of each tool includes an example or case study of tool to show the relevance of green design to quality management. The green design tools we discuss are just a subset of those available from universities and consulting companies for corporate use.

2.1 Life-Cycle Assessment.

Life-cycle assessment (LCA) involves evaluating the environmental effects of a product or activity holistically, by looking at the entire life cycle of the product or process from raw materials extraction through consumer use (See Figure 1). Coca-Cola was one of the first corporations to attempt

LCA when the company analyzed the environmental impacts of beverage containers in the 1960s. LCA criteria are beginning to find their way into environmental labeling schemes such as Germany's Blue Angel and the ISO 14000 environmental management standards (www.iso.ch). Simplified LCAs even pop up in popular magazines; for example, Consumer Reports occasionally comments on the environmental impacts of different product packaging types and chemicals. The use of LCAs as part of labeling systems and management criteria has led to a heated debate about just what criteria should be measured and how impacts should be compared.

A number of efforts involving industry, government agencies and universities are focusing on standardizing the analyses so that LCAs carried out by different organizations are comparable. LCAs have been criticized for a number of reasons:

- Environmental effects for all life-cycle stages are not known.
- It is difficult to compare different types of effects - species extinction versus cancer incidence, for example.
- The amount of data required to analyze even simple products is enormous; the typical cost for a detailed LCA may exceed $200,000.
- Data gathering is difficult as many of the life cycle stages involve proprietary processes.

It is difficult to know where to draw the boundary around the analysis - should raw materials extraction be included? What about the life cycle impacts of the equipment used in the raw materials extraction? What about the environmental impacts of the detergent used to wash the clothes of the people mining the raw materials!

Despite the drawbacks, LCA can show the major environmental problems of a material, product, or process. The act of doing the assessment builds awareness about environmental impacts and focuses improvement efforts. This has led companies, such as AT&T, to develop internal LCA tools for their product lines (Graedel and Allenby, 1995) and Government agencies, such as the EPA, to provide generic guidelines for conducting LCAs (USEPA, 1993).

Most currently used LCA techniques are modifications of the approach developed by the Society for Environmental Toxicology and Chemistry (SETAC). Practical use of the SETAC approach involves drawing a boundary that limits consideration to a few producers in the chain from raw materials to consumers. Researchers at Carnegie

Mellon have found that limiting the analysis in this way may lead to consideration of only a fraction of the total environmental discharges associated with the product or process, and have developed an approach based on models of industrial activity and pollution discharge data (Cobas et al, 1995; Lave et al., 1995; Lave et al, 1996; Hendrickson et al, 1998). The resulting software tool allows for economy-wide aggregate LCA.

Economic input-output analysis is a well-established modeling framework for tracing the flows of inputs and outputs in an economy. Input-output analysis is generally used for economic planning purposes; for example, calculating the resources needed to support an increase in the production of automobiles. The resulting estimates show the increases in production, both for automobiles and for the various sectors that supply products directly or indirectly. For example, an expansion in automobile production would require steel, electricity, petroleum, plastics, etc., and even additional automobiles. We have used the U.S. Department of Commerce input-output tables for the US to develop an economy-wide LCA technique by linking the economic input-output tables with environmental databases (conventional pollutants, TRI, energy use, ore use, fertilizer use, global warming potential, and ozone depleting potential, etc.). The method is called Economic Input-Output Life Cycle Assessment, or EIO-LCA.

The EIO-LCA approach has been used to examine the environmental impacts of various industrial sectors (Cobas et al, 1995, Lave et al, 1995, Hendrickson et al, 1998), automobile use (MacLean and Lave, 1998), automobile components (Joshi, 1997) and construction materials (Horvath and Hendrickson, 1998a, b; Hendrickson and Horvath, 1998). An example of the type of analysis that can be carried out with this tool is included in section 2.

The EIO-LCA tool has advantages over other LCA techniques:
- EIO-LCAs can be used to examine the total direct and indirect economy-wide effects (effects of suppliers) on emissions and energy consumption resulting from changes in production.
- The EIO-LCA model uses the entire US economy as the boundary for the analysis.
- The approach highlights priority areas for reduction in environmental impacts.
- Initial EIO-LCAs can be carried out at a fraction of the cost and time

associated with SETAC-LCAs.

- The EIO-LCA can be used in conjunction with current SETAC approaches, where the SETAC approach is used to analyze specific products and processes in detail and the EIO-LCA explores the indirect economy-wide effects of changes in product mix.
- A transparent, computer-based tool that is easily implemented as part of green design efforts calculates the LCA.
- The LCA has underlying databases accessible by all environmental stakeholders.
- The method provides some environmental information for every commodity in the U.S. economy.
- The method allows for sensitivity analyses and scenario planning.
- The method is complete and the software is available either through the Green Design Initiative at Carnegie-Mellon or via the World Wide Web at www.eiolca.net.
- Results are reproducible by environmental stakeholders

The EIO-LCA approach also has limitations. Even with 500 economic sectors, the amount of disaggregation may be insufficient for the desired level of analysis. EIO-LCA models include sectors of the economy rather than specific processes. Detailed analysis of the environmental impacts of the activities of the individual members of the supply chain will require more traditional SETAC-LCA techniques. The use and disposal phases of certain products may be too difficult to analyze with EIO-LCA. In addition, the data in the model may reflect past practices and imports are treated as if they are U.S. products. Efforts are underway to integrate the EIO- and SETAC-LCA approaches to overcome the inherent limitations of each and allow for a software tool that permits both broad and narrowly defined LCAs. Efforts are also underway to develop models for other countries.

2.2   Environmental Performance Measurement Tools

There are a number of environmental performance metrics that are being used by companies and academia to understand, in a simple way, the impact of a material, product, or even corporation on the environment. In general these tools are fairly crude and aim to handle a number of levels of analysis, simplify complicated multidimensional environmental impacts, prevent sub optimization, and inform decision-makers. Examples include:

- Holistic lists of 'leaders and laggards' (e.g. annual Council on Economic Priorities rankings);
- Simple lists of emissions per facility e.g. the U.S. Environmental Protection Agency's annual Toxic Release Inventory);

- Toxicity- or risk-weighted emissions per facility (e.g. CMU's CMU-ET index, the University of Tennessee's chemical ranking for potential health and environmental impacts);
- Abbreviated LCA methods (such as AT&T's green product design matrices and Volvo Corporation's Environmental product strategies approach -see Graedel and Allenby, 1995).

To illustrate the use of environmental performance metrics, we will discuss the CMU-ET index developed by Arpad Horvath at Carnegie Mellon University.

Facility emissions are often used as a crude initial metric of environmental performance. Most manufacturing firms are required to report the discharges and transfers of over 650 chemicals as a result of Title III of the 1986 Superfund Amendments and Reauthorization Act (The Toxics Release Inventory, or TRI). TRI discharges do not take into account differences in relative toxicity. Yet studies of human health effects suggest that the most harmful chemicals on the TRI list are more than 1,000,000 times more toxic than the least harmful.

Horvath et al. (1995a, b) have developed preliminary environmental performance metrics by applying commonly available measures of toxicity to "weight" TRI emissions by their toxicity. Details of this system are outlined in Horvath (1995a). Data from companies of different sizes can be normalized by dividing the CMU-ET by the pounds of chemicals on the company's TRI discharge list to result in a Toxicity Ratio. This toxicity ratio permits a comparison among companies of different size. This project addresses the second environmental quality goal outlined earlier - minimizing toxic releases into the environment. The project has resulted in an analysis system that may reveal priority areas for emission reduction. The model allows for environmental performance measurements, both temporal and inter- and intra-industry. In addition:
- A transparent, easy to use computer-based tool calculates the CMU-ET ratios.
- The underlying databases are accessible by all environmental stakeholders. Although reliability of TRI data can be a concern, the format of the data is consistent across industries.
- The CMU-ET and Toxicity Ratio can be used on a variety of analysis levels: facility, company, county, state, industry, and product.
- Interpretation of the indices is objective and straightforward.
- International environmental performance measurements may be possible if other nations adopt the TRI approach.

Current follow-up efforts involve studies of the uncertainties and limitations of the current approach and investigation of additional toxicity measures. However, despite these limitations, the software enables the user to think broadly about emissions reduction efforts and prioritize based on toxicity rather than pounds of emissions.

2.2.1   Environmental Management Application – Analysis of the emissions associated with alternative automobile car seat designs.

Lave *et al* (1996) used CMU-ET coupled with the EIO-Life Cycle Assessment model discussed earlier to examine the emissions (both unweighted and weighted) associated with alternative automobile seat designs - a seat made primarily of plastics and one containing a higher proportion of aluminum). The seats were assumed to cost the same amount. The dollar costs of the various materials associated with each seat were entered in the EIO-LCA software and then the electricity and TRI emissions (both direct and indirect) associated with each design were calculated (see Table 2).

From the data in Table 2, it is apparent that the *total* amounts of emissions associated with both car seat designs are similar. However, because different compounds are emitted as a result of the manufacture of each seat type, the *weighted* CMU-ET emissions totals are different. Alternative design 2 has more toxic emissions on average than alternative 1. Design 2 also requires greater amounts of electricity to manufacture.

2.3   Environmental managerial cost accounting and application to a plastics molding facility

Environmental costs are often obscured in overhead accounts and, as a result, overlooked in management decision-making (USEPA 95). Understanding and tracking environmental costs, and incorporating these costs into management information systems (MIS) can provide powerful incentives to reduce such costs. Understanding environmental costs is an important component of any corporate quality management system. Additionally, knowledge of environmental costs provides an incentive for green design and the development of greener products and processes – those with lower environmental costs

The US EPA has issued guidelines describing how to root out all the private environmental costs associated with activities (USEPA 95) in order for these costs to be included appropriately in cost estimations. The EPA discusses three tires of costs: Potentially Hidden Costs (e.g. medical surveillance, record keeping, post-closure care), Contingent Costs (e.g. legal expenses, future compliance costs), and Image and Relationship Costs (e.g. relationships with lenders or host communities). According to the EPA, the environmental costs incurred by a company deserve management attention because:

- Many environmental costs can be reduced or eliminated once they are understood.
- Many environmental costs (e.g. waste) may provide no value added for the company.
- Better management of environmental costs can result in improved environmental performance and benefits to human health and safety.
- Understanding environmental costs provides more accurate information for costing and pricing.
- Accounting for environmental costs and performance can support a company's development and use of an overall environmental management system. In fact, the ISO 14000 environmental management system standard includes discussion of environmental cost accounting.

2.3.1 Environmental Management Application: Environmental Cost Accounting at a Plastics Molding Facility.

Horney (1998) examined the environmental costing procedures at a plastics molding facility located in the U.S. The goal of the research was to go through a step-by-step analysis of the production process to derive the full "private" costs of a specific environmental issue- in this case, scrap production. Understanding the full costs of scrap will provide incentives for scrap reduction, decreasing impacts on the environment and improving the bottom line.

Using mass balance techniques, Horney calculated the waste production and then assessed the current methods of determining the economic impacts of scrap production. She determined the true private costs of scrap production in order to estimate the costs that were being overlooked in the MIS. The results of the analysis are detailed below.

The plastics facility operates an injection molding line. Most of the polycarbonate used in the facility is molded into parts. Rejected or incorrect parts are scrap – in this facility the scrap rate was initially calculated to be 3%. Although some of the scrap is directly recycled back into manufacturing processes, a large amount of scrap must be taken off site and is either sold or disposed at a cost. In all, the annual income from scrap sales amounts to approximately $242,000, while the original material cost of scrap was approximately $852,000 resulting in a calculated net loss to the facility of $610,000.

2.3.2   Additional Cost Impact of Scrap Production on the Facility

The initial materials cost of the scrap was calculated from a mass balance. However, scrap is produced at five main points in the manufacturing process: molding, molding of add-on parts, subassembly, painting, and final assembly. Another 691,000lbs of scrap representing 51,400 parts are recovered from subassembly, painting and final assembly. This scrap represents an additional materials cost of $960,490 and suggests a more correct scrap percentage rate calculation of 6%.

To recover the scrap, there are associated costs throughout manufacturing such as: labor, regrind/rework, loss of parts, and additional electricity usage. The author also estimated these costs to arrive at a better estimate of the scrap costs for the facility as shown in Table 3. The plant managers were omitting much of these costs in their estimates of scrap costs, and thus did not consider scrap to be a significant economic issue. Once the costs were revealed, scrap production became an economic issue in addition to the quality and environmental issues associated with inefficient resource usage.

2.4   Design for the Environment strategies -design for recycling, remanufacturing etc.).

The product design phase typically includes a listing of specific requirements of functionality that the product should exhibit when development is complete, such as strength requirements, tolerances, costs, weight, size, permissible noise levels in use etc. There are also a set of design for "X" parameters which must also be considered, where "X" can stand for manufacturability (M), reliability (R), Compliance (C), etc. A new design parameter has emerged over the last decade, Design for the Environment, or DfE. In the case of DfE, additional design requirements include specific environmental issues such as energy use, recyclability, and remanufacturability. Designing products with environmental concerns in mind will lead to a reduction in environmental costs for both the company and for society.

2.4.1 Environmental Management Application: Example of reuse/remanufacture issues for an electromechanical parts manufacturer

In many countries corporate environmental management has grown to include extended product responsibility issues. Companies are responsible for the post-consumer management of their products. The manufacturer needs to ensure that returned products are being converted to marketable products that meet consumer quality expectations. This results in environmental and quality management concerns for manufacturers, as they need to develop reuse and remanufacturing strategies, which address both waste management and product quality concerns. These strategies need to be incorporated in the design phase as part of quality management and DFE programs.

Klausner et al (1998a,b) recently developed and tested an electronic data log (EDL) circuit for use in small motors. The EDL records data during product use, which is retrieved upon product takeback and used to aid motor reuse and remanufacturing decisions. Klausner and colleagues also describe a cost analysis procedure to determine the potential cost savings that would be achieved through the use of the EDL device.

In this project, an EDL was developed for use in the motors of small hand held power tools – the motor is the most expensive component of these tools and offers the highest potential for component reuse in a product take-back situation. However, it is difficult to determine whether a motor in a post-consumer product is suitable for reuse without detailed and costly analyses. The EDL was designed to record parameters strongly associated with the degradation of the motor to allow for rapid determination of reuse potential. The EDL was designed to be small, cheap, and easy to incorporate into a small power tool. The EDL records information such as: the number of stops and starts of the motor, the accumulated motor runtime, and various sets of sensor information such as temperature and power consumption.

Klausner *et al.* completed a set of economic analyses to determine the motor recovery rates that would be required for profitable motor reuse based on EDL. This information is being used in the development of product return strategies for a power tool manufacturer. The authors also determined that the EDL could facilitate quality management efforts and marketing programs by providing insights into product usage patterns.

3.   SUMMARY

Quality management and environmental management must start with better design, including choice of materials, design for reuse and disassembly, and design for end of life. These tasks are difficult. The choice of materials depends not only on the properties of the materials, but also the specific use that is intended. There are important tradeoffs between designing for reuse-remanufacturing and designing for materials recovery-recycling. Further complicating green design and quality environmental management efforts is the need to make these decisions in the content of the life-cycle implications, rather than focusing on a single life cycle phase or environmental issue.

The last decade has seen a major change in the recognition that environmental management is important, is analogous to quality management, and must be accomplished in large part through green design and industrial ecology. The growing recognition of the need for change among designers, companies, and consumers is a step forward. Some insightful studies have been conducted that reveal the tradeoffs and some of their implications.

We have described some new quality environmental management tools and their applications. While a great deal of work must be done to develop new tools and apply them, we think of the glass as partially full. A long difficult journey has begun, but one that is needed to allow people to enjoy high standards of living in harmony with the environment -a goal summarized in the term 'sustainable development'.

Figure 1. Schematic showing the major life cycle stages.

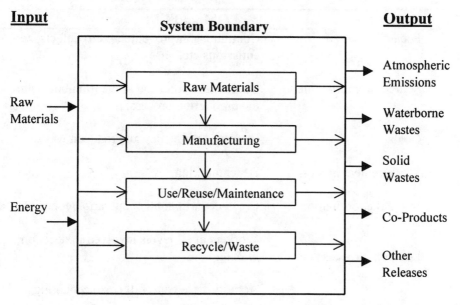

Stages of the Product Life Cycle

Source: Modified from the U.S. EPA

Table 1: Organizational aspects of Total Quality Management and Environmental Management Initiatives

Factor	TQM and EM
Focus	Continuous improvement-zero defects, zero emissions etc.
Source of improvement	Quality and pollution prevention built into the production process.
Goals	Increased efficiency and reduced waste
Measurement process	Benchmarking
Internal coordination	Cross-departmental cooperation/coordination
Decision process	Workers at all levels involved in decision making
Accounting systems	Activity-based and full cost accounting

Source: Modified from the Office of Technology Assessment, 1992.

Table 2: The electricity demand and the toxic releases associated with alternative car seat designs (each seat is assumed to cost $500 in the EIO-LCA model).

	Alternative 1 (plastic)	Alternative 2 (aluminum)
Total Electricity Demand (kwh)	560	1,030
TRI Air Releases (lbs)	2	2
TRI Total releases (lbs)	2	2
TRI Releases and Transfers (lbs)	4	4
CMU-ET Weighted Air Releases (lbs H_2SO_4 Equiv.)	0.1	0.2
CMU-ET Weighted Total Air Releases (lbs H_2SO_4 Equiv.)	1	2
CMU-ET Weighted releases and Transfers (lbs H_2SO_4 Equiv.)	6	9

Adapted from Lave et al., 1996.

Table 3. Summary of the costs associated with scrap production at the automobile parts manufacturing facility (adapted from Horney, 1998).

Scrap Activity	Scrap Costs per Year ($)
Regrind Labor Costs	260,000
Regrind/Rework Costs	592,000
Parts Loss	235,000
Overtime Labor	264,000
Additional Electricity	4,000
Materials Cost	1,499,000
Total	2,854,000

Noellette Conway-Schempf is Executive Director of the Green Design Initiative at Carnegie Mellon University and an Adjunct Assistant Professor of Environmental Management at Carnegie Mellon University's Business School, the Graduate School of Industrial Administration.

Lester Lave is the Director of the Green Design Initiative, a University Professor, and the James Higgins Professor of Economics at the Graduate School of Industrial Administration. He is also on the faculty of the Department of Engineering and Public Policy in the College of Engineering, and the Heinz School of Public Policy.

REFERENCES

BATE, 1999. Business and the Environment's ISO14000 Update, March, 1999, Cutter Information Corporation, Arlington, MA, p.3.

E. Cobas, C. Hendrickson, L. B. Lave, and F. C. McMichael, *"Economic Input-Output Analysis to Aid Life Cycle Assessment of Electronics Products,"* Proceedings of 1995 International Symposium on Electronics and the Environment, Piscataway, NJ: Institute of Electrical and Electronics Engineers, pp. 273-278, May 1995.

N. Conway-Schempf and L. Lave, *"Green Design Tools for Environmental Management"*, In Press, Quality Environmental Management, 1999.

T.E. Graedel and B.R. Allenby, *"Industrial Ecology"*, Prentice Hall, New Jersey, 412pp, 1995.

C. T. Hendrickson, A. Horvath, S. Joshi and L. B. Lave, *"Economic Input-Output Models for Environmental Life Cycle Assessment,"* Environmental Science & Technology, pp. 184A-191A, April 1998.

C. Hendrickson and A. Horvath, *"Resource Inputs and Emissions for Aggregate U.S. Construction Sectors,"* Technical Report Carnegie Mellon University, 1998.

C. Horney, *"Integrating Environmental Costs in a Management Information System: A Full Cost Accounting Case Study of a Manufacturing Plant,"* Cheryl Horney, M.S. Thesis, Department of Civil and Environmental Engineering, Carnegie Mellon University, May 1998.

A. Horvath, C. Hendrickson, L. Lave, F. C. McMichael and T.-S. Wu *"Toxic Emissions Indices for Green Design and Inventory"*, Environmental Science & Technology, Vol. 29, No. 2, pp. 86-90, February 1995a.

A. Horvath, C. Hendrickson, L. Lave and F. McMichael, *"Performance Measurement for Environmentally-Conscious manufacturing"*, ASME International Congress and Exposition, Symposium on Life Cycle Engineering, San Francisco, CA, November 1995, published in MED-Vol. 2-2/MH-Vol. 3-2, Manufacturing Science and Engineering, ASME 1995b.

A. Horvath and C. T. Hendrickson, *"Steel vs. Steel-Reinforced Concrete Bridges: An*

Environmental Assessment" ASCE J. of Infrastructure Systems, September 1998.

A. Horvath and C. T. Hendrickson, "*A Comparison of the Environmental Implications of Asphalt and Steel-Reinforced Concrete Pavements*", Transportation Research Board Conference, Washington, DC, January 1998, and Transportation Research Record, 1998b.

S. Joshi, "*Comprehensive Product Life-Cycle Analysis Using Input-Output techniques*", Ph.D. Thesis, Carnegie Mellon University, The John H, Heinz III School of Public Policy and Management, 1997.

M. Klausner, W. Grimm and C. Hendrickson "*Reuse of Electric Motors of Consumer Products: Design and Analysis of an Electronic Data Log*", Journal of Industrial Ecology, Vol. 2, no. 2, 89-101, 1998a.

M. Klausner, C. Hendrickson, and A. Horvath, "*A Product Takeback Case Study of Electromechanical Consumer Products*", submitted for review to Environmental Science & Technology, 1998b.

L.B. Lave, Cobas-Flores, E., Hendrickson C.T. and McMichael F.C., "*Using Input-Output Analysis to Estimate Economy-wide Discharges*", Environmental Science and Technology (ES&T), 29 (9), pp 420A-426A, Sept. 1995.

L. B. Lave, E. Cobas-Flores. F.C. McMichael, C.T. Hendrickson, A. Horvath, and Satish Joshi, "*Measuring Environmental Impacts and Sustainability of Automobiles*", Sustainable Individual Mobility – Critical Choices for Government and Industry Conference, Zurich, Switzerland, Nov 4-5, 1996.

H. MacLean and L. B. Lave "*A Life Cycle Model of an Automobile: Resource Use and Environmental Discharges in the Production and Use Phases*", Environmental Science & Technology, pp. 322A-330A, July 1998.

OTA, "*Green products by Design; Choices for a Cleaner Environment*". Congress of the United States, Office of Technology Assessment, ETA-E-541, October 1992.

President's Council on Environmental Quality, "*Total Quality Management. A Framework for Pollution Prevention*", Quality Environmental Management Subcommittee, President's Council on Environmental Quality, Washington, D.C., January 1993.

U.S. Environmental Protection Agency "*Life Cycle Assessment: Inventory Guidelines and Principles*", EPA /600/R-92/245, 1993.

U.S. Environmental Protection Agency, "*An Introduction to Environmental Accounting as a Business management Tool: Key Concepts and Terms*", EPA 742-R-95-001, 1995.

AUTHORS

Noellette Conway-Schempf is Executive Director of the Green Design Initiative at Carnegie Mellon University and an Adjunct Assistant Professor of Environmental Management at Carnegie Mellon University's Business School, the Graduate School of Industrial Administration.

Lester Lave is the Director of the Green Design Initiative, a University Professor, and the James Higgins Professor of Economics at the Graduate School of Industrial Administration. He is also on the faculty of the Department of Engineering and Public Policy in the College of Engineering, and the Heinz School of Public Policy.

Chapter 6

Environmental Cost Accounting and Business Strategy

Robert J. P. Gale[1], Peter K. Stokoe[2]
[1]*Royal Roads University, Ecological Economics Inc.*
[2]*Ecostrategies*

1. INTRODUCTION

Firms need information for both financial and managerial accounting. On the financial side, information is required for a range of uses such as corporate financial planning and control, performance evaluation, and to verify credit worthiness and taxes owed. On the management side–the focus of this chapter–the emphasis is on controlling costs. The relationship between the environment and managerial accounting can be seen through the lens of cost control. This is because managerial accounting emphasizes the use of accounting information to serve "business managers in making capit al investment decisions, costing determinations, process/product design decisions, performance evaluations, and a host of other forward- looking business decisions" (EPA, 1996: 28).

We confine our inquiry into environmental cost accounting and business strategy to corporate environmental performance in this chapter. In doing so, we focus on accounting for internal managerial decision -making rather than accounting for reporting to external shareholders and other stakeholders. It is known that environmental costs can be substantial, from five to twenty percent of the total costs of business activities according to Ditz *et al.* (1995:15). Because these costs are likely to rise as pressures for environmental protection measures increase, the purpose of this chapter is to make the case that incorporating environmental costs directly into accounting functions and business strategies can improve a business's competitive position.

2.   THE EXPANDING BASE OF COST ACCOUNTING

While research on social accounting and reporting is rooted in the early 1970s, the validity of corporate environmental accounting in professional practice has only been widely accepted over the last five years. By 1998, many of the major North American accounting organizations had produced at least one publication on environmental accounting. For example, the Canadian Institute of Chartered Accountants published *Environmental Costs and Liabilities: Accounting and Financial Reporting Issues* in 1993, and *Environmental Reporting in Canada: A Survey of 1993 Reports* in 1994. In gathering literature on this emerging area of inquiry, CICA now maintains a twenty-page list of environmental accounting references on its web site.[1] Elsewhere, the Society of Management Accountants of Canada has produced an excellent series of guides including *Tools and Techniques of Environmental Accounting for Business Decisions* (1996) and *Accounting for Sustainable Development: A Business Perspective* (1997). Other accounting organizations have had publications prepared for them by specific experts in the area. These include the Certified General Accountants Association of Canada,[2] the U.S. Institute of Management Accountants[3] and the International Federation of Accountants[4].

A recent survey of environmental management strategies by the Society of Management Accountants of Canada (SMAC) suggests that the majority of major corporations in Canada and the U.S. will begin implementing environmental accounting and reporting practices by the end of 1998 (SMAC 1997). While we will have to await the evidence for this claim, the survey indicated that the three motivating factors to account for this trend in order of priority are:

1. Compliance with standards,
2. A moral commitment to environmental stewardship, and

---

[1] Canadian Institute of Chartered Accountants.1999. On-line Environmental Resources. CICA, Toronto. www.cica.ca/new/pa/environ/envires.htm accessed April 2, 1999.

[2] See Anderson, Elias and Zéghal (1998)

[3] See Epstein (1996)

[4] See Dzinkowski (1998)

3. The desire to promote good relations with the residents of local communities.

Within the corporation, environmental accounting concerns the definition, assessment and allocation of environmental costs and expenditures for the purposes of cost and resource management, compliance reporting, and capital budgeting, planning, and operational decision making. Environmental accounting can be further delineated into two main areas: financial environmental accounting and managerial environmental accounting.

Financial environmental accounting emphasizes the analysis and reporting component of internal costs and liabilities related to environmental matters. This is typically the domain of an accountant who prepares financial reports for lenders and investors. The assessment and reporting of environmental risks and liabilities, capitalization for environmentally related expenditures and the treatment of environmental debt, all fall into this stream of environmental accounting. In these matters accountants are guided by professional accounting standards such as the Generally Accepted Accounting Principles (GAAP).

Managerial environmental accounting has a different focus. It supports the internal management and decision-making process through various techniques of cost allocation, performance measurement and business analysis. This type of environmental accounting is interdisciplinary in scope. On the one hand, scientists, economists, and policy advisors can identify internal and external environmental costs. On the other hand, the management accounting profession can use its expertise to allocate these costs within existing and emerging environmental and sustainability accounting frameworks.

Given the two main areas of environmental accounting and the fact that both accountants and environmental experts are required to delineate and allocate internal and external costs, it is not surprising to find different methods related to environmental accounting in the literature. These include:

- activity-based costing/activity-based management
- total quality management/total quality environmental management
- business process re-engineering/cost reduction
- design for environment/life-cycle design and assessment
- life-cycle assessment/life cycle costing
- total cost assessment

- full cost assessment

In this paper, the focus on corporate environmental accounting is confined to activities where one could reasonably expect the accounting profession to be involved. While this includes all of activities noted above, we will confine our discussion to total cost assessment, full cost environmental assessment and life-cycle assessment.

2.1 Understanding Internal and External Environmental Costs

There are many ways in which environmental costs, losses or benefits may go unrecorded in traditional accounting systems. One broad approach to calculating full environmental costs is to distinguish between internal costs (those borne by the organization) and external costs (those passed on to society), e.g. environmental and health costs. In this approach, internal environmental costs to the firm are composed of direct costs, indirect costs, and contingent costs. These typically include such things as remediation or restoration costs, waste management costs or other compliance and environmental management costs. Internal costs can usually be estimated and allocated using the standard costing models that are available to the firm. Direct costs can be traced to a particular product, site, type of pollution or pollution prevention program (e.g., waste management or remediat ion costs at a particular site). Indirect costs such as environmental training, R&D, record keeping and reporting are allocated to cost centers such as products and departments or activities.

External costs are the costs of environmental damage external to the firm. These costs can be "monetized" (i.e., their monetary equivalent values can be assessed) by economic methods that determine the maximum amount that people would be willing to pay to avoid the damage, or the minimum amount of compensation that they would accept to incur it.

Full environmental costs = (internal + external costs)

Where:

Internal costs = (direct + indirect + contingent)

External costs = the costs of external environmental and health damage

(e.g., the costs of uncompensated health effects and environmental impacts -

Stratospheric ozone depletion; biodiversity loss; climate change)

From the perspective of society as a whole (i.e., the firm and the rest of society), economic efficiency is achieved (i.e., full environmental costs are minimized) when the firm takes internal measures to protect the environment up to the point where the sum of internal and external costs is minimized.

Contingent or intangible environmental costs are costs that may arise in the future to impact the operations of the firm. Contingent costs can fall into both internal and external cost categories, and include:

- changes in product quality as a result of regulatory changes that affect material inputs, methods of production, or allowable emissions;
- an unforeseen liability or remediation cost;
- employee health and satisfaction;
- customer perception and relationship costs; and,
- Investment financing costs or the ability to raise capital.

External costs are typically of less interest to the firm than internal costs, unless the external costs lead to liabilities for the firm. The distinction between internal and external environmental costs is illustrated in Table 1. The total area of the box in the table represents the entire spectrum of environmental costs that can be incurred as a result of the production or existence of a firm.

Within the existing financial reporting framework, the Canadian Institute of Chartered Accountants (CICA, 1993) applies the term "environmental losses" to the category of environmental cost expenditures for which there are no returns or benefits. According to CICA environmental losses are damages that have to be paid to others as a result of damage to the environment that resulted in bodily injury to humans, damage to the property of others, economic damage to others, or damage to natural resources (CICA 1993). CICA also describes another category of environmental cost expenditures as "environmental measures". These are the costs incurred to "prevent, abate, or remediate damage to the environment or to deal with the conservation of renewable and non-renewable resources" (CICA 1993; see also, Judd, 1996).

TABLE 1: Internal and External Environmental Costs

External Environmental Costs
Examples: Depletion of natural resources Noise and aesthetic impacts Residual air and water emissions Long-term waste disposal Uncompensated health effects Change in local quality of life
Internal Environmental Costs

Direct or Indirect Environmental Costs	**Contingent or Intangible Environmental Costs**
Examples: ▪ Waste management ▪ Remediation costs or obligations ▪ Compliance costs ▪ Permit fees ▪ Environmental training ▪ Environmentally driven R & D ▪ Environmentally related maintenance ▪ Legal costs and fines ▪ Environmental assurance bonds ▪ Environmental certification/labeling ▪ Natural resource inputs ▪ Record keeping and reporting	*Examples*: ▪ Uncertain future remediation or compensation costs ▪ Risk posed by future regulatory changes ▪ Product quality ▪ Employee health and satisfaction ▪ Environmental knowledge assets ▪ Sustainability of raw material inputs ▪ Risk of impaired assets ▪ Public/customer perception

Source: Adapted from: Whistler Center for Business and the Arts. Environmental Accounting. Prepared by T. Berry and L. Failing. 1996.

2.2   Internalizing Externalities

The objective of externality costing is to internalize externalities. In other words:

- to allow the external costs a firm imposes on society to be brought to bear in an augmented profitability calculation;
- to bring external costs considerations into the corporate decision making process;
- to ensure future viability of the organization through understanding potential liability and risk scenarios; and,
- to be able to inform stakeholders on the environmental and health impacts of the organization's economic activities.

Externality costing generates monetised estimates of environmental damage created by an organization, either at a specific site or through its activities. There are two widely used approaches for monetising externalities. The only valid approach from the standpoint of economic theory is the damage cost approach, i.e., assessing the value of environmental (and health) damage to those who incur the damage, as described above. The damage cost approach uses the value of loss of use to estimate externality costs. Within the damage cost approach are the following evaluation methods:

1. market price method
2. hedonic-pricing method
3. travel cost method
4. contingent valuation methods (survey questionnaire methods)

However, if firms undertake (or are required to undertake) measures to reduce environmental damage to the "optimal" extent (i.e., the extent which minimizes the sum of internal and external costs), then the marginal external cost (incremental cost of the last unit of harm) will be equal to the marginal internal cost (incremental cost of preventing the last unit of harm). On this basis, marginal external costs are sometimes assumed to be equal to marginal internal costs, and are estimated accordingly; this is generally called the "cost of control approach".

The concern about accounting for external costs is also reflected in the increasingly widespread practice of using "shadow prices" (e.g. dollars per ton of greenhouse gas emissions) in firms' capital budgeting decisions. This reflects the view that, although such costs are not currently imposed on the firm, it is likely that they will be before long.

Alternatively, it might be more pragmatic and realistic for a firm to take account of external costs as impending internal costs. In other words, it might be assumed that each category of external cost would eventually be reflected in internal costs. As external costs become internalized, the internal costs rise from zero (when the costs are purely external) to magnitudes that might meet or even exceed the magnitudes of the initial external costs. Therefore, rather than accounting for external costs directly and immediately, a firm might take account of them in terms of various possible time profiles of (future) internal costs (as external costs become internalized). These time profiles of future costs would still have implications for current capital budgeting (and other) decisions.

3.   IMPLICATIONS FOR BUSINESS STRATEGY: TOTAL, FULL COST AND LIFE-CYCLE ASSESSMENT

Broadly speaking, environmental accounting describes, measures and reports on the allocation of environmental resources, costs, expenditures and risks to various industry groups, to specific firms, or within firms to specific departments, projects, activities or processes. With regard to the expanding base of environmental accounting, three techniques are particularly salient: total cost assessment, full cost assessment, and life-cycle analysis. Each has its basis in activity-based costing (ABC), a technique accountants are familiar with, at least at the theoretical level.

According to the Society of Management Accountants of Canada (SMAC, 1997), the origins of activity-based costing (ABC) can be traced back to 1985. The Society distinguishes between traditional cost accounting and ABC (SMAC (1997: 1):

Traditional cost accounting allocates costs to products based on the attributes of a single unit. A typical attribute is the number of direct labor hours required to manufacture one unit. Allocations therefore vary directly with the number of units produced. In contrast, ABC systems focus on the activities required producing each product or providing each service, based on that product or service's consumption of the activities.

The EPA (1995: 37) states that ABC is,

> *a means of creating a system that ultimately directs an*
> *organization's costs to the products and services that*
> *required these costs to be incurred. Using ABC, overhead*
> *costs are traced to products and services by identifying the*
> *resources, activities, and their costs and quantities to*
> *produce output.*

ABC can be considered as a technique for the economic analysis of a firm's overhead or indirect costs. Unlike total cost assessment or full cost assessment, it is part of conventional cost accounting (Figure 1). Although ABC does not ensure that a broader range of direct, indirect, contingent or less quantifiable costs are included in the analysis, it is a more accurate form of environmental accounting than traditional cost accounting. ABC is also a critical technique for gathering information that is required in TCA and FCA. Ideally, although this is not generally emphasized in the literature on TCA and FCA, costs need to be itemized by using the ABC technique before meaningful results can be generated in a TC or FC assessment. As Foster (1995:296) states:

> *The basic premise of ABC is to "cost" activities, which*
> *then becomes the ways and means for assigning/allocating*
> *costs to products. Subsequently, relevant environmental*
> *costs are allocated on the basis of the individual products'*
> *demand for those services.*

3.1   Total Cost Assessment

Total cost assessment refers to the long-term, comprehensive financial analysis of the full range of internal (i.e., private) costs and savings of an investment. The framework for the total cost assessment (TCA) technique represents an expanded approach to traditional financial analysis. It is a tool for preparing business cases that facilitates identifying and analyzing internal project costs and savings. Total cost assessment builds upon conventional cost accounting models by including:

- direct and indirect financial costs, and,
- recognized contingent costs

Recognized contingent costs include future compliance costs, penalties and fines, relationship costs, release response costs, remediation costs and the time value of money (also a critical concern in "conventional" accounting models).

However, TCA is less comprehensive than full cost environmental accounting in that it necessarily excludes costing for externalities (Figure 3). According to Reid, Fraser and Schoeffel (1997), the primary benefit of TCA is that "it helps 'level the playing field' by allowing projects that generate longer-term savings to compete more successfully for limited capital funds. This, in turn, provides companies with an opportunity to improve their bottom line through captured efficiencies and more accurate costing and pricing."

Investment appraisal techniques such as TCA that incorporate allocations of environmental costs and extended time horizons are appropriate when assessing which investments, including pollution prevention investments, are economically favorable (Schaltegger and Muller 1997). P2/FINANCE, developed by the Tellus Institute, is a spreadsheet software application for conducting financial evaluations of current and potential investments. P2/FINANCE has been designed to capture a broad range of potential environmental costs and savings including internal indirect and less tangible environmental costs and uses profitability indicators and time horizons that capture the long-term characteristics of environmental investments.

There are only a handful of good TCA case studies. For example, White, Savage and Shapiro (1996: 7-9) from the Tellus Institute report on an unnamed paper mill that commissioned a study on the operation of the mill and how to reduce "peak effluent flows, reduce BOD in the effluent and reduce total fresh water intake on a mill wide scale". Since TCA is concerned with a range of costing errors in project financial analysis, a TCA cost analysis is different to a conventional cost analysis. As White, Savage and Shapiro note (1996: 7-10):

> *The extent to which TCA improves a pollution prevention investment's profitability depends on the firm's current cost structure, project evaluation practices, the specific project, and the degree to which less tangibles are significant and quantifiable.*

In the case study these authors report on, results are discussed in terms of three financial indicators: net present value, internal rate of return and simple payback. Their analysis indicates that a proposed investment in new equipment "meets the mill's 2-year-payback rule of thumb" when the costs are properly allocated.

3.2 Full Cost Environmental Assessment

For the purposes of our discussion, full cost environmental assessment is distinguished from total cost assessment and other cost accounting techniques (Figure 1). In addition, while the term 'full cost accounting' is frequently used in discussions of environmental accounting, most FCA studies do not attempt to quantify the social impacts of an organization's activities. For greater clarity, we prefer the terms full cost environmental assessment or full cost environmental accounting when considering the identification, evaluation and allocation of conventional and environmental costs in an organization. The broader term 'full cost accounting' would include these costs plus the external social costs borne by society ((e.g. adjustment costs from lay-offs and involuntary terminations, especially in firms or industries where turnover rates are higher than average and this is not reflected in higher social insurance tax rates)".

Perhaps the best-known case study with full cost assessment is at a major North American power utility. In 1993, with the aid of former Chair and CEO Maurice Strong, Ontario Hydro became one of the few organizations to incorporate full cost accounting (FCA) information into their decision-making process. In particular, Ontario Hydro was a pioneer in taking on external cost accounting as part of its FCA framework.

Prior to developing its FCA framework, Ontario Hydro had conducted assessments of external costs of its activities (and proposed activities) on an *ad hoc* basis, in connection with regulatory reviews and various project assessments. The support of Maurice Strong provided an opportunity to give external cost assessment a greater role. Mr. Strong had been the Secretary General of the first major United Nations environment conference in 1972; he was subsequently the first Director of the United Nations Environment Program (UNEP), and later the Secretary General of the Rio "Earth Summit" in 1992. Upon becoming Chair and CEO of Ontario Hydro in 1992, Mr. Strong appointed a Task Force on Sustainable Energy Development, which received analytical support from several working groups or "teams" within the Company, including a Full Cost Accounting Team. The work of the Task Force and the Full Cost Accounting Team provided the basis for establishing FCA as a principal analytical tool to support decision-making for sustainable development within the company.

Toward this objective, Ontario Hydro conducted and supported [delete: considerable; insert: extensive] work in the mid-1990s on generic assessments of the external (i.e. environmental and human health) impacts of its activities. In 1995, the Environment and Sustainable Development Division (ESDD) developed Corporate Guidelines for FCA. The Guidelines outline the rational for FCA, describe the implementation process, and delineate roles and responsibilities. The guidelines also specified that Ontario Hydro incorporate FCA into evaluations of:

- major integrated resource plans
- operation and dispatch of Ontario Hydro's system;
- investment decisions
- environmental externalities associated with imports and exports of electricity;
- retiring or rehabilitating existing stations;
- benefits and costs of additional pollution control equipment; and
- monitoring environmental performance improvements.

Ontario Hydro applied methods of environmental economics (based on the "damage function approach") to develop monetary values for external impacts, which hence became "monetized external impacts". By monetizing external impacts, they could then be taken into account, like any other monetary value, in economic decision-making. Although the ideal might be to monetize all external impacts, practical difficulties often meant that some external impacts remained as non-monetized external impacts. Where non-monetized external impacts or other considerations (e.g. risk, reliability, and social impact) were considered important, Ontario Hydro resorted to Multi-Criteria Analysis (MCA) to balance incommensurables in decision-making.

In a report (ICF Incorporated 1996) on Ontario Hydro's experience with FCA, it is noted that:

> *Ontario Hydro recognizes that some definitions of full cost accounting include only "internal costs" (also termed "private costs"), which are the costs that affect a firm's bottom line, and exclude "external costs" (also termed "social costs") which is a term used to describe monetized impacts on human health and the environment that currently are not reflected in a firm's bottom line. Ontario Hydro's approach explicitly encompasses both internal costs and external impacts (both positive and adverse), even if the latter cannot be monetized or expressed as external costs*

(i.e., fully monetized in dollars)... Ontario Hydro has explicitly acknowledged that the dividing line between internal and external costs is not static. For example, a cost that Ontario Hydro considers external today may be internalized tomorrow because of new environmental regulations or corporate standards. "

After Ontario Hydro began its FCA work, Ontario regulators also encouraged Ontario's natural gas utilities to adopt FCA principles. At present these initiatives are in abeyance, as Ontario's utilities are currently undergoing substantial restructuring.

3.3 Life-Cycle Assessment

Life-cycle assessment is a third approach to business strategy development because the costs of a product or process throughout its life cycle can ultimately bear on costs for the firm. This is especially so given the increasing expectations on firms to take responsibility for the full life-cycle impacts of products, including through product (and packaging) take-back requirements and "extended producer responsibility". There are also increasing demands for product labeling to take account of "process and production methods" (PPMs), and not only the characteristics of the product *per se.*

There are a number of approaches to Life Cy cle Assessment that reflect its origins in different disciplines (Welford 1995; Epstein 1996; Allen 199x; White, Savage and Shapiro 1996). From an environmental accounting perspective, the focus is on adding a monetary component, that is, to assign a cost to each environmental impact. The sum of all the costs at each stage in a LCA would yield the net environmental costs of a product or process. This means that there is a life-cycle costing (LCC) component to LCA. Epstein (1996: 154) states that life cycle costing is,

... an attempt to identify all the environmental costs (internal and external) associated with a product, process, or activity throughout all stages of its life. The product stages in the life cycle process are raw materials acquisition; manufacturing; use, reuse, and maintenance; and recycling and waste management.

AT&T, in their Green Accounting Glossary, define LCC as:

A costing concept that argues for including all the costs incurred for a product, from its inception to abandonment, as part of its product cost. In Green Accounting, this includes cost of extraction, intermediate manufacturing, manufacturing, transportation, product recycling in take-back, disassembling, reverse distribution, restocking used material, disposing of waste, etc.

Allen (199x) provides case examples of LCA. These studies detail opportunities for energy efficiency, water efficiency, and solid waste reduction. An environmental accounting approach would add costs to this analysis so that the financial impacts are clear.

4.   LINKING ENVIRONMENTAL ACCOUNTING TO BUSINESS STRATEGY

Given the techniques described above, business strategy can be linked to environmental cost accounting in at least three separate ways. Business strategy could focus exclusively on total cost assessment in which case the external environmental costs borne by society are ignored. Or the external environmental costs could be included in the broader full cost environmental assessment framework. Finally, business strategy could be linked to the broad range of external environmental and social costs in a full cost assessment. The later approach ultimately leads to the category of sustainability accounting and reporting which is beyond the scope of this account. Details on this approach can be found in Elk ington (1998).

From a management accounting perspective, the next step beyond Activity Based Costing is "strategic cost management". According to Shank and Govindarajan (1993), strategic cost management "is cost analysis in a broader context, where the strategic elements become more conscious, explicit, and formal. Here, cost data is used to develop superior strategies en route to gaining sustainable competitive advantage:" Strategic cost management thus represents an important link between business strategy and the choice of an environmental accounting tool such as TCA, FCEA, or LCA.

Success in linking strategic cost management to environmental accounting will depend on at least five factors:

1.   The motivation for environmental protection and/or pollution prevention initiatives;

2. A systematic procedure for identifying costs (or to use CICA's terms environmental measures and losses);
3. Achievable but demanding objectives and targets;
4. The integration of various corporate strategies in the organization as a whole; and
5. A reporting system that provides a monitoring and corrective feedback system for the strategy.

First, the motivation to link environmental accounting to business strategy needs to be considered. A concern with compliance, for example, will drive a different choice of management strategies than a concern for the costs of environmental impacts. Compliance-oriented strategies would lead to techniques such as environmental auditing and the development of corporate environmental management systems. In this approach, concern for the costs of environmental impacts would be very general. Whole books are written on environmental topics for business managers without any mention of costs. In contrast, a cost-oriented strategy would go beyond EMS and EA techniques to examine "the costs of production plus the cost of any environmental damage associated with it (Schmidheiny, 1992: 17).

Second, the system for gathering information is critical to the success of an environmental accounting initiative. To this end, the purpose of environmental accounting "is to provide relevant in-house information that will support the making of environmentally compatible decisions by management (Fuller, 1999: 287).

Third, the managerial uses of environmental cost accounting information must be related to achievable but demanding objectives to enhance not only environmental performance objectives but also productivity and profitability objectives for the company. Fuller (1999: 294-295) suggests the following six areas in which environmental cost accounting information can support marketing and managerial decisions: product mix decisions, choosing manufacturing inputs, assessing pollution prevention projects, evaluating waste management options, comparing environmental costs across facilities, and pricing products. We believe that environmental cost accounting could be used in all areas of business decision-making and that these six items these six areas provide a good starting point for analysis. To these six, we would add product (and/or service) *design*.

The forth factor to consider is the integration of corporate strategies. Schaltegger, Muller and Hindrichsen (1996:225) argue in favor of evaluating strategic options on at least three levels: corporate, business and product.

Strategic options include, for example, the choice of new businesses to enter. Business strategy may be concerned with product-mix decisions. Finally, at the product strategy level, options include environmental upgrades or the discontinuation of products on environmental or other grounds. For strategic cost management to be successful, strategy at the corporate, business and product levels needs to be coherent from an environmental accounting perspective.

Rob Gray (1996: 173) best sums up the fifth factor, reporting. He sees accounting and reporting as two sides of the same coin.

> *The two sides are mutually dependent, it is impossible to report until one has something to report, to give account until something is accounted for.*

5. CONCLUSIONS

Managerial environmental accounting provides a comprehensive means for incorporating environmental considerations into business decision-making. The inclusion of internal environmental costs in its accounting will assist a company in working to maximize its current profitability. A firm can further be guided in maximizing its long-run profitability by taking into account external environmental costs, especially to the extent that it may be required to internalize these costs in the future. The adoption of these methods can help put a firm in a stronger competitive position in relation to firms that apply only conventional accounting. The extent of this advantage will depend on how extensively and creatively the firm makes use of these methods in its decision-making.

REFERENCES

Anderson, T., Elias, N. and Daniel Zéghal. 1998. *Environmental Management and Reporting*, Research Study Series, Certified General Accountants of Canada. CGA, Vancouver.

Canadian Institute of Chartered Accountants. 1993. *Environmental Costs and Liabilities: Accounting and Financial Reporting Issues.* CICA, Toronto.

Canadian Institute of Chartered Accountants. 1994. *Environmental Reporting in Canada: A Survey of 1993 Reports.* CICA, Toronto.

Ditz, Daryl; Janet Ranganathan and R. Darryl Banks (eds.). 1995. *Green Ledgers: Case Studies in Corporate Environmental Accounting.* Washington: World Resources Institute.

Elkington, John. 1998. *Cannibals With Forks: The Triple Bottom Line of 21st Century Business.* New Society Publishers. Gabriola Island, British Columbia.

Environmental Protection Agency. 1995. *Environmental Cost Accounting for Capital Budgeting: A Benchmark Survey of Management Accountants.* United States, EPA, Washington, D.C.

Environmental Protection Agency. 1996. *An Introduction to Environmental accounting as a Business Management Tool: Key Concepts and Terms.* United States, EPA, Washington, D.C.

Epstein, M.J. 1996. *Measuring Corporate Environmental Performance: Best Practices for Costing and Managing an Effective Environmental Strategy. A Research Study for the Institute of Management Accountants.* Irwin Professional Publishing and the IMA Foundation for Applied Research: New Jersey. Financial Post. 18 January, 1999. "Skeena not out of woods despite lifeline." 18 January, C3.

Fuller, Donald A. 1999. *Sustainable Marketing: Managerial-Ecological Issues.* Sage Publications Inc., Thousand Oaks, California.

Gray, Rob. 1996. Corporate Reporting for Sustainable Development: Accounting for Sustainability in AD 2000, pages 173-196 in Richard Welford and Richard Starkey (Eds.) Business and the Environment. Earthscan. London.

ICF Incorporated. 1996. *Full Cost Accounting for Decision-Making at Ontario Hydro: a Case Study.* Washington: U.S. Environmental Protection

Judd, Greg. 1996. Environmental Accounting and Reporting Practices, pages 61-84 in Brett Ibbotson and John-David Phyper (Eds). Environmental Management in Canada. McGraw-Hill Ryerson Limited. Toronto.

Reid, Fraser and Schoeffel (1997), [details to follow]

Schaltegger, Stefan, Muller, Kaspar and Hennriette Hindricksen. 1996. *Corporate Environmental Accounting* by Stefan Schaltegger, Kaspar Muller, and Hennriette Hindricksen. John Wiley and Sons. Chichester.

Schmidheiny, Stephan. 1992. *Changing Course: A Global Perspective on Development and the Environment. Cambridge,* Massachusetts.

Shank, John K. and Vijay Govindarajan. 1993. *Strategic Cost Management: the New Tool for Competitive Advantage.* New York: Free Press.

Society of Management Accountants of Canada. 1996. *Tools and Techniques of Environmental Accounting for Business Decisions.* The Society of Management Accountants of Canada, Hamilton, Ontario.

Society of Management Accountants of Canada. 1997. *Accounting for Sustainable Development: A Business Perspective.* The Society of Management Accountants of Canada, Hamilton, Ontario.

White, Allen L., Davage, Deborah and Karen Shapiro. 1996. Life Cycle Costing: Concepts and Applications pp. 7-1 to 7 -19 in Mary Ann Curran (Ed.), Environmental Life Cycle Assessment. McGraw -Hill. New York.

AUTHORS

Robert Gale is a management consultant and presedent of Ecological Economics Inc. He is a faculty associate at Royal Roads University where he teaches on-line graduate courses including a course on sustainability in the MBA program.

Peter Stokoe is an economist and president of Ecostrategies. He specializes in devising strategies for realizing both economic and environmental objectives in businesses and communities.

Chapter 7

Accounting for an Environmentally Conscious Setting

Bin Srinidhi
City University of Hong Kong, Hong Kong, and Rutgers University, New Jersey, U.S.A

Environmental performance, measurement and management issues are becoming increasingly important to many different stakeholders such as investors, creditors, customers, employees, competitors, taxing authorities, regulators, environmental interest groups, Environmental Protection Agency, media, neighborhood communities and general public. All these stakeholders need credible information on environment-related activities of firms to make their own decisions. At the same time, the rate of development-related natural resource consumption has outpaced nature's regeneration rate, making such development unsustainable. A reversal of this trend requires appropriate managerial incentives at the firm level and meaningful policies at the macro level. Consequently, micro-level intra-firm environmental performance measures, environmental performance disclosures by firms and ecological auditing have become increasingly important. This has resulted in an increasing demand for sound commonly accepted internal and external environmental accounting practices. A large part of these practices are in an early stage of development. This chapter presents the current status of environmental accounting, which includes both the accounting for financial impact of environment-related activities and non-financial measurement of environmental performance. It also presents new developments in accounting, which could facilitate environmental performance evaluation. Finally, it briefly addresses the current research in the area of environmental accounting, reporting and management.

1.   ACCOUNTING IN ECO-MANAGEMENT – A FRAMEWORK

Over the years, there has been a growing realization of the close and delicate relationship that exists between the activities of a firm[1] and its environment. Sometimes, the activities of a firm have beneficial effects both for the immediate profit of the firm and for the environment. Often, there is a trade-off between the two. Accounting can provide quantitative estimates and measures of the costs and benefits that can help restore the delicate balance between the two. From its inception, accounting has focused more on the measurement of costs and benefits of the firm's activities in terms of its immediate financial implications but there is nothing in accounting theory that makes such focus inevitable. It is beneficial to society to change the focus of accounting to a broader measurement of costs and benefits over the long term to both the firm and to its environment. The measures that result from such a change of focus can facilitate a more prudent ecological cost benefit analysis.

It is fair to state that in comparison with the magnitude of the task the above change of focus implies, the actual attention given by the accounting practitioners (including accounting standard setters, external and internal auditors, management accountants) and academics (including teachers, educators and researchers) is minuscule. Financial statement disclosure has focused on a limited set of financial effects of environmental performance. The disclosure and accrual of contingent liabilities and certain environmental remediation costs are all that are required of firms. The most prominent technical reasons for this scant attention are outlined below.

The most problematic aspect of measurement is the presence of negative externalities in environmental costs and benefits. By externality, we mean that the party responsible for an activity might not bear all the consequences of such activity. In particular, the party responsible for environmental degradation does not directly bear the cost of such degradation. Those costs are passed on to others. Others might bear these costs contemporaneously or after a considerable length of time. Moreover, there is often uncertainty about who bears these costs, how much and in what form. The accounting profession has not come to grips with either externalities or with uncertainties (an examination of the deliberations of the Financial Accounting Standards Board on accounting for derivatives or on the

---

[1] In this chapter, the term "firm" is used to denote a manufacturing and/or servicing organization rather than the term "corporation". Corporation is a particular way of organizing a firm for legal purposes. Partnerships and proprietorships are other forms of organizing.

estimation of post-retirement benefits confirm this intuition) in an authoritative way. Such a framework of accounting is essential for sound Eco-management.

Figure 1 presents the framework under ideal conditions. Consider a party A who causes environmental degradation and thereby increases the costs for parties B, C and D. A good accounting system, in conjunction with political, legal and management systems, should identify all the affected parties, aggregate the costs imposed on all of them and then impose the total externality costs back on A.

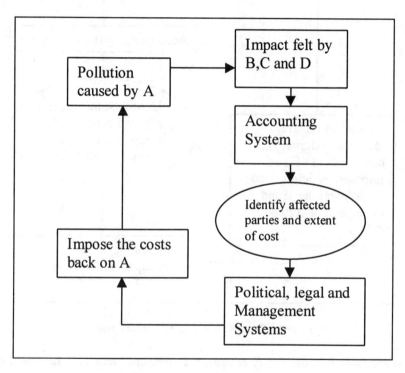

Figure 1. Framework Under ideal Conditions

Under the ideal framework, A equates the total marginal cost of pollution caused to A, B, C and D to the marginal cost of pollution control effort. The resultant equilibrium level of pollution, p_{ideal} will be such that these two marginal costs are equal.

Figure 2 presents the framework in a more realistic setting. It is known that the pollution caused by A has an impact on B and C. There is uncertainty on whether there is or will be an impact on D and E. The accounting system might identify some of the parties who are affected. It is also not clear how much impact the affected parties will feel now and in the future. The costs could be out-of-pocket costs such as clean-up costs or more likely, they could be opportunity costs such as costs of employee health and productivity and quality of life costs.

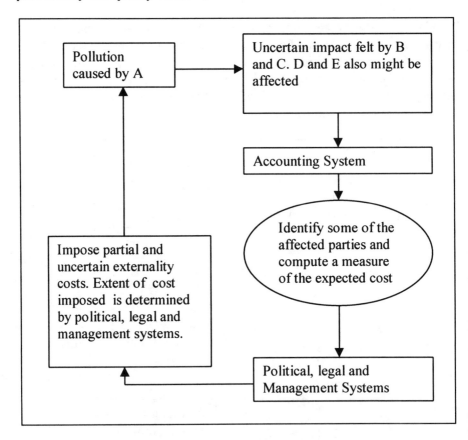

Figure 2. Framework under realistic conditions

Under this framework, A equates the measured expected marginal cost imposed by the conjunction of accounting, political, legal and management systems to the marginal cost of pollution control. Clearly, the resulting extent of pollution, p_{real} will be much higher than p_{ideal}.

The framework presented above provides definite implications on the factors that an accounting system must consider in environmental accounting. These are:

1. Uncertainty in the cost impact of pollution makes it less likely that the accounting system identifies the affected parties and computes the proper measure of environmental cost. The resulting level of environmental cost will be higher than if proper identification and measurement are made.

2. Diffused impact, i.e., small individual impact on a large number of different parties will make it more difficult for the accounting system to identify and measure the proper environmental cost. The resulting level of environmental cost will be higher than when the impact is not as diffuse.
3. The lack of political will, the lack of management leadership and poor co-ordination between the accounting, management and legal systems lead to under-identification of environmental costs and consequently, result in a higher equilibrium level of pollution.
4. Higher complexity and cost of accounting systems will result in reduced effort and consequently under-identification of environmental costs and a higher equilibrium level of pollution.

The above framework gives the technical framework in which environmental accounting needs to operate. It helps us understand the important parameters, the factors to be controlled and the interactions between accounting system and the political, legal and management systems. To complete the framework, we need to identify all the stakeholders of the accounting system. In other words, we need to identify the users of accounting information whose decisions will be moderated by the inclusion of environmental information as part of the accounting system.

Figure 3 gives the dispersion of stakeholders of accounting information. The identification of stakeholders is based partially on the exposition in Schaltegger, Muller and Hindrichsen (1996). Within the firm, strategic decision-makers, operating managers and other employees use management accounting information in their decisions. The inclusion of internal ecological accounting information will therefore influence operating decisions, capital acquisition decisions, strategic decisions and contracting (compensation contracting, labor- management contracting) within the firm. In the business environment of the firm, we identify customers, suppliers, competitors, complementors (businesses like hardware and software which complement each others' products), investors and creditors. These players use the information in financial reports of the firm. Inclusion of ecological information in the external reporting of the firm will influence the contracting and other relationships with these players. In addition, it will also develop new contracting relationships with some regulatory players such as the Environmental Protective Agency and other private interest environmental groups. In the legal and regulatory environment of the firm, we identify additional players such as the Internal Revenue Service, foreign tax agencies, the Federal, State and Local governments, Public Service Commissions and other regulatory agencies such as the Federal Trade Commission. We identify the communities in which the firm branches are located, the media, and the general public in the social and political environment of the firm. The nature and content of external ecological reporting will influence the explicit and implicit contractual and non-contractual relationships with all these agencies. Moreover, institution of changes in any part of the reporting structure will have to consider the potential impact on all these different players in the firm's environment. As an example, if the firm's managers believe that environmental disclosures lead to higher likelihood of legal actions, their incentive to disclose will be greatly reduced. In the absence of proper ecological auditing, their disclosures could be biased as well. This, in turn, will reduce the reliability of the report for external stakeholders. Similar disincentive effects can result if there is significant disparity between the disclosure requirements for U.S. based firms and competing foreign firms.

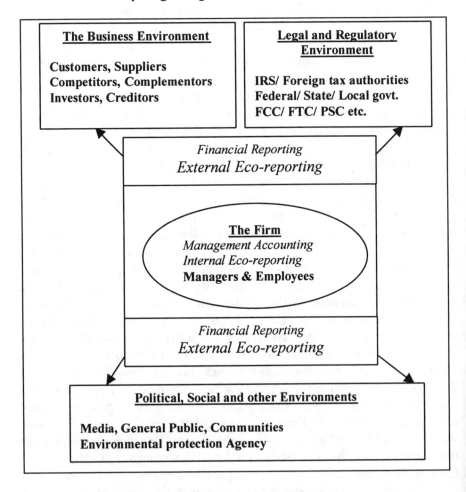

Figure 3. Stakeholders of the Accounting System

The value of any reported information to a decision-maker can be realized only if the information is credible, reliable and relevant. Internal auditors inside the firm and external auditors outside the firm bring such credibility and consequently, reliability and relevance to the information. The inclusion of ecological information in the internal and external reports requires the development of concepts, mechanisms and procedures of ecological auditing.

In summary, the framework of environmental accounting consists of the technical parameters and factors that influence environmental reporting, the various stakeholders who use the environmental accounting information and the institution of auditing which makes such information credible and valuable to them.

2. ENVIRONMENTAL ACCOUNTING SYSTEMS

The focus of this section is to describe the various components of accounting and reporting systems and their relationships[2]. We distinguish between two kinds of accounting systems. *Eco-sensitive accounting systems* collect information and report on the financial impact of environmental issues in monetary terms to the normal stakeholders of accounting information (such as managers, employees, investors, creditors, rate regulators and tax authorities). Schaltegger, et.al. (1996) call such systems "environmentally differentiated" accounting systems. On the other hand, *Ecological Accounting systems* collect and report information on environmental issues *whether or not they have financial impact on the firm* to both the normal stakeholders of accounting as well as to new stakeholders such as the Environmental Protection Agency (EPA) and Ecological Rating Agencies. Moreover, this reporting is not necessarily in monetary units. This terminology is consistent with Schaltegger, et.al. (1996). The term "Environmental Accounting Systems" is used generally to refer to both the Eco-sensitive and Ecological Accounting Systems.

---

[2] This section includes some ideas and concepts that have been developed in Schaltegger, Muller and Hindrichsen (1996)

2.1   A Brief Introduction to Financial and Management Accounting

2.1.1   Financial Accounting

Financial accounting refers to the branch of accounting that deals with the collection, processing and reporting of periodic financial information about the firm to shareholders. The focus of financial accounting is to provide the shareholders relevant and reliable financial information about the performance of the firm over a year (or a quarter, as the case might be), which can help them value the firm. This valuation is based on the present value of future expected cash inflows. A better valuation is possible if the financial accounting information helps predict the future cash flows with greater precision. In other words, financial accounting is value-relevant if and only if the information has predictive ability about the future cash flow from the firm to its shareholders. In particular, information that does not directly or indirectly help in estimating the probability distribution of future cash flows is not considered relevant for good financial reporting.

The above perspective (called the information perspective) is operationalized by a set of accounting principles, conventions and procedures. Every public firm in the United States is required to file a set of financial statements (10 K) with the Securities and Exchange Commission (SEC) every year. This set consists of a balance sheet that gives the accounting values of assets, liabilities and shareholder's equity at the end of the year, an income statement that presents the revenues and expenses for the year and a statement of cash flows that gives a picture of cash inflows and outflows from operations, financing and investing. The generally accepted accounting principles (GAAP) govern the extent of disclosure, the amount and timing of recognition of revenues and expenses as well as the valuation of assets and liabilities. The Financial Accounting Standards Board (FASB) is the principal body that promulgates the standards of reporting. It is a private body that operates under the delegated power of the SEC. The American Institute of Certified Public Accountants (AICPA) and its affiliate bodies, the State Boards of Public Accountancy and other organizations complement the FASB in the standard setting and implementation process. The principles and conventions that have evolved over time are based on the following general criteria:
1. Objectivity and auditability: Objectivity refers to the property that two or more people examining the same transaction will arrive at the same number using the procedure. Auditability refers to the property that an auditor, applying generally accepted auditing principles and

tests of reasonableness will be able to assess whether the disclosed item is a truthful representation. As an example, the historical cost of acquisition is a more objective and auditable number than the replacement cost or liquidation cost.

2. Recognition of revenues and expenses: Revenues are recognized when the probability of cash collection from the transaction is highly probable and the service or product that earned the revenue has been delivered to the customer. Expenses are matched in time with the revenues they help generate, subject to tests of low uncertainty. The matching principle governs such issues as the rules of depreciation and decisions to capitalize or expense costs. However, very long time intervals between the incurrence of cost and the realization of revenues increases uncertainty and lead to periodic, rather than matched, expensing.

3. Conservatism: The standards have generally followed the principle that in the absence of certainty, it is better to report lower income earlier and possible higher income later than to report higher income now and report possible lower report later. This criterion is evidenced in rules such as lower of cost or market and impairment of the value of assets. Loss of value is recognized immediately but gain of value is deferred till a transaction takes place.

The structure and the GAAP principles followed in the United States have had a significant influence on the standard setting process of the International Accounting Standards Committee (IASC). Most of the other countries in the world follow either the standards set by IASC or set their own standards very similar to those of the U.S. or of IASC. Therefore, the arguments presented here based on the workings of the U.S. standard setters are appropriate in most other countries.

The premium that is placed on certainty, objectivity and conservatism in the financial accounting system explains the reluctance of the accounting profession to undertake the difficult task of including the impact of environmental issues in reporting. As detailed in the framework, environmental effects are characterized by a high degree of uncertainty and by a lack of precise timing of the impact. Secondly, the primary notion of the information perspective is that the information should be useful in predicting cash flows. It is often difficult to establish a direct link between say, environmental degradation and specific cash flows.

2.1.2   Management Accounting

Management Accounting is the branch of accounting that collects and disseminates mostly financial but also some non-financial information from the products, departments and other organizational units of the firm to internal managers for the purpose of evaluating and improving their performance. *Cost Accounting* is the part of Management Accounting which is focused on the attribution of costs to cost objects such as products, departments, divisions, projects, customer classes, distribution channels etc. *Cost Management* is the part of Management Accounting, which focuses on the use of cost accounting information to increase the value of the firm by improving efficiency and effectiveness of operations, waste reduction, productivity improvement and inventory and process time reduction. Closely related to Cost management is the field of *Performance management* which deals with budgeting, performance evaluation, transfer pricing, divisional performance evaluation such as economic value added, target analysis and non-financial performance evaluation. The recently evolving field of *Strategic Cost Management* expands the scope of cost and value management to outside the organization by exploiting value chain and value net relationships and strategic positioning of the firm's activities.

Certain characteristics of management accounting stand out in sharp contrast to those of financial accounting. Unlike financial accounting, there are very few standards, principles and procedures that are mandated in management accounting. There is much greater subjectivity than in financial accounting. The value of management accounting to the firm lies in improved decisions made by the firm's managers rather than in improved investment decisions of external investors. There is a much greater scope for introducing non-financial measures of performance than in financial accounting.

The flexibility afforded in management accounting makes it friendlier to the introduction of environmental information. However, the flexibility is a double-edged sword. The firm's management has complete control over the information that gets collected and reported under the management accounting system. Each firm's management will individually evaluate the costs and benefits of including environmental information in their management accounting systems.

2.2 Eco-sensitive Financial and Management Accounting

2.2.1 2.2.1 Eco-sensitive Financial Accounting

The SEC is the only regulator that requires the disclosure of all material effects of compliance with environmental regulations (SEC 101 and SEC S-K 103). However, there are no specific accounting standards of eco-sensitive financial accounting even in the United States. The only direct reference to the impact of environmental issues is Statement of Position issued by the AICPA in September 1996 addresses the issue of remediation costs. It became effective from December 1996. Most of the discussion arising from the exposure draft of this SOP was on whether to include the legal costs of defending against assertion of remediation liabilities. In general, disclosure and accrual of contingent liabilities are covered under SFAS 5. Liabilities arising from potential environmental impact fall under this standard. The guiding criteria under SFAS 5 are (i) the liability to the firm must be probable and (ii) the amount of liability can be estimated reasonably.

From the framework of environmental accounting, the two main characteristics of eco-sensitive financial reporting are (i) negative externality and (ii) uncertainty. The first criterion of SFAS 5 places a stringent filter on the reporting of externalities with uncertainty. The liability must be shown to be probable for it to be reported. The second criterion also places a restriction on the reporting of uncertain amounts of impact caused by the firm's activities. The bias of financial reporting standards is towards the legal defensibility of the reported numbers rather than on full disclosure of the best estimates. This bias acts against the interest of eco-sensitive financial reporting.

In view of the above bias and the lack of specific standards, the actual amount of disclosure of environmental impact is much less than is needed for adequate reapportionment of environmental costs to the party driving those costs. In the rest of this sub-section on eco-sensitive financial accounting, we enumerate the reporting rules that might influence the incentives of firms in undertaking capital investments in the arena of environmental management.

1. *Treatment of environmental expenditures as assets or as expenses*
An expenditure is classified as an asset if the expenditure results in future period benefits and if the future benefits are measurable with reasonable certainty. If it is classified as an asset, the expenditure is written off as depreciation or amortization over the entire period in which the benefits are anticipated. (It may be noted that the actual realization of future benefits is

not the question here – it is essential that the expectation of future benefits must be there at the time of acquisition). If it is classified as an expense, it is written off in the period in which the expenditure is incurred. This is true of the accounting practices of all countries. A clear example of expenditure classified as an asset is the acquisition of a tangible property such as a building or plant and equipment. An example of expenditure classified as an expense is the research and development costs (in the U.S.). The R & D costs clearly result in future benefits but the amount and timing of benefits cannot be estimated with reasonable precision[3].

Similar criteria apply to pollution prevention costs. Property, plant and equipment acquired for pollution control or safety reasons are capitalized and treated as assets under FASB and IASC standards. Both the FASB and IASC (Also FEE – Federation des Expertes Comptables Europeens) agree on the capitalization of costs that extend the life or improve the safety of another asset. They all agree that costs incurred to prevent future environmental impacts should be capitalized but clean-up costs of past damage should be expensed. Costs of voluntary activities that are not needed for legal compliance are generally expensed. Fines and fees are also generally expensed. There seems to be a difference between FASB and IASC in the case of costs, which might prevent or mitigate future contamination that has not yet occurred but which is likely to occur. IASC requires expensing of such costs but FASB allows those costs to be capitalized. There is no need for a separate recognition of environmental assets and expenses as such under any of the standards.

In general, greater ability to capitalize costs will result in greater incentives to incur those costs. This is presumably because it gives managers the flexibility to write the costs off at a future date and such flexibility has value. However, the standard setters are concerned also about the ability of the managers to manage the reported earnings. Greater the ability of managers to manage the reported earnings, less believable the reported numbers become and therefore, less valuable the accounting and reporting function. This trade-off between managerial incentives (for higher environment-related costs) and the need to restrict earnings management has led standard setters to take a position which might not be entirely favorable for environmental improvements.

2. *Treatment of environmental liabilities*

An environmental liability is an obligation to pay to clean up or compensate damage that has occurred due to past events or transactions that

---

[3] Many hold that the rule requiring the expensing of R & D expenditures has caused the United States to under-invest in R & D expenditures and over-invest in the acquisition of tangible capital goods.

can be directly attributed to the firm in question. As stated before, in the absence of any specific standards for environment-related liabilities, the general conditions under SFAS 5 apply. The liability must be probable, measurable and material to be recognized in the balance sheet of the firm. If the obligation is either not judged to be probable (or highly possible, in the words of the standard) or/and not estimable, but material, it needs to be disclosed in the footnotes as a *contingent liability*. If there is a counter-claim against the third party, it cannot be used as an offset under the U.S. GAAP but can be used as an offset under the IASC rules.

3. *Tradable Emission Allowances*

In the United States and in most other countries, the law (such as the Clean Air Act) allows a certain degree of pollution. In other words, the law usually stipulates a level of pollution above which it constitutes a violation of the law. In view of the fact that many of these laws are recent, firms that have been operating under the earlier laws are generally grand-fathered. Firms are given allowances for pollution based on past emissions and as a way of incentive to reduce the degree of emissions, are allowed to trade their pollution rights with others who are less compliant. Stated differently, an artificial market has been set up for pollution rights so as to achieve efficiency in pollution control. Firms that find the marginal cost of pollution less than the marginal cost of pollution control can buy pollution rights from other firms whose pollution control costs are lower. The expectation is that the worst polluters are penalized and the best pollution-controlled firms are rewarded under a market-based pricing scheme that makes the system efficient.

Emission allowances primarily have the characteristics of intangible assets and accounting principles support reporting of these allowances as intangible operating assets. However, there is no authoritative pronouncement in this matter and it is a matter of judgement whether these are treated as intangible assets or as non-current marketable securities.

4. *Management Discussion and Analysis*

In its management discussion and analysis part of the financial statements, the management of every public firm is required to disclose current trends and uncertainties with material effect and is encouraged to disclose anticipated future trends. Some firms disclose their environmental policy in this section. It is in the enlightened self-interest of the firm to present the intended strategy of dealing with environmental issues because this will serve to reduce uncertainty and improve the valuation of the firm. However, very few firms present such an environment strategy. (Appendix

A gives the disclosure of environmental issues by Union Carbide Corporation in their 1998 annual report.)

2.2.2   Eco-sensitive Management Accounting

As discussed in the previous section, management accounting is not a uniform set of conventions and procedures followed by all firms but rather an internal management measurement tool. Therefore, the focus in this section is *not on what should be done* but on *what can be done* under the rubric of eco-sensitive management accounting. First, we discuss **eco-sensitive cost accounting** that involves the identification and collection of environmental costs as well as the allocation of costs to environmental cost objects. Then, we discuss briefly **eco-sensitive cost management** that utilizes this information to evaluate the efficiency and effectiveness of environmental functions and activities in the firm.

2.2.2.1   Eco-sensitive Cost Accounting

The resources that are entirely devoted to environmental functions can be clearly identified as environmental costs. Examples of these costs include emission filters, scrubbers, wastewater treatment plants, incinerators, compensations paid to managers and employees who maintain or work on these specialized equipment, the rent or depreciation of buildings and other space costs attributable to these activities. The equipment is recorded at historical cost at the time of acquisition and depreciated over their useful lives. On a periodic basis, the cost of environmental control will include the depreciation cost. All the costs mentioned above are costs that are direct costs of environmental control.

Indirect costs are allocated to environmental control as well. These include the costs of administration and management that cannot be directly attributed to environmental control, the costs of training and other activities that are not specifically for environmental control, and the costs of material handling, logistics and other activities that are not primarily environmental.

The most difficult-to-allocate costs are those of integrated technology, higher compensation paid to managers who are more conscientious about the effect of environment, and opportunity costs of reduced benefits from competition with firms in other countries that do not have the same environmental standards.

Identification of these costs is the first phase of cost accounting. The second phase is the categorization of these costs. In general, three types of cost are identified:

(i)   costs that are direct to the products and therefore need no allocation (such as materials and direct labor),

(ii)   costs that are direct to operating departments, which handle the product but are indirect to the products themselves (such as depreciation of plant and equipment, salaries of operating department managers etc.) and

(iii)   service costs that are indirect to operating departments but are incurred for service functions (defined as those functions in which the products are not handled) such as the cost of human resources department, maintenance department, and information technology department.

The traditional costing systems are mostly production (or sales) volume-based. The service costs are first allocated to the operating departments based on some measures of service utilization. For example, the cost of maintenance might be allocated to the operating departments based on the maintenance hours utilized by each operating department. These allocated costs are added to the direct-to-operating center costs. For example, the total cost of a fabrication department will be the costs of the fabrication equipment and people that are direct to the department plus the costs allocated from service departments such as maintenance, information technology, and human resources. The total operating department costs are allocated (applied) to products based on volume measures. In such a system, the environmental costs will be fragmented and become part of product costs. No specific environmental cost is isolated. This presents a significant problem in making meaningful and informed decisions on environmental investments, costs and benefits.

The preferred system of eco-sensitive cost accounting is the *Activity Based Costing (ABC)* system. In this system of costing, different operating and non-operating activities are first identified. The resources are attributed to each of the activities based on resource drivers. For example, environmental clean-up would be an activity. All the resources which are directly or indirectly associated with environmental clean-up are tracked. For direct resources, there is no need for allocation (except inter-temporal allocation such as computing depreciation based on the total acquisition cost) but for indirect resources, resource drivers are identified. These could be objective measures or estimated measures. One method of estimation could be a survey of the employees involved in the activity. They could be surveyed on what percentage of their time they spent on environmental clean-up activities. (A full discussion of ABC is given in Turney, 1995). All these costs are aggregated to give the cost of an activity. At this stage, the firm would have aggregated costs of material handling, production scheduling, purchasing, environmental clean-up and other similar activities. In the second stage, the activity costs are attributed to cost objects based on how much of the activity was driven by that cost object. The cost object could be a product in which case, activity costs will be attributed to products based on how much of each activity was driven by that product. The cost object could be a customer in which case, the activity costs could be attributed to customers based on how much of each activity was driven by the individual customers. In this system, it is easy to simply aggregate the costs of all environmental activities and this will give the best estimate of the environmental cost.

Figure 4 gives the costing schemes corresponding to both the traditional and activity based costing.

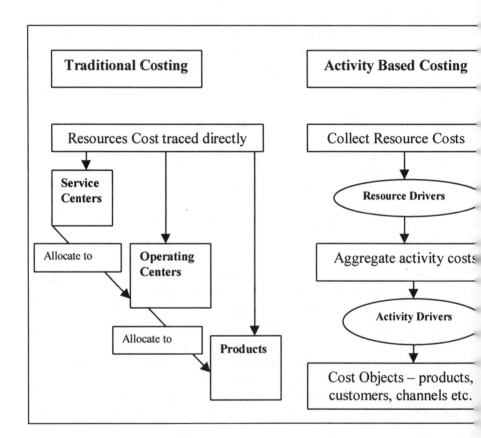

Figure 4. Traditional Costing and ABC

2.2.2.2   Eco-sensitive Cost Management

Cost management is the process in which the cost accounting and other measures are utilized to enhance long-term value and reduce long-term cost. Value enhancement could take the form of quality improvements, technology improvements, customer service improvements, improved safety and working conditions and environmental improvements. Cost reduction can take the form of increased efficiency and productivity, waste reduction, minimization of non-value-added cost, inventory reduction and process time reduction. It is evident that these are neither mutually exclusive nor independent of each other. However, eco-sensitive cost management requires us to define "value" as the total value accruing to society in the long run rather than the value that is accruing to the firm in the short term. An important aspect of cost management is the analysis of costs and benefits of new projects and investments. When these costs and benefits are computed, the most appropriate concept is that if the firm imposes a cost on society, the society will, in turn, impose the cost back on the firm through a combination of political, regulatory and measurement processes. We consider some important decisions of cost management below and discuss how environmental issues impact on these decisions.

First, consider the investment appraisal process. Firms take any material investment or disinvestment decisions after appraising the expected costs and benefits as well as the uncertainties in the project. The well-documented methods of investment appraisal are the payback, accounting return and the discounted cash flow methods. The discounted cash flow method requires the firm to compute the net present value of the expected future cash flows using a discount rate that incorporates the risk premium of the investment. In other words, the discount rate used for assessing riskier investments is higher than the discount rate used for assessing less risky investments. In reality, however, most firms use a common rate to compute the present values of different investment based on the Weighted Average Cost of Capital (WACC). This is often intuitively unappealing (as well as technically flawed) and reduces the credibility of the system. A number of firms use the simple payback method in which the time for payback using undiscounted cash inflow is computed and benchmarked against a subjective and often arbitrary time target. Environmental considerations influence the decision in many different ways. First, it is necessary to incorporate in the cost of the investment the estimated minimum amount needed for legal compliance of the environmental laws. Second, the uncertain costs such as possible future regulations, future legal liabilities, future political and social costs of environmental failures as well as the business costs of tainted reputation from possible environmental mishaps, need to be considered in selecting the appropriate technology and the proper skill and training levels of the personnel. In other words, technology and human resources should be evaluated in terms of their incremental effect on these costs. The estimation of these costs and their attribution to the project in question is far from trivial. Often, the lack of information about such costs has precluded the firm from taking up some investments. In some cases, after investment has been made, some of these costs have been realized but the phase-out has been delayed because of the difficulty involved in the evaluation of these costs. (Example: Phasing out of Chlorofluorocarbon production at Du Pont after its adverse environmental effect came to be known.)

Now consider the waste reduction process. Under Activity Based management (ABM), activities are first classified into three categories: (i) Value-added Activities that add value to the customer and therefore, are instrumental in deriving revenues from the customer, (ii) Essential Support activities which do not add any value to the customer but are needed for the operations of the firm and (iii) waste activities which are either unnecessary or have become necessary only under the current process. The process is then taken up for changes that will make the waste activities redundant. The essential support activities are examined for possible cost reductions. The value-added activities are taken up for possible value enhancement. There are many environmental activities that might not add value to the customer. There may also not be a specific law that requires the activity for compliance. Would it still be a value-added activity? Would it be an essential support activity or would it be a waste activity? It is possible to classify it as any of the three but we recommend classifying it as a value-added activity that needs to be examined from the viewpoint of not only customer value added but also social value added.

2.3   Ecological Accounting and Auditing

Unlike eco-sensitive accounting that measures the impact of environmental issues on financial statements, ecological accounting focuses on environmental measurement *per se*. It is a measurement in physical units and is not meant for reporting in financial reports. Similarly, ecological audits expands the scope of traditional financial auditing by including audits for compliance of environmental laws and audits of corporate environmental policy implementation. There are stakeholders such as the investors, governments and regulatory bodies who are interested not merely in knowing how the firm has assessed the financial impact of its environment-related activities but also in the direct information on the activities of the firm. This can help them assess the financial impact with greater precision than eco-sensitive accounting and reporting. There are other stakeholders such as environmental pressure groups, neighborhood communities and EPA that are not primarily concerned with financial impact assessments but are very concerned about the environmental impact *per se*.

Ecological accounting does not rest on widely accepted conventions and practice. It has not yet evolved to a stage where it can be discussed in as structured a manner as eco-sensitive accounting. However, ecological accounting has attracted greater attention in recent years because of the twin effects of decreasing marginal costs of ecological accounting and increasing marginal costs of environmental impact reduction. The first aspect, namely, decreased marginal cost of ecological accounting is related to the generally decreasing costs of information collection and dissemination. The second aspect, namely, the increasing cost of environmental impact reduction comes about because most firms have already carried out the low cost environmental impact reduction and they are now attempting to reduce environmental impact further.

2.3.1   Ecological Improvement Concepts and Evaluation

Evaluation of environmental performance of firms based on ecological accounting information rests on a few important concepts. They are: (i) *Sustainability* which is that the rate at which society is depleting natural resources must be equal to or less than the rate at which nature can regenerate those resources; (ii) *Sustainable Development* which is the development of the present generation without compromising the development of future generations. A society whose activities are sustainable and which achieves sustainable development can, in theory, continue the process for an infinite number of generations.

Much of the problem in achieving sustainability and sustainable development can be traced to the general managerial attitude that environmental efficiency and economic efficiency are substitutes. This attitude needs to be replaced by the belief that the two forms of efficiency are, in fact, complementary. For example, Schaltegger and Strun define a performance measure, Economic-Ecological Efficiency as the ratio of economic value added to the environmental impact. In effect, this assumes a trade-off between the two (See Schaltegger et.al., 1996). This might very well be true in the short run but in the long run, improved technologies and systems can be developed which moves the economic-ecological efficiency frontier to a higher level. In a manner similar to eco-sensitive accounting, we classify ecological accounting into (i) *internal ecological accounting*, a system that seeks to inform internal managers about environmental activities, issues and impacts and (ii) *external ecological accounting*, a system that seeks to report to the external interested stakeholders on the environmental activities and impacts. We also discuss in this section the notion and procedures of *ecological auditing*.

2.3.2   Internal Ecological Accounting

Internal ecological accounting consists of the following stages –
1. *Identification of environmental interventions.* This is similar in spirit to the identification of resource costs. Often, an environmental intervention is an emission into air or water or soil. Each such emission can be measured in terms of ponds or gallons etc. As an example, emission of CO_2 into air is an intervention.
2. *Tracing and allocation of interventions to activities*: The same structure as in the ABC can be used. If an environmental intervention is the direct consequence of an activity, it is traced to that activity. If it is the consequence of two or more activities, a reasonable measure is found to allocate the environmental intervention to the activities. For example, there might be a mixing activity followed by a boiling activity that results in the emissions. The extent of CO_2 emission is dependent on both the mixing and the boiling activities. It will be allocated between the two activities.
3. *Aggregation and assessment of interventions*: For each activity, all interventions are listed. A severity index is associated with each type of intervention. For example, emission of a toxin into water might have a much higher severity index than the emission of CO_2 into air.
4. *Allocation to assessment objects:* Depending on how the activities are "consumed" by each assessment object, the interventions associated with each activity are allocated to the assessment objects. A manager of a department could be an assessment object. Similarly, divisions, projects, plants, sites, products and even customers could be assessment objects. At this stage, if a department manager is an assessment object, it should be possible to identify the different types and extent of interventions with each department manager.

Intervention Management is a concept similar in spirit to cost management. The objective is to improve the environmental efficiency of each assessment object. For example, each departmental manager will prepare a trend chart of each type of intervention attributed to him and explain the actions taken by him to reduce the intervention or to increase the production at the same level of intervention on a periodic basis. Similarly, a trend sheet of interventions is prepared for each product, project, division and site.

When managerial decisions are made, the effect of the decision on different activities and consequently on the nature and extent of interventions can also be computed.

An estimate of the incremental (could be negative) intervention will be useful information in deciding about undertaking a project, product modification, new equipment purchase, or site selection.

Ecological Investment Appraisal is a process similar in spirit to financial investment appraisal. Unlike financial investment appraisal, discounting will not be used because environmental impact is not measured in financial terms. One measure could be the Ecological Payback Period that measures how long it will take to compensate the environmental impacts caused by an investment. For example, consider a pollution-reducing device. The production and the disposal of the device result in an environmental impact. The annual reduction of pollution caused by the device is used in the denominator to compute the number of years for payback. If the payback period is less than the expected life of the device, the investment is ecologically effective. It is evident that eco-sensitive accounting helps in the financial appraisal of the investment but ecological accounting helps in the ecological appraisal of the investment.

2.3.3 External Ecological Accounting

External ecological accounting primarily refers to the disclosure of ecological information that is useful to the EPA, regulators, ecological pressure groups and neighborhood communities. Currently, there are no accepted standards of ecological reporting other than for the compliance of the environmental regulations and laws. In the United States, the laws that bear on ecological performance are the following:

- Clean Air Act
- Clean Water Act
- Resource Conservation and Recovery Act
- Comprehensive Environmental Responsibility, Cleanup and Liability Act (generally known as "Superfund")
- Safe Drinking Water Act
- Toxic Substances Control Act
- Federal Insecticide, Fungicide, and Rodenticide Act
- Surface Mining and Reclamation Act
- Community-Right-to-Know Law
- occupational Safety and Health Act

Unlike GAAP, the laws and regulations in this area do not have general disclosure requirements. Acts such as the Clean Air or Clean Water Act require the firms to apply for permission to emit chemicals into air or water as the case might be. Emission without permit will be considered a criminal act. Emission above the limit set by the permit will constitute a violation. Once the permit is given, the EPA has the right to inspect and measure the amount of emission. Ecological audit is employed to check the compliance with the laws.

It is not at all clear that requiring environmental disclosures similar to the financial disclosures will result in greater benefits to society than the cost that needs to be borne by the firms to prepare the reports. This is an area where research is needed to establish the need (or the lack of it) for periodic disclosures.

2.3.4   Ecological Auditing

Cahill (1987) lists the typical U.S. corporate objectives for ecological auditing as follows:
- Assurance of compliance with the environmental laws
- Definition of liabilities
- Protection against liabilities for company officials
- Fact finding in acquisitions and mergers
- Tracking and reporting of compliance costs
- Information transfer among operating units
- Accountability of Management

From the above list, it is evident that the focus of ecological auditing is on compliance with the environmental laws. The EPA has also strongly encouraged the use of ecological auditing. Firms view ecological auditing as signals of serious effort to reduce environmental impacts, which will reduce the amount of oversight and inspections by the EPA. In other words, firms consider the economic trade-off between incurring the additional costs of ecological auditing and the reduced expected costs of penalties by EPA resulting both from a reduction in the violation detection likelihood and from a possible reduction of the amount of penalty if an infraction is found.

A structure of evaluation of environmental performance has been developed in the United States by the Coalition for Environmentally Responsible Economics (CERES). These principles were developed after the Exxon Valdez oil spill. Compliance with the principles is purely voluntary and is not monitored. Even though these principles have the potential of eventually becoming the generally accepted ecological principles (GAEP?), there is not much movement towards it at present.

A list of the methods of ecological auditing given in Callenbach et.al. (1993) is reproduced here.
- ***Environmental indicators***: These are levels of specific pollutants identified in the regulations. These indicators are quantitative and

objective but do not consider the interactive effects of all the emissions.

- *Auditing of Eco-sensitive external reports:* This is part of financial auditing. This audit might extend beyond the compliance with GAAP and provide an audit of estimates of costs and benefits.

- *Environmental Impact Assessment:* This is the backbone of auditing for compliance with the regulations. It includes the environmental indicators that are quantitative. It usually includes other qualitative and subjective assessments as well which will help determine compliance not only with the form but also with the substance of the regulations.

- *Technology Assessment:* This is the audit of management representations of the predicted social, political and ecological consequence of new technologies.

- *Material and Energy Auditing:* This involves the auditing of processes within the firm that result in emissions. Process auditing is likely to be more useful than output auditing in helping the management control the flow of pollutants.

- *Surveys:* These are surveys of customers on the premium they might be willing to pay for environmentally superior products. This serves as an input to the cost-benefit analysis of ecological expenditures.

- *Social /Ecological Balance Sheet:* This is the development of a "snapshot" of the firm's social and ecological assets and liabilities. Presumably, the difference between the two is the "social and ecological equity" that has been built up by the firm.

- *Environmental Data banks:* Data banks on pollutants, waste management and various regulations act as the basis for ecological audit planning. Some of the data banks identified by Callenbach (1993) are UMPLIS (general information and document system), AWIDAT (waste management information) and DABAWAS (water pollutant data).

- *System Dynamics:* This is a method based on simulation of processes through computer analysis.

- *Scenario Building:* This is a technique in which discussion groups are asked to build different scenarios (status quo, optimistic, realistic etc.) for the future.

Both external ecological auditors as well as internal auditors have used these methods. Some methods such as Scenario Building and Systems Dynamics are more suited for internal auditing whereas some other methods such as environmental impact analysis are equally useful for both internal and external auditing.

David Owen's essay in Smith (1993) provides some cases of ecological audits. As can be expected, they vary widely between firms and between countries. As an example, he refers to Elkington's (1990) discussion of current practice at British Petroleum that includes (i) compliance audits with both statutory obligations and with voluntary industry codes, (ii) activity audits with a focus on activities that cross business boundaries, (iii) corporate audits of entire BP business sectors, (iv) associate audits of companies in other countries acting as agents of BP and (v) Issues audits that focus on specific key environmental issues. He also talks about "cradle to grave" audits at Norsk Hydro that undertook a comprehensive eco-balance review of all its manufacturing companies with many stages.

3.   ACCOUNTING AND ENVIRONMENTAL MANAGEMENT SYSTEMS

This section presents two recent developments in Management Accounting and discusses how environmental management might get aligned with them. The first is the *Balanced Scorecard* that is a performance evaluation system that includes both the financial and non-financial measures of performance. The second is the *Strategic Cost Management* system in which the locus of control extends beyond the boundary of the firm and incorporates other players in the value chain.

3.1   The Balanced Scorecard and Environmental management

Kaplan and Norton (1996) state that a Balanced Scorecard "translates an organization's mission and strategy into a comprehensive set of performance measures that provides the framework for a strategic measurement and management system".

Increasingly, the Balanced Scorecard has supplemented the traditional financial budget as the performance evaluation criterion in many companies. Kaplan and Norton (1992) introduced the concept of Balanced Scorecard to complement financial measures with operational measures on customer satisfaction, internal processes and the organization's innovation and improvement activities. They suggested four perspectives in evaluating any entity in an organization: (i) The *financial perspective* is the evaluation of the actions and decisions of managers from the shareholders' point of view, (ii) *customer perspective* is the evaluation from the customer's point of view, (iii) *internal business perspective* is the evaluation of actions and decisions from the view of the core competency and the vision of the organization and

(iv) *innovation and learning perspective* is the evaluation of the actions and decisions from the long-term value creation perspective. Initially, the purpose of the system was mainly operational, just coordinating the activities of different managers so support the defined strategy.

However, the experience of many firms implementing the method was that they could use the information collected in these Balanced Scorecard reports to formulate and revise strategy. Kaplan and Norton (1996) build on this experience and suggest four processes in the management of the strategy: (i) translating the vision, (ii) Communicating and linking, (iii) Business planning and (iv) Feedback and learning. The actual experience of one company is described in that article. Among the operating managers, the discussions and exchanges after the introduction of Balanced Scorecard are based on the focus provided by the strategy in place rather than on the diffused foci resulting from the disconnect between the exhortations in the mission statement and the formal performance evaluation measures.

The Balanced Scorecard concept is broad enough to incorporate more than the four dimensions that Kaplan and Norton have recommended. They provide the following example of a chemical company that wished to create a new dimension on environmental management. Kaplan and Norton challenged the management on the need for a separate dimension and the reply of the CEO and other senior executives is noteworthy:

"We don't agree. Our franchise is under severe pressure in many of the communities where we operate. Our strategy is to go well beyond what current laws and regulations require so that we can be seen in every community as not only a law-abiding corporate citizen but as the outstanding corporate citizen, measured both environmentally and by creating well-paying, safe, and productive jobs. If regulations get tightened, some of our competitors may lose their franchise, but we expect to have earned the right to continue operations." (Kaplan and Norton, 1996 – page 35)

In this case, the firm has recognized environmental management as a dimension of strategic interest. The firm is not satisfied in merely complying with the regulations. It wants to be a leader in environmental management and wants to be known as an outstanding corporate citizen. In this case, the use of the environmental management dimension means that environmental impact becomes a criterion for assessing the actions of every manager in the firm. If it is a dimension in Balanced Scorecard, it would be possible to justify an investment using positive environmental impact even if it cannot be justified in the financial dimension. In such a firm, ecological accounting has a very real effect and the managers can be expected to be very interested in environment-friendly investments.

Not every firm that uses Balanced Scorecard will have an environmental dimension. However, eco-sensitive accounting will link the financial perspective to environmental concerns. Customers are often willing to pay a premium for environmentally friendly products. Using Quality Function Deployment (QFD), the voice of the customer will become the criterion for evaluating the decisions of managers in the customer dimension. This aspect of ecological accounting and auditing will link the customer perspective to environmental concerns. Activity-based ecological accounting described in the previous section will be useful in making the processes environment-friendly and thus links the internal business process with environmental issues. Finally, the learning and growth perspective includes the quality-of-life issues of employees and their families. Environment-friendly work atmosphere and work objective can give both pride and satisfaction to the employees. Ecological accounting can therefore link the learning and growth dimension with environmental issues.

In summary, the Balanced Scorecard facilitates the elevation of environmental management to a strategic level. Whether the firm chooses to treat environmental management as a strategic Balanced Scorecard dimension or not, eco-sensitive and ecological accounting can help link environmental issues to the dimensions of the Balanced Scorecard.

3.2 Strategic Cost Management and Environmental Management

The conventional cost management framework is aimed at maximizing "value added". In other words, activities are examined to determine whether they add value to the customer or not. If they add value, improvement of the efficiency and effectiveness of such activities are undertaken. If they do not, their cost is minimized. In all cases, the analysis is confined to the boundary of the firm.

In a similar vein, environmental management as specified earlier seeks to determine the environmental value of each activity and process in the firm. By maximizing the environmental value and minimizing the adverse environmental impact caused by each of the activities an processes, the environment is "managed".

Unlike the conventional cost management framework, Strategic Cost Management (SCM) framework seeks to find the drivers of value and cost both inside and outside the firm. It is quite possible, for example, that the suppliers who are outside the organizational boundary are the primary drivers of a non-value-added activity (like inspection). The best way of managing this cost is to manage the supplier's production process or to change the vendor selection process.

In a similar vein, Strategic Environmental Management (SEM) is a framework that identifies the primary drivers of environmental impacts both inside and outside the organizational boundaries. For example, it is conceivable that the primary determinant of the emission level is the supplier's technology. In such cases, the best way to manage the environmental impact could be to help improve the supplier's technology or change the supplier to a more environment-friendly supplier (who could command a price premium).

In order for SEM to work effectively, the eco-sensitive accounting as well as the ecological accounting/auditing systems must extend beyond the organizational boundaries to include the suppliers of the firm a and the customers of the product. Porter (1985) has introduced the term *Value Chain* to denote this external focus. For every product, value is added in separate but sequentially linked processes starting from the basic raw materials to the final use or consumption of the product. The identification of these linked processes, supporting activities for each process and the determination of the amount of value added in each of these linked processes constitutes the creation of the value chain of the product. This value chain then becomes the framework of strategic management including SEM.

It is unlikely that any one firm spans the entire value chain. In this value chain, the firm will perform a subset of processes. In general, a firm spans only a subset of the larger set of linked processes in the value delivery system. The value chain will include the suppliers, the customers, the customers' customers, some firms that could be both suppliers and competitors and also some firms that could be both competitors and customers etc.

2   Environmental Accounting Research and Concluding Remarks

Research in environmental accounting is considerably sparser than research in other areas of accounting. The current research can be classified into three general categories:

1. *Management and Incentive issues:* This area of research examines the incentives that managers and other players face in measuring, communicating and disclosing environmental accounting information. Katz(1991) talks about the worry of many managers that information from ecological auditing (which is not mandated) could be used against them in lawsuits. Zachry, Gharan and Chaisson (1998) examines environmental costing issues and the need to apply environmental costs to work centers. The authors also expect the pressure to price a product low because other countries might have lower environmental standards, to decrease, as global need for environment management gets recognized.

2. *Measurement and Disclosure Issues:* This area of research deals with improved measurements and disclosures by firms. Ilinitch, Soderstrom and Thomas (1998) use both theoretical and empirical approaches to compare process measures with measures of public reaction and advocate the need for explicit environmental performance metrics in the accounting framework. Bennett and James (1998) examine how management accounting can help business better understand and manage environmental issues. Neu, Warsame and Pedwell (1998) uses environmental disclosures in Canadian firms to examine the influence of external pressure on environmental disclosures, the comparison of environmental vs. other social disclosures and the association between environmental disclosures and actual performance.

3. *Compliance and other issues:* This area of research essentially deals with the compliance with the laws and regulations as well as with voluntary standards and industry codes. For example, Lally (1998) discusses how cost accounting can help compliance with ISO 14000, which is a voluntary international environment management standard. There are also a number of case studies that examine specific cases of environmental accounting.

Environmental accounting is a rich area for future research. There is no known theory that helps us in valuing the usefulness of environmental accounting in a way similar to financial accounting theory. Financial accounting theory suggests that accounting and reporting derive value from the additional information that the investors get in valuing their securities and this helps a more effective allocation of resources in a capital market setting. Compared to this structure, there is not much theory that guides public goods, social or environmental accounting. Would mandating certain disclosures provide more value to the stakeholders than allowing voluntary disclosures by firms? Voluntary disclosures have the benefit of signaling that gets reduced when disclosures are mandated. What would be the cost of collecting information to comply with mandated disclosures? What would happen to the possibility of litigation based on such disclosures? These are some of the questions that researchers need to address in this area.

In conclusion, it is evident that more effort will be made in identifying, collecting, applying and disclosing both financial and non-financial impact of environment-related activities by firms. There is an increasing global awareness of the need to grow sustainably in future. This needs better measurement techniques, a clearer understanding of the negative externalities involved and a deeper examination of the incentives of various parties involved in the process of environmental accounting.

REFERENCES

Bennett, Martin; James, Peter, "The green bottom line: Management accounting for environmental improvement and business benefit", *Management Accounting-London*, Vol: 76 Iss: 10 Date: Nov 1998 p: 20-25

Cahill, L.B. Editor, "Environmental Audits", 5th Edition, Published by *Government Institutes*, 966 Hungerford Drive #24, Rockville, Md. 20850, 1987

Callenbach, Ernest; Capra, Fritjof; Goldman, Lenore; Lutz, Rudiger; Marburg, Sandra, "Ecomanagement – The Elmwood Fuide To Ecological Auditing and Sustainable Business" Published by *Berrett-Koehler Publishers, Inc.*, 155 Montgomery Street, San Francisco, CA 94104-4109, 1993

Elkington, J., "The Environmental Audit: A Green Filter For Company Policies, Plants, Processes and Products", Published by *SustainAbility Ltd.*, 49, Princes Place, London W11 4QA, U.K., 1990

Ilinitch, Anne Y; *Soderstrom*, Naomi S; Thomas, Tom E, "Measuring corporate environmental performance", *Journal of Accounting & Public Policy* Vol: 17 Iss: 4,5 Date: Winter 1998 p: 383-408

Kaplan, Robert S.; Norton, David P, "The Balanced Scorecard", Published by *Harvard Business School Press*, Boston, MA, 1996

Kaplan, Robert S.; Norton, David P, "The Balanced Scorecard - Measures That Drive Performance", *Harvard Business Review*, Boston; Jan/Feb1992; Vol. 70, Iss. 1; pg. 71-79

Katz, David M, "Risk managers seek EPA cover on audits", *National Underwriter [Property & Casualty/Risk & Benefits Management]* Vol: 103 Iss: 5 Date: Feb 1, 1999 p: 1, 32

Lally, Amy Pesapane, "ISO 14000 and environmental cost accounting: The gateway to the global market", *Law & Policy in International Business*, Vol: 29 Iss: 4 Date: Summer 1998 p: 501-5

Neu, D; Warsame, H; Pedwell, K, "Managing public impressions: Environmental disclosures in annual reports", *Accounting, Organizations & Society*, Vol: 23 Iss: 3 Date: Apr 1998 p: 265-282

Porter, Michael. "Competitive Advantage", Published by *The Free Press*, New York. 1985

Schaltegger, Stefan; Muller, Kaspar; and Hindrichsen, Henriette, "Corporate Environmental Accounting" Published by *John Wiley and Sons*, Baffins Lane, Chichester, West Sussex, PO19 IUD, England, 1996

Smith, Dennis; Editor, "Business and The Environment", Published by *St. Martin's Press inc.*, 175 Fifth Avenue, new York, NY 10010, 1993

Turney, Peter B.B., "Common Cents", Published by *Cost Technology*, Hilsboro, Oregon, USA, 1993

Zachry, Benny R; Gaharan, Catherine G; Chaisson, Michael A, "A critical analysis of environmental costing", *American Business Review*, Vol: 16 Iss: 1 Date: Jan 1998 p: 71-73

APPENDIX

Union Carbide (1998 annual report)
Discussion and Analysis part of the report
Costs Relating to Protection of the Environment

Worldwide costs relating to environmental protection continue to be significant, due primarily to stringent laws and regulations and to the corporation's commitment to industry initiatives such as RESPONSIBLE CARE, as well as to its own internal standards. In 1998, worldwide expenses related to environmental protection for compliance with Federal, state and local laws regulating solid and hazardous wastes and discharge of materials to air and water, as well as for waste site remedial activities, totaled $91 million. Expenses in 1997 and 1996 were $100 million and $110 million, respectively. In recent years, such environmental expenses have decreased as the corporation has made progress toward completing major remediation projects. In addition, worldwide capital expenditures relating to environmental protection, including those for new capacity and cost reduction and replacement, in 1998 totaled $57 million, compared with $68 million and $43 million in 1997 and 1996, respectively.

The corporation, like other companies in the U.S., periodically receives notices from the U.S. Environmental Protection Agency and from state environmental agencies, as well as claims from other companies, alleging that the corporation is a potentially responsible party (PRP) under the Comprehensive Environmental Response, Compensation and Liability Act and equivalent state laws (hereafter referred to collectively as Superfund) for past and future cleanup costs at hazardous waste sites at which the corporation is alleged to have disposed of, or arranged for treatment or disposal of, hazardous substances. The corporation is also undertaking environmental investigation and remediation projects at hazardous waste sites located on property currently and formerly owned by the corporation pursuant to Superfund, as well as to the Resource Conservation and Recovery Act and equivalent state laws.

There are approximately 118 hazardous waste sites at which management believes it is probable or reasonably possible that the corporation will incur liability for investigation and/or remediation costs. The corporation has established accruals for those hazardous waste sites where it is probable that a loss has been incurred and the amount of the loss can reasonably be estimated. The reliability and precision of the loss estimates are affected by numerous factors, such as the stage of site evaluation, the allocation of responsibility among PRPs and the assertion of additional claims. The corporation adjusts its accruals as new remediation requirements are defined, as information becomes available permitting reasonable estimates to be made, and to reflect new and changing facts.

At Dec. 31, 1998, the corporation's accruals for environmental remediation totaled $220 million ($264 million in 1997). Approximately 53 percent of the accrual (55 percent in 1997) pertains to estimated future expenditures for site investigation and cleanup, and approximately 47 percent (45 percent in 1997) pertains to estimated expenditures for closure and postclosure activities. See Note 17 to the financial statements for a discussion of the environmental sites for which the corporation has remediation responsibility. In addition, the corporation had environmental loss contingencies of $121 million at Dec. 31, 1998.

Estimates of future costs of environmental protection are necessarily imprecise, due to numerous uncertainties. These include the impact of new laws and regulations, the availability and application of new and diverse technologies, the identification of new hazardous waste sites at which the corporation may be a PRP and, in the case of Superfund sites, the ultimate allocation of costs among PRPs and the final determination of the remedial requirements. While estimating such future costs is inherently imprecise, taking into consideration the corporation's experience to date regarding environmental matters of a similar nature and facts currently known, the corporation estimates that worldwide expenses related to environmental protection, expressed in 1998 dollars, should average about $110 million annually over the next five years. Worldwide capital expenditures for environmental protection, also expressed in 1998 dollars, are expected to average about $50 million annually over the same period. Management anticipates that future annual costs for environmental protection after 2003 will continue at levels comparable to the five-year average estimates. Subject to the inherent imprecision and uncertainties in estimating and predicting future costs of environmental protection, it is management's opinion that any future annual costs for environmental protection in excess of the five-year average estimates stated here, plus those costs anticipated to continue thereafter, would not have a material adverse effect on the corporation's consolidated financial position.

Footnote Disclosure:

Environmental - The corporation is subject to loss contingencies resulting from environmental laws and regulations, which include obligations to remove or remediate the effects on the environment of the disposal or release of certain wastes and substances at various sites. The corporation has established accruals in current dollars for those hazardous waste sites where it is probable that a loss has been incurred and the amount of the loss can reasonably be estimated. The reliability and precision of the loss estimates are affected by numerous factors, such as different stages of site evaluation, the allocation of responsibility among potentially responsible parties and the assertion of additional claims. The corporation adjusts its accruals as new remediation requirements are defined, as information becomes available permitting reasonable estimates to be made and to reflect new and changing facts. At Dec. 31, 1998, the corporation had established environmental remediation accruals in the amount of $220 million ($264 million in 1997). These accruals have two components, estimated future expenditures for site investigation and cleanup and estimated future expenditures for closure and postclosure activities. In addition, the corporation had environmental loss contingencies of $121 million at Dec. 31, 1998.

The corporation has sole responsibility for the remediation of approximately 37 percent of its environmental sites for which accruals have been established. These sites are well advanced in the investigation and cleanup stage. The corporation's environmental accruals at Dec. 31, 1998, included $169 million for these sites ($197 million at Dec. 31, 1997), of which $65 million ($79 million at Dec. 31, 1997) was for estimated future expenditures for site investigation and cleanup and $104 million ($118 million at Dec. 31, 1997) was for estimated future expenditures for closure and postclosure activities. In addition, $66 million of the corporation's environmental loss contingencies at Dec. 31, 1998, related to these sites. The two sites with the largest total potential cost to the corporation are non-operating sites. Of the above accruals, these sites accounted for $39 million ($36 million at Dec. 31, 1997), of which $19 million ($17 million at Dec. 31, 1997) was for estimated future expenditures for site investigation and cleanup and $20 million ($19 million at Dec. 31, 1997) was for estimated future expenditures for closure and postclosure activities. In addition, $44 million of the above environmental loss contingencies related to these sites. The corporation does not have sole responsibility at the remainder of its environmental sites for which accruals have been established. All of these sites are in the investigation and cleanup stage. The corporation's environmental accruals at Dec. 31, 1998, included $51 million for estimated future expenditures for site investigation and cleanup at these sites ($67 million at Dec. 31, 1997). In addition, $55 million of the corporation's environmental loss contingencies related to these sites. The largest of these sites is a non-operating site. Of the above accruals, this site accounted for $10 million ($14 million at Dec. 31, 1997) for estimated future expenditures for site investigation and cleanup. In addition, $3 million of the above environmental loss contingencies related to this site. Worldwide expenses related to environmental protection for compliance with Federal, state and local laws regulating solid and hazardous wastes and discharge of materials to air and water, as well as for waste site remedial activities, totaled $91 million in 1998, $100 million in 1997 and $110 million in 1996.

Chapter 8

The Development of Eco-labelling Schemes
An Economic Perspective

S. Salman Hussain[1], Dae-Woong LIM[2]
[1]*Department of Natural Resource Economics, SAC, West Mains Road, Edinburgh, UK*
[2]*Eco Management Consulting Co., Dongdaemoon-Ku, Seoul, Korea*

1. INTRODUCTION

Eco-labelling might be considered as an extension of conventional marketing practices, a profit-driven response by industry to the commercial pressures of green consumer-consciousness. Peattie defines green marketing as 'the management process responsible for identifying, anticipating and satisfying the requirements of customers and society, in a profitable and sustainable way' (1992:11). If this definition is adopted then eco-labelling is simply a form of media that communicates information to a (receptive) user on the impact of a firm's product on the natural environment compared with those of its competitors. An externally verified eco-labelling scheme should then serve to validate such a marketing claims, therein protecting the user from deceptive and/or false environmental information, a phenomenon known as 'greenwashing' (Rockness, 1985). The Organization for Economic Cooperation and Development (OECD) interprets the goals of environmental labelling as follows: improving the sales or image of a labeled product; raising the awareness of consumers; providing accurate information; directing manufacturers to account for the environmental impact of their products; and protecting the environment (OECD, 1991).

This is very much a 'business-as-usual' interpretation of what eco-labelling implies; it has been criticized by Welford who describes it as a 'marginalist' approach, which 'invites companies to pay lip-service to environmentalism,' (1995:152). Welford proposes a more radical alternative stakeholder approach: 'if our ultimate aim is sustainable development all other demands must be considered as secondary to this and profit-cantered strategies replaced by more holistic and integrated approaches' (1995:154). It is noteworthy that such 'marginalist' approaches have been endorsed and propagated by many influential corporate bodies. Blaza (1992) presents the view of the Confederation of British Industry: 'One of the other problems of 1990 which hindered progress in corporate environmental performance was the tendency in some quarters to promote the idea that the environment as a business issue was different and required a special approach. This is clearly untrue' (1992:32-34). Although we have some sympathy with Welford's position, our primary aim in this chapter is to present the status quo in green marketing, as opposed to a vision of what may or may not be ideal. Further, although it might appear crass, a firm that is bankrupted through competition owing to Welford's proposed relegation of profitability is not going to have any further real influence on the sustainability debate.

This chapter adopts the following structure. Section 2 reviews the extant eco-labelling schemes, charting the historic development and applications of schemes in different places. We provide a synopsis of the attributes and conditions of the competing schemes. Section 3 is a detailed investigation of one particular scheme, which has not been well documented to date, the Environmental Mark scheme in South Korea. The lessons that were learnt in the implementation procedure are elucidated. We then turn to the appraisal of the incentives for participation in voluntary eco-labelling schemes; section 4 provides the theoretical perspective for consumer (green) decision-making and green consumerism. Section 5 concludes.

2.   A REVIEW OF ECO-LABELLING SCHEMES

According to Kuhre (1997), there are two categories of environmental labelling: mandatory and voluntary labelling. The former is usually used to provide the users with warning and caution about handling and treating specific substances such as radioactive waste and hazardous materials, which are regulated, under various sections of legislation. Voluntary labelling is generally considered as a symbol of the environmental superiority of products and services. It is a form of information provision to consumers to encourage them to take the environmental credentials of the product or service into account.

This section focuses on the voluntary environmental labelling simply because this type of scheme is what many commentators, including the US Environmental Protection Agency (EPA, 1993), consider to be the most significant in terms of achieving environmental policy goals. Voluntary schemes are also known as environmental certification programmes (ECPs).

According to the International Organization for Standardization (ISO), the fundamental goal of voluntary environmental labelling is to provide information about the environmental superiority of products and/or services to consumers in order to support their (green) purchasing decision. If consumers are willing (and able) to demonstrate this green preference in the market through purchasing then providers, i.e. manufacturers, retailers, brand holders etc., might be motivated to attempt to achieve the eco-label voluntarily in order to achieve a competitive advantage (ISO, 1997). The implication of the ISO argument is that participation in eco-label schemes demonstrates that the product/process is environmentally friendly; whether this is justified is considered below. Some commentators (e.g. North, 1992) have proposed that improvements in the environmental quality of products will become a prerequisite for the survival of firms in the future; although we believe that this may be overstating the case, there is little doubt that there is some momentum for industrial eco-change, and merely attaining a (significant) competitive advantage is a sufficient argument for accreditation.

Generally, voluntary environmental labelling is divided into three types: environmental labels, environmental self-declaration and certified scientific information (Table 1).

- *Type 1 Environmental labels:*

This refers to general voluntary eco-labelling schemes with external accreditation; they are a type of ECP. There are two principle types: seal-of-approval schemes, which categorize the product and/or service as more environmentally-benign than similar non-accredited competing products and/or service with the same function. A common methodology is applied: product categories (e.g. 'solvents') are selected; criteria are set based on environmental impacts through the application of life cycle assessment (LCA), life cycle management (LCM) and/or life cycle consideration (LCC)[1] of the selected products and/or services in each category; the labels are awarded to products and/or services that satisfy the criteria. The second type is single attribute certification, wherein the consumer is left to decide whether accreditation implies improved overall environmental performance.

---

[1]LCA is an objective process used to evaluate the environmental burden associated with product, process or activity by identifying and quantifying energy and material usage and environmental release, to access the impact of those energy and material uses and release on the environment and to evaluate and implement opportunity to effect environmental improvement (Fava *et al*, 1990)

Life cycle management integrates environmental occupational safety, with recycling as part of the overall management decision making or government policy making process (Fox and Singh, 1997).

Life cycle consideration (LCC) is a process to review environmental impact of cradle to grave of product and/or service qualitatively as well as quantitative.

- *Type 2 Self-declaration of environmental claims*

This type of eco-claim is made by firms that want to declare the environmental superiority of their products and/or services based on such environmental issues as recyclability, energy recovery, design for dismantlement etc. Since the firm determines which aspect(s) of environmental performance it reports on, it can select those aspect(s) that put its product and/or service in a good light. Thus, such self-declaration should be monitored appropriately by government authority or consumer organizations to protect consumers from 'greenwashing', that is false or misleading environmental marketing claims.

- *Type 3 Quantified product information labels*

This eco-label-type provides scientific information on the products and/or services. This is a relatively new development in eco-labelling. Generally, this is likely to be the results of life cycle analyses. For instance, Volvo published *Environmental Product Declaration : Volvo S80 2.9* (Volvo, 1998). This included quantified and qualified environmental information about the car. This was sub-divided into four principal areas of environmental performance; manufacturing, operation, recycling and environmental management. This type of eco-label differs from types I and II in that the information disclosed might be negative, as opposed to only neutral or positive.

Table 1. The Types of Environmental Labelling

	TYPE I	**TYPE II**	**TYPE III**
	Environmental labels	Self-declaration of environmental claims	Quantified product information labels
Characteristics	Independent certification	Independent certification, Self-declaration	Independent certification, Self-declaration
Product-category	Selected categories	All categories	All categories
Methods	LCA*, LCC** or Scientific Data	Scientific Data	LCA, LCM***, etc.
Validation body	Independent	Independent, Dependent	Independent, Dependent

* LCA : Life Cycle Assessment, ** LCC : Life Cycle Consideration *** LCM : Life Cycle Management
Source: Modified from "1998 Environmental Mark Scheme" of Ministry of Environment, Korea, 1998

Type 1 environmental labels are arguably the most familiar type to the general public. The United Nations (UNEP, 1991) has set out guidelines for the efficient operation of seal-of-approval schemes. These include: a legally protected logo or symbol for the scheme; governmental endorsement; operated by institutions (including government) which do not have a commercial interest in the awards; periodic review of the process, categories and criteria.

There is considerable debate as to what categories ought to be included in an eco-label scheme. Accreditation sends signals to both consumers and producers, and therein lies the problem. Take the automobile industry: if a customer is going to purchase a car anyway (as opposed to using more ecologically-benign transport options) then that car should have as low an environmental impact as possible from production through till disposal. An eco-label might facilitate this in that the environmentally-conscious purchaser would be able to express his or her preferences accordingly, and car manufacturers would fill this 'green' niche market. Although the (combustion-engine-based) car that was awarded the eco-label is likely to be more eco-friendly relative to competing cars, it is still likely to be environmentally harmful in absolute terms. An eco-label might provoke a consumer to buy a car (which he or she would not have done otherwise) as it has been given an environmental seal-of-approval.

Another contentious debate in eco-labelling is the question of how high to set the bar; the scheme's organizers must trade-off the benefits of the momentum generated by widespread approval of the scheme in a particular industry and the positive impact on green consumerism which is implied by limiting the percentage of applicants that achieve the criteria. If all products in a category pass then green consumerism is stifled.

The German Blue Angel was the first scheme to become operational, starting in 1977. As of 1999, there are about 30 environmental label schemes extant in the world. A synopsis of some of the features of the major eco-labelling schemes is given in Table 2. Each is considered in more detail below.

Table 2. The Characteristics of Major Environmental Labelling Schemes

Name of Scheme	Blue Angel	Eco-Labelling	White Swan	ECP[*]	Green Seal	Eco-Mark	Environ-mental Mark
Country	German	EU	Nordic Council	Canada	USA	Japan	Korea
Operation Organization	UBA[**]	EU countries	Nordic Council	Terra Choice	Green Seal	JEA[***]	MOE[#] KELA[##]

Established Year	1977	1992	1989	1988	1989	1989	1992
No. of Product-Categories	About 70	12	43	90	37	69	29
No. of Awarded Products	920	About 200	n.a.	n.a.	18	About 1300	145
Methods of Criteria Devel.	LCA	LCA	LCA	LCA and LCC	LCC	LCC	LCA and LCC
Level of selectivity	20-30%	5-30%	20-30%	20%	15-20%	50-70%	10-20%

*ECP: Environmental Choice Program   #MOE: Ministry of Environment
**UBP: Germany's Federal Environmental Agency ##KELA: Korea Environmental labelling
    Association
***JEA: Japan Environmental Association   n.a.: not available
Source: Modified from "1998 Environmental Mark Scheme" Ministry of Environment, Korea,
1998

- *Germany's Blue Angel*

As mentioned above, this was the first Type 1 scheme in the world. The scheme was established by Federal Ministry for the Environment (BMU) in 1977.   Currently, this is operated by the BMU, Federal Environmental Agency (UBA) and German Institute for Quality Assurance and Labelling (RAL).   In the past, the criteria for participation in the scheme focused on single specific environmental aspects of the products and services such as recyclability, recycled content, etc. However, the scheme's coordinators are presently conducting a transition from single issue to multi-issues based on LCA.

- *The European Union's Eco-label*

This was established in 1992, based on Council Regulation (EEC) No. 880/92. The EU Eco-label was designed in line with the principles, goals and priorities selected by the Fifth Community Environmental Action Programme and its Review, as well as with Agenda 21. Each EU member country operates their own schemes based on the guidelines and criteria that are agreed at Commission level.   A 'lead country' evaluates product categories. The criteria are determined by both quantitative and qualitative research using life cycle methodologies in feasibility studies and market structure analyses.   The Groupe des Sages prepared Guidelines for the Application of Life-Cycle Assessment in the EU Eco-label Award Scheme to determine the appropriate use of LCA in developing ecological criteria. Initially there were relatively few applications, both because potential applicants were not familiar with the LCA techniques and the criteria were quite difficult to meet in the short time. However, the rate of applications did increase, as consumer awareness of the label and the purpose of the scheme grew.

- *The Nordic Council's White Swan scheme*

This was introduced by the Nordic Council in 1989 and has been operated by the Nordic Eco-labelling Board with Sweden, Norway, Finland, Iceland and Denmark applying this scheme. The protocol and methodologies applied by the White Swan scheme were influential in the development of the EU's Eco-label. Initially, the aim of the scheme was to prevent consumer confusion arising from the co-existence in the Nordic countries of several different schemes. However, the scheme's remit was subsequently extended to consider the development of sustainability through the ecology/industry interface. Further, social implications have augmented the environmental criteria. In 1996, the Board announced new strategies to improve the credibility of the scheme to consumers, institutions and providers simultaneously.

- *Canada's Environmental Choice Program*

This was established in 1988 by the Canadian government. The Ministry of Environment has commissioned TerraChoice Environmental Service to operate the Scheme since 1995. Currently, TerraChoice manages all operations and they consult with government to issue product licenses. The scheme has about 90 product categories and approximately 1700 products from 165 companies have been licensed. The LCA methodology is applied. Product category selection can be initiated by industry as well as the eco-label board; in such cases, the firm must corroborate its application for a new product category with a submission outlining the environmental performance and impact of the product type from 'cradle to grave'. An expert panel then assesses this evidence, although the Programme organizers conduct no original research. This procedure is deemed as an important incentive to encourage voluntary environmental awareness and utilization of the environmental label as a business strategy for Canadian companies.

- *Environmental labels in the USA*

The US government only certifies one environmental label - the Energy Star program. However, it is different from general labelling schemes in that it is a single-issue scheme. There are two other major eco-labelling schemes in the US. The Green Seal is a Type I scheme run by a private organization, Green Seal. The other is a Type III scheme, Certified Eco-Profile, which is run by Scientific Certification System Inc.

- *Japan's Eco-Mark*

This was established by JEA (Japan Environmental Association) in 1989. As of June 1996, 2023 products had been licensed in 69 product categories; this is a significantly high number of accreditations. One reason for this is that the designation of product categories under the Eco-Mark are determined by 'guiding principles' as opposed to more rigorous, in-depth analysis. The scheme is in transition. Previously, the ecological criteria of most product groups were based on single issues. However, JEA has tried to revise the scheme based on ISO 14020s. JEA has launched two revised product categories, which have multi-issue criteria applying quantified targets using life cycle environmental impacts.

- *Summary*

The schemes discussed above have been developed for different purposes and forms. Several schemes are operated by governments and some by private agents or companies. Several criteria are developed for a single-issue and some for multi-issues. Even in the same product category, the criteria for each scheme are often quite different. Some schemes are used independently whilst others are coordinated with other policy instruments, such as institutional purchase or energy saving scheme.

It is, therefore, quite important to harmonize the schemes through international standardization. As well as effecting efficiency and improving transparency, such a harmonization would ameliorate one of the problems associated with eco-labelling schemes, viz. protests that eco-labels act as a barrier to free trade. . Currently, the ISO (International Standardization Organization) has commissioned the Technical Committee 207 to establish guidelines for international standardization for the three types of voluntary environmental labelling schemes (Table 3).

Table 3. Environmental Labelling and International Standardization Organization (Dec.1997)

Subjects	Document	Status
Under TC207/SC3[*]		
WG[**]1	ISO 14024	DIS[#]
Environmental Labelling Type III	ISO 14025	WD[##]
WG2 Self-Declaration Environmental Claims	ISO 14021	DIS
WG3 General Principles	ISO 14020	DIS

* TC207/SC3 : Technical Committee207/Subcommittee3
# DIS : Draft of International Standard
** WG : Working Group
WD : Working Draft

3. SOUTH KOREA'S ENVIRONMENTAL MARK SCHEME

The environmental mark scheme was introduced in South Korea in 1992. The scheme's organizers state its purpose to be to encourage environmentally friendly production and consumption and to increase consumers' environmental awareness. The South Korean Ministry of Environment has commissioned the Korean Environmental Labelling Association (KELA) to operate this scheme. As of April 1999, KELA had 29 product-categories in operation, with 121 products from 93 companies under license.

Initially, most product categories had specific environmental performances and functions, such as recycled paper, empty can collector etc. This naturally made the designation of environmental criteria a single issue. In addition, the targets set were easily achieved by the industrial applicants. In many cases, most of the products in a given product category were awarded certification. Consumers could not express their preference for the green option owing to this inefficient lack of selectivity in the scheme.

The Ministry of Environment made a decision to renew the scheme and since 1997 has based it on the International Standard ISO14020s. Under this renewal, various methodological changes have been investigated: the labelling mechanism; methods for product-category selection and criteria development; benchmarking with foreign schemes; integration of the three ecolobel scheme types in South Korea. The project has focused especially on operational transparency, participatory procedures, and scientific approaches.

3.1 Transparency in the operation of the scheme

The designation of the methodologies and practices applied in the scheme required the participation of various stakeholders and interest groups such as individual consumers, consumers groups, environmental groups, industrial associations, manufacturers and retailers. In setting up the framework for the renewed scheme, their participation was considered at various stages although there were some limitations imposed by operational costs. All interest groups had the opportunity to participate directly in the public hearing during the review process concerning the selection criteria. There was also participation (but to a lesser extent) in the selection of product categories and the decision-making process for certification.

3.2   The selection method for product categories

The fundamental goal of the revision to the scheme was to maximize the competition and voluntary environmental improvement in the context of inter-firm competition in the same product category. The assessment procedure was in two stages: effectiveness and efficiency. Both assessment checklists are given in Tables 4 and 5.

Table 4. The assessment checklists for product categories: South Korean Environmental Mark

Questions in the first assessment	Assessment	
	Positive	Negative
1. Is the product category related to the foods, pharmaceutical dispensing, agricultural medicines and so on?	No	Yes
2. Are the accessibility and measurability of the criteria developed for the category acceptable?	Yes	No
3. Is there a possibility that the designation of the category might cause interested parties to misunderstand other labels or symbol?	No	Yes
4. Is it possible to consider the environmental aspects or impacts of the life cycle of the products in this category, and develop multi-issues criteria?	Yes	No
5. Is the manufacture and consumption of the products in the category over a short-run (instant) or long-run (continuous) period in the market?	Continuous	Instant

Table 5. The efficiency assessment checklist for product categories: South Korean Environmental Mark

	Questions in the second assessment	
Categories	Sub-categories	Assessment
Potential & imminent environmental improvement (40 points)	The significance of the environmental burden imposed by the products in the category	/10
	Technology level required for environmental improvement	/10
	The imminence of environmental improvement	/10
	The accessibility of life cycle data	/10

Potential market force (30 points)	The size of consumer demand	/10
	The environmental elasticity* of consumers	/10
	The market structure (number of providers)	/10
Interest level of the product (30 points)	The level of interest of consumers for an eco-label	/10
	The level of interest of product suppliers	/10
	The level of interest displayed by foreign label programmes	/10
	Total	/100

3.3   The development of assessment criteria

Three types of criteria have arisen from this revision: ecological, quality and consumer information. The ecological criteria are developed through life cycle studies and benchmarking. The latter two criteria are developed by expert advice, experimental results and benchmarking. Although quantitative life cycle studies are quite time and cost-intensive, they do establish the scientific impacts of product systems in markets where consumers and providers are sensitive to the environmental performances of products, such as laundry detergent, refrigeration etc. Several life cycle databases are needed to conduct life cycle studies, including life cycle environmental inventories on materials, energy, processes, transportation and so on. The South Korean Ministry of Environment has prepared these at national level. These can be used for Types I and III environmental label schemes, as well as in various policy simulations. The Ministry has determined that quantitative life cycle studies shall be used more and more in the development of criteria.

The Ministry plans to facilitate the integration of the three types of eco-labelling scheme. The main purposes of this project are to avoid consumers' confusion about the meaning of the three different types and to share operational resources in order to reduce the operational costs of three schemes. It is thus possible to share life cycle studies and databases, operational organization, and advertising costs for the various schemes.

---

* If a product-category is selected for the Environmental Mark, consumers are expected to buy more products from that category that have achieved the award. The elasticity measures the degree of responsiveness of consumers (the change in aggregate quantity demanded) to the product being certified.

3.4   Consumers' and providers' current response

Although the scheme is in a trial period, there is evidence that consumers in Korean markets have begun to perceive the purposes and importance of the scheme; according to a 1997 survey by the Ministry of the Environment, 80 per cent of consumers recognize the eco-label mark.   Although the evidence is only anecdotal, on the supply side the major Korean companies, such as Samsung, Hyundai, Daewoo and LG, appear to be very positive about the scheme, and small and medium sized companies consider the scheme as an opportunity to advertise their environmental credentials to the public. Thus the scheme appears to have been relatively successful to date.

One of the principal lessons to be learned from the South Korean implementation are that success is dependent upon facilitating a forum for the various stakeholders to express their opinions and misgivings. Further, a central unifying force (in this case the Ministry of Environment) is required to optimize the applications of co-existing eco-labelling schemes such that energies are not wasted and the consumer does not suffer from misinformation and/or confusion. The latter is a key issue and is explored further in the next section.

4.   GREEN CONSUMERISM

The fundamental driving force of eco-labelling is green consumerism. It is therefore important to evaluate how it is that consumers make decisions in general, and 'eco-decisions' in particular. Williams (1982) breaks consumer decision-making down into four sequential steps: problem perception, deliberation, solution, and post purchase review. These various steps are well documented in the marketing literature (e.g. Howard and Sheth, 1969; Engel et al., 1978). Information-provision in the form of eco-labelling is used in the deliberation phase to evaluate purchase alternatives.

The Lancaster-Rosen approach is the principle economic theory related to consumer decision-making. A product/service is conceptualized as a 'bundle' of attributes or qualities, both positive and negative, which the potential consumer perceives the product/service to have at varying levels. Thus if the product was, say, a car then these attributes might include: shape and style; engine specifications; reliability; options; safety features; storage capacity; warranty; insurance group etc.   Under Lancaster-Rosen, an individual consumer has a well-defined set of preferences/tastes, which determine the weighting applied to each attribute. These preferences can be modified by marketing, or alternatively the marketer can try to appeal to an individual's tastes.   Green consumerism generally falls into the latter category. One of the 'attributes' then is environmental responsibility or stewardship. Callenbach et al (1993) report that, in a survey of consumer preferences conducted in the US, 25 per cent of respondents claimed that they had changed their spending patterns owing to negative perceptions formed of particular firms.

The notion of consumers' perceptions of the attributes is of significance; these perceptions might not correspond with the 'true' levels of the attributes for a particular product. Thus, the consumer might perceive the product as 'green' but it might have a high environmental impact (in applying LCA methodology) relative to competing products. If asymmetric information exists between firms and consumers vis-à-vis the environmental characteristics of a commodity, i.e. the firm knows that its product is not green but the consumer does not, then the market does not operate efficiently. There is an incentive for self-interested, profit-maximizing firms to make false green marketing claims, i.e. 'greenwashing' (Rockness, 1985; Wiseman, 1982). The incentive that the profit-maximizing firm has for making only 'true' green marketing claims depends on monitoring and enforcement in advertising standards, the associated penalty, and the firm's preference for risk. 'Greenwashing' hinders environmentalism in that the consumers who weight the product attribute of 'greenness' cannot distribute their income efficiently. The costs required for a firm to reveal environmental credibility represent a loss to society (Klein and Leffler, 1981). Societal welfare would rise if 'greenwashing' could be ameliorated.

An eco-labelling scheme is one way to achieve this. If green marketing claims could only be made if externally-verified under an eco-labelling framework then consumers' confidence in green marketing would rise and the payoff to real eco-design and environmentally-friendly production processes would rise concomitantly. The higher the (perceived) proportion of greenwashing claims in all green marketing claims, the lower is the likelihood that the green consumer (who cannot discriminate) will choose a product because it is marketed as 'green', and vice versa. Once the market is functioning properly, with greenwashing ameliorated, there is an enhanced financial profit-driven incentive for corporate environmentalism. Further, the green 'consumers' can be firms as well as individuals. This type of green consumerism is increasingly becoming a dominant driver for small to medium-sized enterprises (SMEs) who are being forced to meet 'supplier-challenges' from larger firms producing final goods for sale to the public.

The European Commission (EC) has expounded the argument for compulsory validation of eco-marketing claims. ENDS (23.4.99) reports that the EC has brought together experts from the public sector, industry and NGOs to examine how to stop greenwashing. The Commission's consumer affairs directorate (DGXIV) presented a report on 21.4.99, which stated that

EU action was needed to harmonize the regulation of green product claims. The report highlighted such practices as claiming that a product is free from some substance that it would never have contained anyway; highlighting that the product is CFC-free, even though all products in the EC are also CFC-free by law etc. The European Consumers' Association's spokesperson, Barbara Moretti, called for a reversal in the current status of legislation, placing the onus on firms to prove the validity of their claims rather than on regulators/NGOs to prove them to be greenwashing as is the status quo. She also called for harmonization of such terms as 'reduced chlorine', which at present does not inform the green consumer adequately. The EU directive on misleading advertising (84/450) is to be revised this year, leaving the opportunity for green claims to be considered under its remit. The EU is planning to follow the ISO regulations used by the eco-labelling schemes mentioned above.

One issue that remains to be addressed is the validity of the fundamental premise of seal-of-approval eco-labelling schemes, viz. an accredited product/service is likely to be more environmentally-friendly than a competing one, which is not accredited. This need not necessarily be the case, in that the minimum conditions required to meet the criteria set might not require a change in industrial production and design, merely the documentation of the extant industrial processes. Is it then the relatively eco-friendly producers that seek certification, or merely those with a sufficient budget to finance the accreditation procedure? These two sub-sets of industry are unlikely to be identical. Further, just because a product is not accredited does not necessarily imply that its production is relatively environmentally-harmful; nor does it imply that any green marketing claims associated with it are necessarily greenwashing.

Although we would agree with the European Consumers' Association given the arguments in this chapter, the benefits of making the validation of green marketing claims compulsory are dependent on how well the market for green goods can be operationalized. If such validation does not weed out greenwashing then it only serves to add a layer of bureaucracy to the marketing process, an outcome that would actually reduce social welfare. Further, if the validation process were to be costly, this would decrease competition, as smaller firms might not be able to meet these marketing overheads. On the other hand, eco-labelling could facilitate an integrated approach by industry to environmental management in reducing the environmental impacts of production throughout the product life-cycle.

5.   CONCLUSIONS

This chapter has considered the three extant types of eco-labelling - environmental certification programmes, self-declaration, quantified product information labels - as a subset of green marketing. As such, eco-labelling is intended to facilitate green consumerism, wherein a consumer who values the product attribute of 'environmental quality' significantly can express this preference through product selection and purchase. Green marketing should be informative and accurate, therein raising consumer awareness, improve the sales of green alternatives and thereby improve overall environmental quality. However, with the status quo of voluntary participation in eco-labelling schemes and no legal requirement for firms' green product claims to be externally-validated, the momentum of green consumerism has been stifled. There is evidence of 'greenwashing', i.e. false or misleading green marketing. This goes against the public interest, as consumers cannot differentiate between truly green eco-design and greenwashing. This in turn hampers the greening of industry.

Externally accredited seal-of-approval schemes could ameliorate this problem. There are about 30 major schemes in operation around the world. There is some commonality in their methodologies, in particular the use of life cycle assessment methodologies as set out in various ISO standards, but there is a need for harmonization. Eight extant schemes (Blue Angel; EU Eco-label; Nordic White Swan; Canada's Environmental Choice Program; Green Seal; Japan's Eco-Mark; Korean Environmental Mark) were discussed and compared. Experience from the development and operationalizing of Korean label suggests that a central controlling body is required to coordinate eco-labelling activity to avoid duplicating efforts and misleading the public. Further, it is necessary to engage the various stakeholders - industry, regulators, NGOS etc. - in the developmental phase in order to achieve the goals that are set out for eco-labelling.

Eco-labelling can assist in the greening of industry. Eco-labelling schemes could provide a framework for the mandatory external validation of green marketing claims, thereby ameliorating greenwashing. However, if such validation becomes prohibitively expensive then competition might be reduced, and thus prudence is required in the administration of eco-labelling schemes if the public interest is to be served.

REFERENCES

Blaza, A.J. (1992) Environmental Reporting - a View from the CBI. in Owen, D.(Ed) *Accounting and the Challenge of the Nineties.* London: Chapman & Hall.

Callenbach, E., Capra, F., Goldman, L., Lutz, R. and Marburg, S. (1993) *EcoManagement: The Elmwood Guide to Ecological Auditing and Sustainable Business.* Twickenham, UK: Adamantine.

Engel J.F., Blackwell R.D. and Kollat D.T. (1978) *Consumer Behavior*, 3rd ed.. Hinsdale: Dryden Press.

EPA (1993) *Status report on the use of Environmental Labels Worldwide.* Report prepared by Abt Associates for the US Environmental Protection Agency, Washington D.C.

Fava, J.A., Curran, M.A., Denison, R., Jones, B., Vigon, B., Selke, S. and Barmum, J. (1990) *Technical Framework for Life Cycle Assessment*, SETAC.

Fox, M.A. and Singh, M. (1997) Life Cycle Management : Status of Concepts and Techniques. *Proceeding of the 1997 Total Life Cycle Conference – Life Cycle Management and Assessment.* SEA:1-7.

Howard J.A. and Sheth J.N. (1969) *The Theory of Buyer Behavior.* New York: John Wiley and Sons.

ISO (1997) *ISO/DIS 14020 Environmental Management/Environmental Labelling : Environmental Labels and Declarations.* International Standards Organization.

Klein,B. and Leffer, K.B. (1981). The role of market forces in assuring contractual performance, *Journal of Political Economy*, **89** (4): 615-641.

Kuhre W.L. (1997) *ISO 14020s Environmental Labelling and Marketing.* Prentice Hall.

Ministry of Environment, Korea (1998) *1998 Environmental Mark Scheme.* Ministry of Environment, Korea.

North, K. (1992) *Environmental Business Management.* Geneva: International Labor Office.

OECD (1991) *Environmental Labelling in OECD Countries.* Paris: Organization for Economic Cooperation and Development.

Peattie, K. (1992) *Green Marketing.* London: M&E Handbooks, Pitman Publishing.

Rockness, J.W. (1985) An Assessment of the Relationship between US Corporate Environmental Performance and Disclosure. *Journal of Business Finance and Accounting*, **12**(3): 339-354.

UNEP (1991) *Global Environmental Labelling: Invitational Expert Seminar, Lesvos, Greece.* New York: United Nations Environment Programme/IEO Cleaner Production Programme.

Volvo (1998) *Environmental Product Declaration Volvo S80 2.9.* PR/PV 981003, Volvo Car Corporation.

Welford, R. (1995) *Environmental Strategy and Sustainable Development. The corporate challenge for the 21st century.* London: Routledge.

Williams T.G. (1982) *Consumer Behavior : Fundamentals & Strategies.* New York: West Publishing Company.

Wiseman, J. (1982) An Evaluation of Environmental Disclosures. *Accounting, Organization & Society*, **7** (1):53-63.

AUTHORS

Syed Salman Hussain gained a MA Hons in Economics in 1994 and a postgraduate MSc in Ecological Economics with distinction in 1995, both from the University of Edinburgh. He was employed in 1995 as a lecturer and researcher in resource and environmental economics at the Scottish Agricultural College. He won the *Journal of Agricultural Economics* prize essay competition in 1995 with an article considering the economic optimality of making participation in environmental auditing schemes mandatory for certain industrial sectors. He has written and published extensively in the subject of the greening of industry, with a particular focus on policy issues and economic incentives. His current research models the adoption of green initiatives by smaller firms. He teaches on and coordinates modules for two Masters programs as well as undergraduate teaching in environmental economics and management, and has supervised seven Masters dissertations.

Dae-Woong Lim gained a BSc in Business Administration from Hankuk Aviation University in 1996. He then proceeded to the MBA program at the Graduate School of Hankuk Aviation University, which he completed in 1998. He is currently studying for an MSc in Environmental Sustainability at the University of Edinburgh. Since 1995, he has been a researcher and consultant at the Korean Eco-Management Consulting Co. As well as being involved in various Life Cycle Assessment projects, he was part of the team responsible for setting up the ecological criteria in 27 product-categories for the Korean ecolabelling scheme, developing operation procedures and a type 3 quantified environmental information scheme. In 1998, he was awarded the best paper prize by the Korean Construction Industry Association for research in the construction industry.

Chapter 9

Green Advertising
Issues and Guidelines

Dr. Stephen J. Newell
Associate Professor, Department of Marketing,
Bowling Green State University

This chapter examines the ISO 14000 environmental standard, the basic components of Environmentally Conscious Manufacturing, and gives some examples of how they may be integrated in support of an organization's environmental policy

1. INTRODUCTION

"The marketplace is greener than ever before - and will become even more responsive to products and services promising environmental responsibility well into the 21st Century." -Jacquelyn A. Ottman 1998

"The ecological friendliness of products was never of primary importance to consumers, and its importance is waning. The general public has become lukewarm about the environment in the past few years..." - Tibbett L. Speer 1997

Businesses who are considering an environmentally-based promotional strategy are getting mixed signals from experts in the field. Some argue that consumers are more focused on the environment than ever, while others seem to think that there are much more important issues that drive buying behavior for most consumers. Whatever side one agrees with, recent polls indicate that the general population consider themselves less environmentally concerned than they were a few years ago. This diminished concern may stem, in part, from numerous companies making vague and often misleading environmental claims about the environmental benefits of their products and packaging. It may also be related to the general feeling that society, as a whole, is more environmentally friendly and companies, in particular, are now more environmentally conscious than ever before. Whatever the reason, marketers are becoming more cautious about using the environment as the centerpiece of their advertising campaigns.

How does a firm decide whether it should incorporate environmentally related product information into their advertising and promotions? The answer depends on a number of factors. One of these is whether there are significant and salient environmental advantages to the products and/or materiels that they are packaged in. In addition, companies must determine the degree of concern for the environment among the consumers they are targeting and whether these consumers base their purchase decision, at least in part, on the perceived environmentally friendliness of the product and/or its packaging.

2.   HISTORY OF THE ENVIRONMENTAL MOVEMENT

It has been argued that the modern environmental movement began in 1962 with the publication of Rachel Carson's *Silent Spring* in which she reported that the use of dangerous chemicals had caused enormous harm to the environment and to the health of those living in affected areas. The government, in response to a small, but growing, vocal minority, passed a number of regulatory acts including the Federal Water Pollution Act in 1966 and the National Environmental Policy Act in 1969, and in 1971, established the Environmental Protection Agency. Also during the 1970's, a few companies, such as IBM, developed corporate environmental policies. Although, the establishment of environmental agencies, the development of environmental policies within a few prominent companies, and events such as the first Earth Day, created more public awareness about pollution and the environment, a relatively small percent of the population seemed concerned enough to warrant a change in the general product and promotional strategies of most companies. However, events in the later part of the 1970's and through the 1980's altered the thinking of many businesses concerning their environmental practices.

From the late 1970's through the 1980's, one highly publicized environmental catastrophe happened after another. In 1976, the Niagara Gazette newspaper had a series of articles describing reports of numerous illnesses of residents in an area called Love Canal. This led to the eventual evacuation of many families from their homes and increasing public outcry about abuses by business toward the environment and their lack of concern for the general public's health and well being. Other disasters such as Union Carbide's toxic leak in Bhopal which killed thousands, the Chernobyl nuclear power plant meltdown that wrecked havoc in Russia, and the Exxon

Valdez which spilled millions of gallons of oil off the coast of Alaska all triggered world wide outrage and demand for environmental reforms.

These events helped to shape the environmental thinking of the early 1990's. By this time, a large percentage of Americans considered themselves environmentally conscious and started to demand that environmentally-friendly corporations develop environmentally-friendly products that were packaged with environmentally-friendly materials. Consequently, the 1990's was plagued by companies falsely promoting their products as "earth-friendly" and their companies as "environmentally-focused" (Mendleson 1995). The use of terms such as "biodegradable," "environmentally-friendly," and "green" were found to be vague and ambiguous to consumers and lawmakers alike (Carlson, Grove, Laczniak, and Kangun 1996). This onslaught of misleading information brought about a backlash among consumers who, almost overnight, became skeptical about any environmentally related product claim. Between 1990 and 1996, the percentage of people who considered themselves the most concerned about the environment declined, and those who were the most environmentally apathetic rose dramatically.

Today consumers continue to be cautious about any environmentally-related product or package claim. This customer wariness has resulted in tough questions that need to be answered before a company decides to develop environmentally focused promotions.

3. PAST MARKETING RELATED ENVIRONMENTAL RESEARCH

There have been a number of environmentally related marketing studies in the past twenty-five years. In general, past "green research" in marketing can be broken down into two categories: 1) research profiling the environmentally concerned consumer; and 2) research concentrating on environmentally focused advertising (Shrum, McCarty, and Lowrey 1995).

3.1 Defining the Environmentally Concerned Consumer

Academic studies that have focused on defining the environmentally concerned consumer have tried to give marketers a better understanding, at least demographically, of those individuals that are the most concerned about the environment (Kinnear, Taylor and Ahmed 1974; Murphy, Kangun, and Locander 1978; Newell and Green 1997; Shrum, McCarty and Lowrey 1995; Tucker 1980; Tucker, Dolich, and Wilson 1981). In general, this research has revealed limited information that is useful to marketing practitioners when developing environmentally focused promotional strategies.

Practitioner-oriented studies have also tried to define the environmentally concerned consumer. The Roper Organization developed one of the most widely used categorizations for S.C. Johnson & Son, Inc. The environmentally related consumer segments that they developed are as follows (Ottman 1998):

True-Blue Greens - This group makes up ten percent of the U.S. population. They are the most environmentally focused and active group of consumers. They are more likely to contribute to environmental organizations. They are politically active and have made lifestyle changes to accommodate their environmental beliefs. They are the most likely to search for products that are environmentally friendly. In general, their behavior matches their strong concern for the environment. About twenty-nine percent have graduated from college, fifty-seven percent are female, ninety-one percent are white, one third of them hold a professional job position, and geographically they are spread out relatively evenly in all regions of the country.

Greenbacks Greens - These are environmentally focused consumers who are willing to pay more for green products, however, they are less politically active than True-Blues. This group represents five percent of adults. They are the most educated (forty-four percent are college grads), are relatively young, and a significant number are males (63%). Fifty-six percent hold white collar or professional jobs and like True-Blues, Greenbacks tend to be evenly dispersed throughout the U.S.

Sprouts - Sprouts are willing to engage in environmental activities when it requires minimum work. They are price sensitive and may purchase environmentally friendly products only if they are comparably priced to other similar products. One third of the population are considered Sprouts. Fifty-six percent are women, twenty-six percent have a college degree, and forty-two percent hold professional or white collar positions. Approximately, seventy-five percent consider themselves either "conservative" or "middle-of-the-road," politically.

Basic Browns - These individuals just don't care that much about the environment and don't participate in environmentally positive activities. Thirty-seven percent of the population are considered Basic Browns. Fifty-two percent of this group are males. Browns earn the lowest average income of all the groups ($22,000). Fifteen percent have graduated college, forty-three percent are from the south, and seventeen percent are African-American.

Grousers - Grousers feel that they are too busy to get involved in environmental matters and that even if they did, it wouldn't have any effect. They make up fifteen percent of the population. Nineteen percent have a college degree, fifty-four percent are female, and twenty percent are African-American. A third work in professional or white-collar positions and forty percent live in the South.

The percent of adults who fall into each of the above categories has changed somewhat dramatically since 1990 (Ottman 1997). From 1990 to 1996 those in the *True-Blue* category, the most environmentally active, have dropped from 11% to 10%. Those willing to spend more money for environmentally friendly products, the *Greenbacks*, have decreased from 11% to 5%. The *Sprouts*, the middle of the road environmentalists, have increased from 26% in 1990 to 33% in 1996. Those uninvolved with the environment, the *Grousers*, have dropped from 24% to 15%, and the *Basic Browns*, the least environmentally friendly group, has increased from 28% to 37% of the population.

Overall, these categorizations are useful as a means to gain a broader understanding of who the environmentally concerned consumer is. Unfortunately, the consumers that a business wants to target will probably not fit neatly into one of the proposed categories. Thus, a more detailed analysis of a companies target market is necessary to thoroughly understand the viability of an environmentally focused ad campaign.

3.2   Research in Environmental Advertising

Another topic researchers have investigated is environmental advertising. Within this area there have been a number of studies that have analyzed the content of environmental ads. Banerjee, Gulas, and Iyer (1995) analyzed the content of green print ads in magazines between 1987 and 1991, and in TV ads produced between 1991 and 1992. They found that promoting a green corporate image was the predominant theme (31% for print and 40% for television). They also found that most of the ads tended to be relatively "shallow" in terms of content and that majority of the "deep" ads were sponsored by environmental or nonprofit organizations. Emotional appeals were used in 27% of the TV ads and 17% of the print ads. Generally, the authors found that green advertising can be segmented in three parts: sponsor (profit or nonprofit), ad focus (corporate or consumer related environmental behavior), and the depth of information in ad (shallow, moderate, or deep).

Carlson, Grove, and Kangun (1993) classified environmental ads into categories and found ads that make claims that are product-oriented (the claim focuses on a specific environmentally related product benefit) or image oriented (associates the organization with a general environmentally attitude or ideal), are more often thought to be misleading and/or deceptive by consumers than other environmentally oriented ads such as process related ads (which focus on environmentally sound production) or ads that present recycling or package disposal benefits.

Schuhwerk and Lefkoff-Hagius (1995) found that for those consumers who were less involved with environmental matters, advertising that presents environmental benefits prominently (green appeal) had significant positive effects on attitudes toward the ad and purchase intentions. However, those consumers who considered themselves more environmentally involved, had no more favorable attitudes toward the ad or indicated greater purchase intentions toward products that used environmentally focused ads versus those ads that did not.

Shrum, McCarty and Lowrey (1995) found that women, who are the most environmentally concerned, tend to be more skeptical of advertising than eco-conscious men. In addition, they found that, in general, green consumers were less brand-loyal than their non-green counterparts. Also, green focused consumers were more receptive to environmentally related print ads than they were to similar ads on television.

Newell, Goldsmith, and Banzhaf (1998) looked at the effects of misleading environmental claims on the perceptions of advertisements. Not surprisingly, they found that if consumers perceived the green claims in an ad to be deceptive, this translated into lower perceptions of corporate credibility, less favorable attitudes toward the ad, less favorable attitudes toward the brand, and decreased intentions to purchase the product. However, more interestingly, they found that those who consider themselves the most environmentally conscious were no more able than consumers who were less environmentally concerned, to distinguish between an ad that was considered misleading and one that was not. Thus, even an ad that was not deceptive was often viewed as such and elicited negative company, ad, brand, and purchase intention reactions. The authors conclude that even if a product makes environmental claims that are true, many consumers, both those highly concerned and those less concerned with the environment, may mistakenly feel that the company is misleading them and consequently perceive the company and its products more negatively than they would if they had made no claims at all.

Overall, there have been a number of studies that may help to guide marketers when they develop promotional strategies for their products (see Table 1). However, research in this area is relatively new and a note of caution is in order before developing environmentally based promotional strategies solely around the findings of these studies.

4.   SHOULD COMPANIES USE ENVIRONMENTALLY - FOCUSED ADVERTISING?

Should companies use green promotions? Before marketers can answer this question, they must first determine: 1) whether their product does, in reality, have environmentally friendly benefits and whether these benefits are significant; and, 2) to what degree the consumers they are targeting with their promotions make purchase decisions based on environmental issues.

4.1   Is The Product Really Green?

With the current levels of distrust among consumers about any environmental claim, marketers have to think long and hard about the possible positive and negative effects of environmental advertising. Before firms develop environmental related advertising they must first ask themselves, "Does our product have environmentally beneficial benefits?" According to Ottman (1998) and Coddington (1993) these benefits may come from three different areas:

1) *The environmentally sound acquisitions and processing of raw materials used in manufacturing products.*

Have we used renewable resources? Are the materials used in production transported in a safe manner? Are recycled materials being used?

2) *The environmentally friendly manufacturing and distribution of products.*

Does the manufacturing process create significant reductions in the amount of both hazardous and non-hazardous waste that is being produced? Has the use of water in the production process been reduced? Is the use of raw materials being minimized? Is energy consumption at minimal levels? Are emissions to air, land, and water minimized?

3) *The environmentally sound packaging, use, and eco-sensitive disposal of the products.*

Is the product or packaging recyclable? Can the product be reused? Is the product long lasting? Are the ingredients of the product environmentally friendly? Is the product energy efficient? Is the product or packaging biodegradable? Can the package be refilled? Is the product or packaging safe when incinerated or landfilled?

If the product has one or more of the above attributes then it may be considered "green."

However, companies must determine whether these are significant or relatively minor benefits. For example, a company that manufactures a hair spray that is made from natural products, in a package that is made from 100% recycled materials, is in a non-aerosol dispenser, and whose package is also recyclable, may have significant environmental benefits, however, a hair spray whose only claim of environmental friendliness is that its package can be recycled, may not.

4.2   Who are Your Target Market and Do They Really Care About Green Benefits?

For companies to develop effective marketing strategies they must first define the target market (or target markets) that will be the focus of their promotional efforts.   This target market usually consists of current and potential customers.   Companies often use the Roper Organization classifications as an indication of the degree to which their target market is environmentally concerned. Though these classifications offer some insight about the environmental views of the general population and gives some limited demographic and geographic information, it fails to provide the detailed information marketers need in order to make informed strategic decisions as to the most appropriate advertising strategy to use for their specific target markets.

When marketers are determining the appropriate target markets they want to focus their promotional efforts on, they must gather demographic, geographic, psychographic, and benefit related information so they not only know *who* to present the information to, but *how* to present it. The following are brief descriptions of the segmentation variables.

- Demographic Variables - These involve defining the target market by issues such as age, gender, marital status, income, race, religion, etc.
- Geographic Variables - These focus on where the target market lives, works, and purchases products. Country, state, city, county, region, etc are all geographic based segmentation variables.
- Psychographic Variables - These describe the personality of individual in the target market. These variables provide a richer portrait of the target market and often are used developing the theme or tone of the ad.
- Benefit Variables - These are the benefits that the consumer looks for when they buy a product. These would include issues relating to the environment.

Once the company gathers sufficient information defining their target, a determination must be made as to which, of any, of the environmentally-related benefits of the product are salient to the target market. In addition, they must determine whether consumer purchase decisions are positively affected by these environmentally related product benefits.

If environmental benefits are important to a large portion of the target market, then a promotional strategy incorporating environmental issues may be appropriate. If, however, only a relatively small portion of the target market express concern about the environment or feel that environmental benefits play a minor or non-existent role in the purchase decision, then marketers must decide whether to include any environmental-related information in their advertisements.

5. DEVELOPING ENVIRONMENTALLY FOCUSED ADVERTISING

"Clearly, both opportunity and danger await makers of everything from automobiles to consumer products. To sell goods and services to an interested-but wary-marketplace, it is critical to bear in mind that eco-marketing and eco-advertising are very different from any other type of marketing campaigns. Whatever you say, in any media format, will be carefully scrutinized by both the law and environmental watchdog groups."
- Bennet, Freierman, and George (1993)

Once a company decides that their products have environmental benefits that are important to a significant portion of their target markets, advertisements must then be developed that incorporates these benefits along with other salient consumer issues into their promotions. Marketers, however, need to understand how best to present this information. First, marketers should be aware of the guidelines the FTC established for companies who are informing their target market about environmentally related product information.

5.1 FTC Guidelines

In 1992, the Federal Trade Commission (FTC) issued guidelines about environmental claims of products and companies. Though these are not legal constraints, they offer advertisers some suggestions on presenting truthful, environmentally related, product information to consumers. Specifically, they state that environmental claims should:

1) Be sufficiently clear, substantiated and prominent to prevent deception
2) Be clear as to whether any assumed environmental advantage applies to the product, the package, or a component of either
3) Not overstate the environmental attribute or benefit. In essence, claims should not concern trivial matters; and,
4) Be clear if comparing the environmental attributes of one product with those of another.

In general, the FTC gives marketers some basic direction in developing ads that present environmentally related product information. The following section presents further advice for marketers to consider when developing eco-centered ads.

5.2   Other Suggestions for Environmentally Focused Advertisements

Because of possible consumer mistrust about environmental claims, marketers need to be cautious when developing environmentally focused advertising. Expanding on the FTC guidelines, the following are general suggestions for marketers to consider:

1.) Understand the Relative Importance of the Environmental Benefits
      You Are Promoting
"When the average American is making a purchase decision, a brands environmental record ranks below product attributes such as price, quality, and the consumer's personal experience with the brand" - (Stisser 1994)

Make sure you understand the importance of the environment to your target market in relation to other benefits they are seeking, then put the appropriate level of emphasis on the environmental claims in the ad. If the environment is a major issue that drives the purchase, then make it the central theme of the ad. If, however, it is of less importance than other issues, present the environmental information as a secondary benefit.

2.)   Be Specific About Your Environmental Claim
Coddington (1993) says that marketers who wish to position their products as eco-friendly must communicate two things. First, what environmental problem they are addressing and second, how the product contributes to the solution. Make sure this information is straightforward and easy to understand. In addition, make sure you provide complete information about the claim and provide any proof that is appropriate. Finally, don't overstate the positive environmental impact of your product. According to Ottman (1988):

"Avoid vague, trivial, or irrelevant claims that can create a false impression of a product's or package's overall soundness. Broad labels such as "environmentally safe" or "ozone friendly," if used at all, should be qualified to prevent consumer deception about the specific nature of the environmental benefit being asserted."

3).   Avoid Catchy Phrases or Slogans
"Today's consumers are not likely to be swayed by image pieces with the general message 'we care about the planet.' Instead, many consumers are

in 'action mode' and are not about to be swayed by mere talk or hollow slogans." - Bennett, Freierman, and George (1993)

Don't use grand slogans or catchy phrases about your product or company's commitment to the environment as consumers may consider these bragging or misleading. Be straightforward and specific about any claim you make.

4.) Don't Create An Environmentally Friendly Company History of It Isn't True

"You can't rewrite or reinterpret history to paint your company as an environmental hero.... The public and environmental critics will acknowledge positive action in the future but won't simply erase bad memories because the company has announced an "about face" with regards to the environment." Bennett, Freierman, and George (1993)

If your company has a long history of environmental stewardship, then this may be information that is appropriate to include in your advertisements. If, however, your company does not have a history of being environmentally conscious, don't invent one. Focus on environmentally related things that your company and products are currently striving to achieve.

5.) When Possible, Get An Endorsement by an Environmental Organization

If you can gain an alliance from a well-respected environmental organization it will give your environmental message more credibility. Make sure you choose independent organizations with a good reputation with your target market. Mendleson and Polonsky (1995) state that:

"Strategic alliances with environmental groups can provide five benefits to marketers of consumer goods (they can): (1) increase consumer reliability in green products and their claims; (2) provide firms with access to environmental information; (3) give the marketer access to new markets; (4) provide positive publicity and reduce public criticism; (5) educate consumers about key environmental issues relating to the firm and its product(s)."

6.) Use an Integrated Communications Strategy to Promote Your Products and Company

Marketers must understand that environmentally focused advertising is most effective when it is accompanied by other environmentally focused promotions (public relations, publicity, personal selling, etc.). This unified strategy is referred to as Integrated Marketing Communications, or IMC. Carlson, Grove, Laczniak, and Kangun (1996) explain that:

"...the marketing of green products would seem to be a prime candidate for IMC due to the abundance of plausible promotional tools that might reasonably be used (e.g., public relations, brand image advertising, direct response advertising, etc) to target the highly involved environmentally concerned consumer segment."

Overall, if companies want their target markets to believe they are environmentally sound, advertising alone may not be enough. An entire promotional campaign incorporating all the promotional elements may be only way to create a lasting eco-friendly impression about your company.

6.   CONCLUSION

With consumer environmental concern waning, companies have to determine whether it is worthwhile to pursue an environmentally based ad strategy. Companies who do decide to go ahead with environmentally focused ads need to adhere to a few basic guidelines. First companies must determine if there is a significant environmental benefit of their product. Next, companies must decide whether this benefit is important to the consumers they are targeting. If the benefit is not salient to those consumers, then the environmental information should not be included in the advertisement. However, if the environmental benefit is important to the target market, then any claim within the ad should be truthful and specific. Finally, in order for environmental advertising to be most effective, a complete and integrated communications strategy needs to be developed.

Table 1. [Findings in Green Advertising Research]

Study	Results
Banerjee, Gulas, and Iyer (1995)	Green advertisements, in general, tend to be categorized in three distinct ways: 1) sponsor (profit or nonprofit); 2) ad focus (corporate or consumer); and, 3) depth of information (shallow, medium, or deep).
Carlson, Grove, and Kangun (1993)	Product oriented or image oriented environmental ads are thought to be misleading more often than environmental ads that present production or recycling related environmental benefits.
Schuhwerk and Lefkoff-Hagius (1995)	Ads that prominently presented environmental benefits of the product were

	more affective for those consumers who were less environmentally involved than for those consumers who considered themselves more environmentally concerned.
Shrum, McCarty and Lowrey (1995)	Women who were the most environmentally concerned were more skeptical toward environmentally focused ads than their eco-conscious male counter parts. In addition, they found that green consumers were less brand loyal overall, and that environmentally focused print ads were more effective with environmentally concerned consumers than similar green television ads
Newell, Goldsmith, and Banzhaf (1998)	Environmental ads that were considered misleading/deceptive resulted in less favorable attitudes toward the ad, the brand, and lower purchase intentions, even if the ad was not, in reality, misleading or deceptive. In addition, environmentally concerned consumers were no more able to distinguish between misleading ads and non-misleading ads than those consumers who considered themselves less environmentally concerned.

REFERENCES

Banerjee, Subhabrta, Charles S. Gulas, and Easwar Iyer (1995), "Shades of Green: A Multidimensional Analysis of Environmental Advertising," *Journal of Advertising*, XXIV (2), Summer), 21-31.

Bennett, Steven J., Richard Freierman, and Stephen George (1993),*Corporate Realities & Environmental Truths: Strategies for Leading Your Business in the Environmental Era*, John Wiley and Sons, New York.

Carlson, Les, Stephen J. Grove, and Norman Kangun (1993), "A Content Analysis of Environmental Advertising Claims: A Matrix Method Approach,"*Journal of Advertising*, 22 (3), September, 27-39.

Carson, Rachel (1962), *Silent Spring*, (Re-released 1994) Houghton Miffin, Co. NY.

Coddington, Walter (1993), *Environmental Marketing: Positive Strategies for Reaching the Green Consumer*, McGraw Hill, New York.

Kinnear, Thomas C., James R. Taylor, and Sodrudin A. Ahmed (1974), "Ecologically Concerned Consumers: Who Are They?", *Journal of Marketing*, 38 (April), 20-24.

Mendleson, Nicola and Michael Jay Polonsky (1995), "Using Strategic Alliances to Develop Credible Green Marketing," *Journal of Consumer Marketing*, 12(2), 4-18.

Murphy, Patrick E., Norman Kangun, and William B. Locander (1978), "Environmentally Concerned Consumers - Racial Variations," *Journal of Marketing*, 54 (October), 61-66.

Newell, Stephen J., Ronald E. Goldsmith and Edgar J. Banzhaf (1998), "The Effect of Misleading Environmental Claims on Consumer Perceptions of Advertising," *Journal of Marketing Theory and Practice*, (Spring), 48-60.

_____ and Corliss L. Green (1997), "Racial Differences in Consumer Environmental Concern,", *Journal of Consumer Affairs*, 31(1), 53-69.

Ottman, Jacquelyn A. (1998), *Green Marketing: Opportunity For Innovation*, Second Edition, NTC Business Books, Lincolnwood, IL.

Schuhwerk, Melody E. and Roxanne Lefkoff-Haqius (1995),"Green or Non-Green? Does Type of Appeal Matter When Advertising A Green Product?," *Journal of Advertising*, 24(2), Summer, 45-54.

Shrum, L. J., John A. McCarty and Tina M. Lowrey (1995), "Buyer Characteristics of the Green Consumer and Their Implications for Advertising Strategy," *Journal of Advertising*, 24 (2), Summer, 71-82.

Speer, Tibbet L. (1997), "Growing the Green Market," *American Demographics*, (August).

Stisser, Peter (1994), "A Deeper Shade of Green," *American Demographics*, (March).

Tucker, Lewis R. (1980), "Identifying the Environmentally Responsible Consumer: The Role of Internal-External Control of Reinforcements," *The Journal of Consumer Affairs*, 14 (Winter), 326-340.

_____, Ira J. Dolich, and David Wilson (1981), "Profiling Environmentally Responsible Consumer-Citizens," *Journal of the Academy of Marketing Science*, 9 (Fall), 454-478.

AUTHOR

Stephen Newell joined the Bowling Green State University Marketing faculty in 1993 and currently teaches Professional Selling, Marketing Principles and Marketing Communications. His prior business experience includes work as a sales representative and as a sales manager. He holds an undergraduate business degree from Michigan State, an MBA from Indiana University, and a PhD. from Florida State University.

Chapter 10

Recycling as Universal Resource Policy

Paul Palmer

Paul Palmer received his PhD. in Physical Chemistry from Yale University in 1966 and taught Chemistry in Ankara Turkey, Copenhagen Denmark and Indiana University.

The universal recycling of all products is an achievable reality if attitude change to accept it but it will take hard, academic research. The Garbage Industry today dominates and determines public attitudes and policies. The insidious notion that recycling should pay for itself through the mere recovery of materials assures the perpetual deferment of progress toward a recycling society. Industrial designers are capable of designing in total recycling of industrial products, independently of bioorganic methods e.g. composting. Chemicals are the most prominent and likely candidates for universal recycling. The five inescapable laws of recycling are: 1) Reuse function, 2) Provide for recycling in advance, 3) Exclude the Garbage Industry totally, 4) Remove all subsidies for dumps and 5) Make sure that recycling is profitable and determine how they may be integrated in support of an organization's environmental policy

1. INTRODUCTION

Recycling, as the term is understood today, is a fraudulent program thrust forward by the Garbage Industry to protect its profits by preventing true recycling from replacing dumping. The universal recycling of all products is an achievable reality if attitudes change to accept it but it will take hard, academic research that has never been allowed to get started. The Garbage Industry today dominates and determines public attitudes and policies. The insidious notion that recycling should pay for itself through the mere recovery of materials assures the perpetual deferment of progress toward a recycling society. Effective recycling consists of recycling functions, not materials. An example is refilling bottles, not breaking them. On the contrary, recycling must be designed for and paid for **before** a product is allowed to be sold. Industrial designers are capable of designing in total recycling of industrial products, independently of bioorganic methods e.g. composting. Chemicals are the most prominent and likely candidates for universal recycling. The five inescapable laws of recycling are: 1) Reuse function, 2) Provide for recycling in advance, 3) Exclude the Garbage Industry totally, 4) Remove all subsidies for dumps and 5) Make sure that recycling is profitable.

1.1 Why Is Recycling So Boring?

Consider:
- It's only about packaging.
- All you need to do is put bottles, cans and paper in baskets. The garbage company handles the rest.
- It comes out of garbage, which is boring by definition (it is that which you no longer want responsibility for).
- It seems to lead nowhere for the future. It's an environmental issue bereft of activism or discussion.
- Recyclers are seen as leftover hippies in dirty jeans (unlike garbage men who are union members and upstanding citizens).
- It is associated with minimum wages, employment programs for the homeless and the demeaning collection of bottles and cans.

My thesis is that this is no accident. Recycling has been corralled, boxed in and robbed of all its vigor because that serves the purposes of those powerful interests that thrive on digging up and stripping the earth of all its resources, to use them once and then discard them.

Take hope. Real recycling is an exciting and revolutionary way to reorganize the way we use all material resources. It will require changes at every stage of economic life.

The longer that a meaningful examination of a resource policy for a technological society can be delayed, the longer that the profits of those powerful interests can be maintained.

Obviously these industries include the extractive industries - mining, logging and monoculture - but the biggest player is elsewhere. The industry that most directly benefits by discard is - the Garbage Industry!

The Garbage Industry (the "GI") is a fascinating industry, which deserves a book of its own. With immense skill at public relations, they have fashioned a public image that is seemingly mild and unobtrusive, while they actually control every feature of the public attitude concerning the fate of those materials and articles of commerce that they call waste, which are of such vital importance to their continued profits.

While pretending to be mildly environmentally positive, they are actually one of the most perniciously anti-environmental industries you can find.

While positioned as simple folk, providing a public service, they are a worldwide, integrated, politically sophisticated, amazingly well connected industry that exploits political influence and the lobbying process as effectively as any industry can. Finally, while posing as one hundred percent private sector entrepreneurs, their luxuriant profits are actually based on abundant public subsidies of every description. To name just a few:

1.1.1 Dump Pollution Subsidies

A dump is an enormously destructive facility. It solves no problem except how to defer the problem to a later generation. Forty percent of Superfund sites are municipal garbage dumps. The chemical variety has directly been responsible for unknown thousands of illnesses, deformities and deaths, not to mention the destruction of wildlife and neighborhoods. Yet the government supports these facilities to the hilt, often winking at their role in causing cancer, birth defects and other illnesses, as well as wildlife habitat despoliation.

When a dump is closed, the former operator usually monitors it for a few years, taking responsibility for finding and stopping any major belowground leaks. The environmental damage actually evidences itself for hundreds of years if not millennia. Any problems that emerge later on will be remediated, or maybe made bearable, by large infusions from the public purse, including sometimes buying out the surrounding town and moving everyone to a new location. The local garbage company will have already banked its profits based on the bizarre bookkeeping of cheap, unwanted land, cheaply and irresponsibly destroyed.

1.1.2 Dump Siting Subsdies

Governments normally find and site a new dump whenever an old one is filled up. This process involves geological searching to find the site where the environmental damage that can confidently be expected from any dump, can be sold to a trusting public. Local governments may seize land for dumps by eminent domain as though a dump were a social good.

In the small California county where I reside, the official in charge of siting a new dump told me the county would spend ten million dollars of public money in just the geologic phase of siting a municipal dump. Other costs include hiring public relations firms and defending lawsuits to force protesting citizens into submission, passing enabling legislation, enduring endless hours of regulation by elected and appointed officials and enormous inputs by public activists and volunteers. The best-connected garbage company takes all of this as a monopoly, a gift.

1.1.3   Dumping Promotion

By suppressing all meaningful efforts to study and apply recycling for the last thirty years, the government, especially the federal government, has maintained a vicious circle whereby no viable alternative to dumping and incineration is ever on the horizon. The public is repeatedly forced to embrace dump construction as a short-term solution to garbage generation. Then, once the enormous public funding of a new dump is committed, the public is forced, like sheep to the slaughter, to make use of the dump, further dooming development of any long-term alternative. This circle has been actively maintained since the EPA's first Report to Congress in 1972 and still serves well to argue against any research into recycling. It is particularly galling that that report identified recycling as the number one, most favored response, to the number one problem, that of chemical waste.

With that Report to Congress freshly in mind, in 1975, at an EPA public meeting, I testified by scanning a two-inch thick EPA bibliography of existing environmental research. It included many hundreds of expensively contracted research projects into the mechanics and economics of garbage collection and dumping. Many of the entries were themselves long bibliographies of other EPA publications about garbage. There was not one single entry under recycling. Under reuse, there was a single article about recapping tires. [i] This imbalance where it counts, in money spent for research, has never changed.

The land under, over and near the dump has effectively been removed from the surface of the earth. When purchased, the land is invariably a rural ranch, sometimes state or federal pristine land, worth little in the calculus of development. Later on, it may be quite otherwise. It may become a borough of New York City or a wildlife preserve. But when starting out, the land is priced in the same way that it would be priced for agriculture, ranching, recreation or parkland. All of these other uses are in stark contrast to the land's use as a dump - they are all temporary. None of them actually destroy the land's future utility as the dump does. The dump will at best be covered with a few feet of dirt and the dump operator will move on to his next location. No excavations will be allowed from then on. Residential or commercial uses will be very dangerous, due to the threat of contamination. Trees and plants must not put down deep roots. Yet none of this depredation was paid for by the dump owner or his customers. The GI invariably promotes the cover story that the "closed" dump can become a park.

It would be appropriate for the destruction of the nation's land surface to be subjected to a new kind of tax designed to reimburse the citizenry for such a unique assault on its patrimony. No such compensation has ever been mentioned by our government as it implicitly accepts, and distributes among the taxpayers, the damages of dumping.

The reader may be noticing by now my use of the word dump. It makes no sense to sweeten the image of garbage by substituting "landfill" for the too descriptive, Anglo-Saxon, "dump".

All of the above subsidies spring from only the active pollution and visible land destruction of the GI. Yet this is not even their biggest impact on our planet.

1.1.4 Subsidies For Resource Depletion

We live on a finite planet. Just as no one is making new real estate, whether for houses or for the watershed and crops to support those houses, no one is making more beryllium, coal, iron, soda ash, gallium arsenide, lithium niobate, silver, platinum or sulfur needed for industry. Petroleum, which can be best used to make a wealth of specialized organic molecules, is being wastefully burned for fuel. Clean water, air and soil are in critically short supply and no extra-planetary sources loom on the horizon.

In light of these problems, many people have begun to think of sustainability as a critical element to build into all of our activities, both industrial and personal. This means that unless we can run our lives in such a way that we leave an undamaged planet to our children, we all have no future. The root meaning of sustainability is to leave a planet behind us just as rich and complex as when we received it.

It has been fashionable of late to substitute the oxymoron "sustainable development" for the more pure idea of sustainability but I do not follow this path. When discussing the sustainable development of fisheries for example, it is common to accept the loss of, let us say, only 25% of a species annually as being so immensely preferable to the loss of 50% per year that the former is anointed by this distorted badge of "sustained" development even while the latter is rejected on environmental grounds. In both cases, the fisheries will soon be destroyed. I do not see the utility of this distinction if we are trying to design a world that preserves a fully functioning, ecologically rich planet for our progeny. I prefer to set up sustainability as a target of zero destruction/full conservation. Then we have a meaningful, fairly fixed objective to measure our progress against.

I am not suggesting that no development is possible, though I do think it is important to be able to take that stand in a world gone mad with growth.

I draw a distinction between extensive and intensive forms of development. Extensive development is that which gobbles up resources in direct proportion to the amount of development. If 10 houses cover 2 acres and 20 houses cover 4 acres then, as more residences are needed, houses will soon cover all the available land. On the other hand, if residences are redesigned so that 20 dwelling spaces are now able to occupy only 1.5 acres, the development is intensive. It packs more use into the same resources. It is extensive development that has traditionally been the darling of the developers, presumably since they can continue to do that which they already know, in one place after another, with no risk of change. We are challenged today to eliminate extensive development in favor of intensive development.

Would that such a convenient distinction were ironclad, affording us the luxury of ever more population and resource extraction without any penalty in overall environmental destruction. Of course this is not so since even an additional residence, cunningly built on existing residences, represents, on a simple level, the extensive use of more water, more wood, more food and all other consumable resources. Population will overwhelm any and all conservation. But intensive growth, based on intense recycling of resources, offers some temporary hope at least for semi-sustainable "smart" growth.

A dump, and the mentality of destruction promoted by the GI are the most unsustainable, extensive forms of resource usage that can be devised. Yet our government embraces and legitimizes them as the resource tool of first choice in the past and in the future. Were the governmental and social subsidies to be withdrawn, the GI would quickly collapse of its own contradictions.

The subsidies enjoyed by the GI are essential to an understanding of the economics of recycling. It is too common to discuss the economics of recycling as though it should float free of any support, as though all recycling should be self-financing, and as though garbage provides some counterexample of an unsubsidized alternative. None of this is true.

1.1.5 Subsidies Thru Law And Regulation

In 1976, Congress passed the Resource Conservation and Recovery Act. Notwithstanding the upbeat name, this was primarily a way to institutionalize garbage. Suddenly unwanted chemicals had an official name - "hazardous waste". Valuable, usable, desirable, excess chemicals were no longer contemplated. New definitions insured that any chemical that had ever been an excess in the mind of any owner, no matter how fleetingly, was forever declared to be hazardous waste and could never be reclaimed from then on[ii]. Only one fate was allowed. Such a chemical was forcibly required to be buried in a chemical dump. Suddenly the stock prices of any company that owned a chemical dump skyrocketed. Many industry people swore that they would never go along with this insane kind of lawmaking, and continued to recycle solvents and other brand new chemicals anyway. Many of them were vigorously persecuted by state and federal governments. No government agency would ever admit that its target was recycling. It was always more acceptable to brand such bold souls as "polluters". A naive public, including, I am sorry to say, the environmental movement, never dug far enough to understand the difference. The EPA began to shift the public view of "recycling" by embracing the general term "sham recycling" in its stead. It was as though there was no longer any concept of genuine recycling, only sham recycling.

For the next 5 to 10 years, the EPA mounted what can only be termed a campaign of terror to drive industry into the nets of the garbage industry. The penalties for making an unacceptable disposition of "hazardous waste" were set at $25,000 for every day it went on. Later they added a year in jail for corporate officers for every day of illegal management of hazardous waste. These unbelievable penalties were hammered home in thousands of public meetings, informational meetings and orientation sessions designed to frighten corporate officers into submission to the new garbage oriented framework.[iii] The campaign was incredibly successful. Corporations were terrified of these penalties. Talk of recycling had to be almost carried on in a hushed whisper. Any activity that failed to put chemicals into dumps when they were no longer needed was simply too dangerous. Hundreds of times I made a bid to buy brand new or reusable chemicals only to be trumped by the legal department who insisted that the only safe place for an unwanted chemical was underground.

Through the end of the decade, the entire recycling staff of the EPA consisted on one single staff person with no resources or power.[iv] By his own description, he worked on municipal garbage recycling in the morning and chemical recycling in the afternoon. At this point the EPA had a budget of multiple billions of dollars, tens of thousands of employees and their official position was that recycling was "the number one" (their own emphasis) response to the problem of hazardous waste.

On May 19, 1980, the EPA published a new definition of "solid waste" (which included all chemicals, even liquids and gases) as "any material which has served its original intended use and is sometimes discarded".[v] The mischief done by this absurd definition was indescribable. Since everything is sometimes discarded, and since recycling works with materials that have served their original intended use, recyclers had nothing left to work with. The governmental prescription to bury anything and everything now had the force of law. One might think that a foolish definition like this would be pooh-poohed, even ignored, but that was not the case at all. Courts struggled, wrongly I believe, to give meaning to this construction and recycling was all but outlawed.

Throughout the history of all this harmful legislation, it is a curious fact that the term "waste" has never been defined, even when it was used in such constructions as: "solid waste is a waste which ...etc."

As I have pointed in my paper <u>Definitions of Waste,</u> on my website[vi], the only meaningful definition of waste is this: "A waste is any item which its owner does not wish to take responsibility for". Any other approach misses the essence of waste. Not only is this the common meaning, it correctly expresses the critical relation between waste and owner, that leads to every problem, every solution and every property of waste which so concerns our society. Had a common sense and truthful definition such as this been adopted early on, recycling would have flowed naturally, since the simplest form of recycling would have consisted of changing the owner of the former waste. But the government was adamant instead that waste-ness had to be considered an intrinsic and inalienable property of a lump of matter, making it intrinsically unusable.

2.   THE TECHNOLOGICAL BASIS OF GARBAGE

The permission for the assault on the planet by the GI that I have described above flows from the huge public subsidy and toleration extended to the GI. It is not surprising that the GI has had to become as politically entrenched as an invasive tumor. After all, in a technically sophisticated modern world, they actually provide a "service" based on no technology worth mentioning. The most pretentious garbage company actually consists of little more than a fleet of trucks and a hole in the ground.

Years of searching, using unlimited funds, for a high tech way to manage garbage, has resulted in the introduction of only such primitive programs as:

2.1   RDF - Refuse Derived Fuel

An abortive attempt in the early nineteen eighties to create fire logs from paper based garbage. As the plants repeatedly exploded, the concept was abandoned.

2.2   MORF's - Material Only Recovery Facilities

A shortsighted attempt in the early nineties to view all garbage as merely a collection of glass, steel, aluminum and burnables, ignoring all the complex assembly of those materials into articles. All garbage was smashed and then poorly paid pickers were used to remove shards of glass, steel, aluminum and paper.

2.3   Incinerators

An ongoing misguided attempt to treat all garbage as merely fuel.

2.4   Dump liners

A futile attempt to prevent dumps from polluting underground water by lining them with a plastic sheet.

Abundant funding and years of work into developing a technical base for the management of garbage have yielded nothing. The reason is simple. The very concept of "garbage" is too simplistic to lead to a modern resource policy. In the final analysis, no advanced society can produce garbage in the sense the GI hopes for. Had the same amount of research been expended into the kind of social, technical, marketing, logistical and juridical answers needed to make widespread recycling a productive reality, we could be today enjoying the fruits of a resource policy for a modern world. Instead, thirty years of so-called environmental progress have brought us no closer to a rational policy.

The basic problem is that the GI has the mission of treating all garbage as a single, undifferentiated (fungible) item in much the same manner as ... nothing else actually. Investors, for example, treat all commodities as equally saleable and priceable - but not fungible. Pork bellies and action toys are bought and sold similarly - but they are recognized as different items. Not so when they are discarded.

Whether it is a mattress, a baseball glove, a glass bottle, a drum of radioactive thorium, a ton of benzoyl peroxide or a carton of tree leaves, it is all the same to the GI. It has only two characteristics of interest - weight and volume. How much can fit into a truck, or a dump, and how much will it weigh down the truck delivering it. In recent decades, one more trait has been spottily taken note of, namely toxicity. But this sits ill in the GI's mission and wherever possible, it has been ignored, or simplified into meaninglessness, usually with dreadful results. Pollution is not an ineluctable consequence of resource usage, not even of chemical processing. It is the result of an attitude. That attitude is the attitude of garbage – "I don't want this, I don't want responsibility for it, I won't treat it with respect, I just want to get rid of it any way I can".

In my work, this comes up time and time again. Tens of thousands of chemical plants are designed the same way - to produce byproduct solvents, which are then all put into a single tank for disposal. Sometimes they are put into separate tanks that are so small that they cannot store up enough solvent to ship to a recycling facility economically. Sometimes expensive and eminently reusable solvents are burned internally with no thought given to recycling.

For example, in the early days of microchip manufacturing, in the 1970's, xylene and butyl acetate were used separately in the etching process. Every single plant I ever dealt with then plumbed both solvents into the same tank, thus cross-contaminating them. Yet these solvents began life as extremely pure electronic grades and were contaminated so little in use that they remained clean industrial grades after use. Not one single process engineer in Silicon Valley considered it his responsibility to provide for their subsequent reuse. Even worse, water used at the plant was added to the same tank. Anyone who has ever seen a process blueprint has seen that pipe ending in space marked "To waste". This can no longer be accepted.

Garbage has only three fates of interest to the GI.

- *Destroy the garbage uniformly*, without any nod to the separate components it is actually made up from, by burning, grinding or compressing.
- *Store it in the cheapest long-term storage* that can be found, usually a pit in the ground.
- *Drop it into some environmental niche that no one is watching* at the moment, such as the deep earth (for liquids and chemicals), the ocean, outer space or a third world country.

Time has shown that these are destructive, wasteful and dangerous approaches. "Garbage" is not uniform. It consists of every material, every assemblage, every device, and every chemical that a modern society produces. The sobriquet "garbage" is used to cover up these differences as though an unwanted kitchen table could somehow be equated with a truckload of unwanted isopropyl acetate. There are millions of old computers that are forcibly discarded - they are required to be broken up - even though there are millions of eager users who could wring more years of use from them. Millions of tons of reactive chemicals now sit underground undergoing unknown and unpredictable reactions due to the foolishness of the GI over thirty years and the cowardice of the bureaucrats that required underground burial, under pain of law, for decades.

3. THE LEGAL RIGHT TO DESTROY VALUABLE RESOURCES

In the seventies, the Navy had a truckload of benzoyl peroxide to dispose of. As was the norm in those days, they insisted that it be buried in a dump. Only afterwards did it strike home that benzoyl peroxide (a common polymerization initiator) is a high-energy oxidizer that can even explode. What prevents it from exploding is an inhibitor that slows down the free radical reactions it undergoes. But when that inhibitor is exhausted, you have a potential bomb. So far as I know, that benzoyl peroxide is still buried underground, waiting...

Many times I have offered a price to companies for a reusable item only to be told that they were going to bury it instead. I offered once to buy 80 drums of unused thionyl chloride from a large chemical company in Menlo Park California. They decided to bury it in the Kettleman Hills dump. Thionyl chloride is an extremely energetic oxidizer and chlorinating agent with many uses in synthetic chemistry. Sitting underground, it will have many years to find a suitable partner for its corrosive and energetic reactions. The drums have long since rusted through and the thionyl chloride is right now underground, surrounded by acres of unknown other chemicals, searching for a reaction.

In the case of the thionyl chloride I applied to the official regulatory agency for help. I was told in writing that the owner had a complete right to bury this oxidizer in a dump if they so wished, and the agency could not interfere.

We live in a society that grants an implicit right of property that is never discussed but strictly enforced. Anyone who owns any item is granted the absolute right to make a socially damaging decision with no oversight or hindrance. To wit: he can arbitrarily decide to destroy, or waste, or reduce that item to what we call garbage and then insist that that item, or its remains, be subjected to any of the treatments mentioned above.

In Novato California a few years ago, a software company rented a dumpster for a large inventory of unsaleable computer programs. Many of the diskettes were blank. A local recycler activist discovered what he considered an unconscionable waste and removed a small quantity of the blank diskettes from the dumpster for reuse or resale. The police were called in to investigate and the recycler was charged with theft. The right of the company to destroy usable diskettes could not be questioned.

In Los Angeles in the early nineties, a chemical recycler learned of a few hundred drums of a plastic resin that had been rejected by a customer for a minor problem. The manufacturer, a major oil company, tested and recertified the material as being perfectly within its manufacturing specifications. The recycler was able to obtain the resin cheaply and sold it to a plastics manufacturer in India. The material was shipped to India. A zealous District Attorney arbitrarily decided that this was a case of illegally dumping hazardous waste in a third world country. He reached out to India and forcibly returned the ship, with all its valuable cargo, to Los Angeles, at great expense. This absurd case served a private political campaign to prove the zealous public servant was "cracking down on pollution". In a world dominated by pressure to "get rid of" rather than "find a way to make use of", this kind of public campaign can be sold. Ultimately, after hundreds of thousands of dollars of unnecessary money spent for legal defense, storage and shipping, the recycler received an apology. The District Attorney was able to hide behind Sovereign Immunity - he could claim to be honestly doing his job. There was no effective recourse for the recycler.

Any chemical garbage company could have destroyed or incinerated or buried the resin and no official could ever have opposed that fate for it. Only the manner of destruction would be subject to oversight.

I propose that this particular right to destroy property is outmoded - in a modern industrial society it is a legalized insanity. There are four main reasons why it is nonsensical, of which the first is mother to the rest:

- *We live on a finite planet. Any reasonable ratio of available resources to the demand for them by a swollen population is approaching zero rapidly.* This situation is inherently unsustainable and is leading to social and economic collapse. We must think intensely about resource conservation.

- *We are producing millions of separate articles with unpredictable and unknown public and planetary impacts.* This characteristic is partially summed up in the term "toxicity". It includes not only the obvious chemicals, which act on a molecular level, but also biotoxins acting on a cellular level. Or six-pack holders and aluminum pop-tops that act on an organism level. Even radioactives that act on an atomic level and ozone destroyers that act on a global level.

- *Our modern industry produces immensely complex articles, far removed in intricacy and investment from the basic materials of which they are constructed.* A computer is made (largely) of metal in complex, wired connections, employing silicon, boron and phosphorous in incredibly complicated chips requiring entire factories to create. All that production in turn requires enormous inputs of resources (buildings, workers, vehicles, housing, and food), which are expended solely to create the chips and the computers. The computers contain, and represent, all of those inputs. To discard a computer or microchip is not to discard three grams of silicon or two kilograms of iron. It means the discard of a portion of the entire industrial complex, which produced that item. To arbitrarily destroy complex items, as though the impact were limited to a few pounds of material, misses the whole point of recycling in a complex society. The analysis of the true impacts of recycling is called Life Cycle Analysis.

- *Even as we extract identified resources, we destroy allied civilizations* such as Indian tribes, ecosystems such as rain forests, information sources such as archeological or fossil repositories, and potential resources as yet unrecognized, with barely a notice. For a variety of reasons, these lost treasures, held in common, are, and will be, sorely missed.

Underlying the above four reasons, is the fact that this mentality of discard is not necessary in order for us to enjoy a resource rich life. We have enforced public sanitation, building codes, toxic disposal limitations and hosts of other behaviors. We have the power to withdraw the right of arbitrary and unlimited property disposal and to replace it with a legal obligation to take personal responsibility for the planet and its parts, with respect to any object owned by a person, particularly when that object is no longer wanted.

Responsibility - this is the key term that cannot be avoided. There will be no sanity in managing resources without it. There will be no such thing as sustainability without it. Our planet, including ourselves, will have no future without it. We cannot continue to delude ourselves that we can foul our nest indefinitely and there will always be a new, unspoiled cave to move into. We are out of planetary real estate, finally and forever. We can no longer talk and act as though things can be thrown "away" because "away" is in someone else's backyard.

Responsibility is not an easy mantle to wear. Historically we have shunned it for its opposite, a laziness, a set of rose colored glasses that make it look as though we can continue to generate garbage willy-nilly forever, or at least until a devastating crisis forces us to adopt a thoughtful resource policy.

4. UNIVERSAL RECYCLING AS RESOURCE POLICY

The reason that a modern resource policy must consist of universal recycling is that there is no other basis for proceeding. There is no untapped lode of resources. There is no safe place to put once-used resources. And there are manifold ways to keep resources turning over. Is there any other possible policy for a modern, technologically based industrial civilization?

Current resource destruction is almost entirely done for greed, laziness, or lack of infrastructure. Our society refuses to make more than the most superficial recycling infrastructure available. The Yellow Pages are our best roadmap to available consumer services but each individual, at the time he faces a discard situation, finds no help in recycling there. He can however always turn to the ubiquitous, commodious, welcoming garbage can. Recycling will never be real until the phone book, at least, provides help in recycling every conceivable excess item.

Opposition to universal recycling is based on several presumptions:

- *We have grown up knowing easy discard and we assume that any other way of life would be so incredibly difficult as to be intolerable.* We know from experience the pressing need to "get rid of waste". We know the need as biological organisms needing to vacate our own wastes, as workers clearing off desks, as food preparers, and myriad ways throughout our lives and work. This appears to be an imperative that brooks neither delay nor frustration - as indeed it is. Yet the dice are loaded in favor of simplistic, short-term solutions.

Consider the obverse problem - that of obtaining the resources we need. Through thousands of years of economic development, winning, distributing and utilizing them has been designed for, built in, financed, promoted and mandated. We have purchasing departments, roads, distributors, brands and retail stores. Complicated? Yes. Unworkable? Obviously not. What better model could there be for a universal re-collection and reprocessing scheme?

- *We have seen recycling and it is inadequate.*

In fact, all that we have seen to date is a purposely emasculated pretense at recycling. The GI has called their self-serving plan "recycling" precisely in order to make it appear that no such plan can ever work. The basis of a true universal recycling plan will be presented below. It does not resemble current, so-called "recycling".

Such a resource policy will require study and research. Not one penny has been spent by the federal government in the last thirty years to study this issue though many billions have been spent studying garbage management (another major subsidy of garbage).

Creating, adjusting and integrating a functioning resource policy will require trained researchers and technicians. Yet not one single department of Recycling Research exists in one single university in this country. There is not one professor or one graduate student or one course. Anywhere! There has never been one dollar available from the federal government to support this research.

- *Recycling is not economic.*

In fact, it is the absence of recycling which is not economic, not profitable and not sustainable. Garbage generation could never pay its own way if it had to pay for all of its external costs.

Although the GI enjoys enormous subsidies, they plead for recycling to pay its way entirely through the paltriest of revenues and they elevate that financing method to a means test for recycling. It is but a ploy, to put forward the notion that recycling should be financed by the sale of recovered materials. This would insure that no funding for adopting a modern resource policy will ever exist.

Not only would the GI and a pliant federal government have recycling be financed in this inadequate manner, it would have this done in the face of a heavily subsidized alternative (cheap dumping), universally offered to a public that is told garbage is convenient, historically proper, legally required and the sole reasonable program.

On the contrary, universal recycling is but a part of a grand, circulating loop of resources. All the monies, which circulate in that loop, must be available to insure that the loop is closed. When a computer is manufactured, its return to reutilization following its immediate, first-time use, must be provided for and paid for *AS AN INTEGRAL PART OF THE SALE PRICE*, not as some afterthought based on extracting and reusing the least important part of the article - its materials. A computer serves as a useful example because its value lies entirely in its complexity, *ITS FUNCTION*, not its materials. Most articles share this characteristic in varying degrees.

If research into recycling, its development and application, received even a small portion of the luxurious subsidies enjoyed today by garbage dumping, we could build a smoothly operating, high capacity, effective program for recycling virtually every article produced by our industrial society. The reason for doing this is that we must.

5.   UNIVERSAL RECYCLING - OPPOSITION

Over many years of discussing universal recycling, virtually without exception, every listener or reader immediately responds, or imagines, that it is impossible. It cannot possibly work; usually because it would be *TOO INCONVENIENT.*

No matter that such commentators have probably never spent more than the merest moment considering why they are so certain of this point, what assumptions it is based on and how that set of assumptions might be proven false upon further analysis.

It is therefore essential that I provide some descriptions of how a universal recycling program would work and what it would feel like, to show that most of the objections are shortsighted. But first, I must discuss the underpinnings of the widespread dubiousness concerning recycling.

First, most popular assumptions about the inevitability of discard are based on familiarity. We are burdened with a major social behavior affecting goods that is essentially unchanged since Neanderthal times. Can any other behavior be so described? Is this something that we can actually hope to continue unchanged in a world that now produces five million chemicals, genetically altered organisms and radioactive bombs? *That doesn't make sense.*

Beyond that, we are immersed in a propaganda war supporting garbage generation at all costs and we are all affected by it.

We are told that we have seen successful recycling programs - when ten percent of, let's say, glass bottles are collected for remelting. That is a ninety-percent failure rate! Nowhere else would an audience accept that as a success. Of course recycling will appear laughable where even a "successful" program fails to put a dent into the flood of garbage heading for the dump. Declaring victory this way is purposeful nonsense, designed to provide a false sense that the battle is over. Let us adopt a standard of 98% recycling for success, not 10%.

Making use of its political dominance, the GI has introduced putative recycling programs that are actually designed to fail and irritate the public into a dismissal of recycling. For example, in California, when a bottle deposit bill was being debated, the GI intervened to insure that the deposit on bottles was set at so low a figure that it would not seem worthwhile to an average shopper to jump through the hoops set up for collecting the deposits. Many shoppers became angry and then opted out of the system entirely. Their future attitude toward recycling suffered predictably.

We have already dispatched the claim that recycling will happen all by itself when "it is economic". The effect of this canard is to deny all funding to research into recycling and consequently all serious discussion of an end to the Age of Garbage. The phrase "modern resource policy" is needed to begin discussion. Yet the very concept must be newly invented and promoted, since no such commonsense notion has yet circulated in public discourse.

The fact is that instituting a policy of universal recycling will take a great deal of thought, study and analysis. Had that analysis begun twenty years ago when I first started urging it on the US EPA, it could have been complete today. But not one dollar has been spent on it to date. The result has been the intended one - the public has no concept that recycling is a valid enterprise, capable of implementation and worthy of serious study. We have all seen with what remarkable celerity environmental issues such as pesticide residues, asbestos and chemical pollution can flash up and become household words. The same could happen for the need to create a new resource reuse policy. But so long as all discussion is squelched, it is not surprising that the average person has absorbed the notion that recycling is not realizable - expressed as "too inconvenient".

Of course for all these recent years, garbage has continued to pour unchecked into the world's dumps and gigantic profits have continued to pour into the GI's coffers. This has been aided particularly by federal government policies, which force industry to pay prices out of all proportion to the actual cost of the treatment or fate being purchased. The strategy of delaying recycling at any cost is an impressively successful one.

No social change or even important industrial progress was ever made in this country because it immediately made gigantic profits and never stopped. The history of industrial change is one of initial government subsidy, until an industry can achieve profitability. The government provides unremitting subsidies of all of the basic industries of the United States. Embassies and trade missions and international conferences and tax supports and enormous quantities of research in publicly supported universities support agriculture, manufacturing, arms, airplanes, the Internet and computer chips.

It is only recycling which is somehow supposed to be entirely self-supporting with no initial support or capital investment. This argument is just a device for starving recycling theory of all research funds, so it can never get started. In fact no progress will be made until recycling becomes an academic discipline with federally funded graduate programs, professors testing and reporting on new methods and students going out into industry bearing the latest research.

6. UNIVERSAL RECYCLING - NUTS AND BOLTS

6.1 First Law Arguments

Recycling will not be possible so long as a common fallacy continues to be accepted, namely that recycling is about materials. *Recycling must concern the highest function that any article embodies.* That is usually its constructed function. "What does it do?" not "what is it made of?"

A glass bottle's value consists 5% of common chemicals and 95% of a carefully designed ability to contain a liquid. In creating the bottle, 95% of the money was spent in crafting the shape, the integrity, the look, the sealable rim and the brand logo and in affixing a label and placing it into a distribution system.

The value of the function is significant because that value depends on the amount of predecessor resources used to craft that finely designed bottle - the factory, the furnace, the design, the research and all the human work and all the resources those humans in turn consume on their own.

The GI's version of recycling breaks the bottle, discarding 95% of its value, and remelts the glass at an enormous waste in heat energy. Although all of the glass is conserved, 95% of the net resources consumed in making the bottle must be reinvested. The net result is hardly better than putting the bottle into a dump.

The obviously superior way to recycle a bottle (or any container) is to refill it, reusing both the materials AND the function. The United States once had a thriving bottle cleaning and refilling industry. Just as the emerging automobile industry took pains to destroy public transportation, the bottle and garbage industries took pains to destroy the refilling industry.

The technology needed to recycle functions is not always known. Paper is universally used to present written information. The current view of its reuse dictates that it be ground up back to fibers and perhaps reassembled into a new form of paper. But the only reason for doing so is to remove the ink presently used on its surface. How much better it would be if that ink could be made to vanish so the paper would be available for multiple uses. Could a special ink be easily bleached by heat, by oxidation, by irradiation or some other way? A few research groups have reported preliminary results on paper, which is not written on per se but is manipulated to become black where ink would have been applied. The principle involves light transmission in the body of the "paper". The clean paper can then be regenerated by electrical impulses or even by air pressure. The research is in its infancy.[vii]

Paint is normally sold in standard cans with standard rims and covers. The containers are admirably suited for retail packaging, handling and delivery but they are poorly suited to protect the contents for long periods of storage, especially once rusting starts. Most discarded paint is probably discarded because of its container's deterioration, resulting in drying and oxidation of the paint, rather than because of any intrinsic lifetime of the paint. The way to reuse partial cans of paint is to force the initial marketer to give strong weight to the requirements for long-term storage by changing to a container, which is designed to protect the paint until it is used up, and then be refillable. Purveyors of certain jug wines use special plastic containers, which might be good candidates for paint.

When doing the accounting for manufacturing, raw material costs are generally distinct from *added value*[viii]. The added value roughly measures the cost of creating a final product and thus represents the cost of creating its functionality. This provides a crude way to estimate the planetary impact of function as compared to materials.

This analysis is universally applicable to industrial products. This leads us to our first law of universal recycling.

6.1.1 First Law Of Universal Recycling

Recycling consists of reusing both materials and function.

Defining the exact meanings of function and assigning useful units to these quantities will be a subject of future research. The relation of function recycling to Life Cycle Analysis and the Second Law of Thermodynamics (concerning Entropy) will also be explored.

6.2 Second Law Arguments

It is not an entirely new idea that *the recycling of an article should be built into the original design*. This concept has been used in Germany and elsewhere to require that all packaging used to deliver appliances must be returnable to the shipper.

This is particularly important when working with chemicals. For example, if a chemical process produces reasonably clean acetone and methanol, the acetone is frequently easy to find a new user for and the methanol can sometimes be reused (under current hostile conditions). But if the process is plumbed to mix the two together in one tank, the mixture is practically useless, without a good deal of specialized processing. Many a mixed output currently goes into an incinerator or worse, just because the process engineer never considered recycling as the normal fate for the byproducts.

Many byproduct acids, bases and salts are produced by neutralization, separation from reaction mixtures or washing. The processes are tuned for the main product. The byproducts are, as a result, mixed together, diluted or contaminated.

Bromides and iodides for example are entirely recyclable in simple ways, yet dilute byproduct streams cannot economically be shipped to a reuser if the only return is the value of the bromine or iodine. The same is true for salts of ammonia, aluminum, magnesium, and other metals. Huge amounts of these salts are put into rivers, wastewater treatment plants, and injected underground or otherwise disposed of, in spite of their inherent value. Were the byproducts designed to be reused right from the start, these salts could be kept sale.

A manufacturer may be left free to design any recycling scheme he wishes, such as the current scheme for automobiles involving the availability of replacement parts, the availability of schematics and manuals and the ultimate sale to a scrap yard. Though this scheme has many gaps, it shows the outlines of a recycling scheme, which is anticipated and provided for before sale. This leads to:

6.2.1 Second Law Of Universal Recycling

No article of commerce shall be allowed to be placed on the market for sale, unless and until, the recycling of that article shall have been provided for, including complete funding, after its next intended use.

Future academic research may devise various social and commercial schemes for making these anticipatory arrangements.

Some articles, such as food or fuel, are consumed in use and so have no apparent residuals. But even these are occasionally in excess, resold, contaminated, over shelf life or rejected, resulting in a need for recycling.

For example, since many small items cannot have the kind of horizontally and vertically integrated aftermarket arrangements used by automobiles, there would probably spring up general purpose recycling enterprises that would receive funding by manufacturers, as well as easy access to manuals, schematics and spare parts, and which would perform recycling functions for hundreds of related products. Such businesses would fill the Yellow Pages with easily found listings, making it a simple job for the owner of an unwanted couch, shed or bicycle, working or not, to pass it on to its next owner. This is how convenience will be reinterpreted and built into a new paradigm of responsible ownership.

A direct result of this approach to marketing would be that items would no longer be allowed to be specifically designed for discard. For example, although a common challenge to universal recycling takes the form of "Well how could you possibly recycle a disposable polystyrene cup", the obvious answer is not to struggle with a way to process that cup (that is a logical dead end) but to observe that such an item would no longer be on the market. Here is where defenses (if not hackles) are raised in the belief that a modern industrial society cannot exist without polystyrene cups (or name your favorite one-time-use item). It is important to point out that no research into drinkware recycling, based on an enforceable mandate for reuse, has ever taken place. I cannot believe that the limitless ingenuity of the American industrial designer is incapable of tackling a new challenge and coming up with, not merely a single disposable product, but a system of services, supports and multiple products which do not result in wanton, irresponsible discard. In any case, this is but one more example of the need for continuing research into new ways to juggle resources responsibly.

6.3 Third Law Arguments

Consider the scurrilous role played by the GI up to now in misdirecting and stalling progress in recycling. This is not an accidental effect but is an essential and structural aspect of the GI's main mission. *No conceivable progress toward universal recycling can be made if the garbage industry is allowed to control or even influence it.* Recycling must not be considered a subset of garbage management.

6.3.1 Third Law Of Universal Recycling

Large scale recycling cannot succeed if an organization pursuing the mission of the garbage industry, namely to treat used articles as lacking differentiation, is allowed to participate at any level in the recycling industry.

6.4 Fourth Law Arguments

We have seen that *dumps are non-economic, wasteful enterprises, requiring public subsidies* to operate. Such subsidies also subsidize various social harms detailed above.

6.4.1 Fourth Law Of Universal Recycling

Recycling will only succeed when no dump receives a subsidy of any form.

All public subsidies related to resource management must be directed at closing resource loops.

It is anticipated that this provision alone will provide so much pressure on dumps through the free market that virtually all dumps in the entire country will be forced to close. A few highly specialized dumps may continue to exist for a short time, until recycling schemes are devised for items, which were formerly dumped there.

Should dumps stay open longer than socially desired, it may be necessary to mandate their phaseout via legislation. Between the third and fourth laws, dumps can be expected to fade away naturally.

6.5 Fifth Law Arguments

Over the past 25 years, it has become obvious to me that anything can be recycled if it is valuable enough. The economic argument is after all the bedrock of American commercial ideology.

Much of the above discussion is aimed at making discard too expensive and insuring that recycling, paid for in advance if necessary, is profitable.

In working with chemicals for 25 years, I have frequently encountered a misunderstanding, to wit: only materials which can be introduced into natural organic cycles e.g. composting, bacterial degradation, atmospheric burning can be recycled. The environmental community tends to suffer from this misconception. [ix,x]

It is important to realize that toxics – chemical, biological, and radioactive to name the prominent ones – are extremely valuable because of their toxicity. At all stages of their manufacture and use, they must be treated with great respect. Their use and storage is controlled and monitored. This is expensive. Therefore toxics are prime candidates for 100% recycling. The only special requirement is that they must be managed by trained specialists who understand them intimately.

6.5.1 Fifth Law Of Universal Recycling

The economics of recycling must be manipulated to insure that recycling is profitable. The more valuable an item is, the more easily it is recycled. Toxic materials and articles are prime candidates for easy and early recycling.

7. CONCLUSIONS

It is unfortunate that so much time had to be expended here on the negatives of the history of chemical recycling and in dissecting the operations of the garbage industry. I would have hoped that all of that reporting would be the stuff of history by now, but such is not the case. Hopefully future writings will now be able to focus more strongly on an analysis of future recycling.

In providing this preliminary look at a society enforcing universal recycling, I have left many holes to be filled in either by the reader or by future generations. The treatment has been somewhat theoretical as befits an early treatment but I believe detailed prescriptions can flow from the precepts presented here.

The reader may well ask what would be the first step in proceeding to implement a change to universal recycling. In the USA, a country awash in free market rhetoric, I would suggest trying to apply the third and fourth laws, namely get the garbage industry out of recycling and remove all subsidies for garbage. If these changes were ever to take effect, the action of the market would force all the rest to follow. Of course putting any of the five recycling laws into effect will entail a major political battle. So the actual first step is to build a political constituency for these changes.

Perhaps the public could buy up dump development rights of candidate tracts in the same manner in which overall development rights are now purchased.

Much academic research, most of it highly technical, will be spent in studying ways to integrate industries so that no unusable garbage is produced and all byproducts can be reused in the simplest possible ways. Perhaps the current excess of chemists and physicists can find meaningful employment in building a more environmentally responsible industry for a change, rather than in finding ways, as heretofore, to either enforce or evade ill devised regulations fated to collapse under the weight of the contradictions of the garbage model. Rather let them build a new, rational system for integrating our society into a sustained and healthy planet. Just creating an active discussion of ways to create a better model than the garbage-discard model will change the reigning mentality and make it harder for the public to be deceived into an automatic rejection of universal recycling.

AUTHOR

Paul Palmer received his PhD. in Physical Chemistry from Yale University in 1966 and taught Chemistry in Ankara Turkey, Copenhagen Denmark and Indiana University.

In 1972, he founded Zero Waste Systems Inc. in Berkeley California. The company's mission was to find new homes for the thousands of chemicals then being discarded, especially by the nascent microchip industry. Though solvent distillers were already widely distributed, any other kind of chemical recycling was almost unknown. Zero Waste embodied the conviction that any chemical can be reused if there is enough value in it. Zero Waste put no restrictions whatsoever on the type of chemical it would work on. In this sense, it was the first and only broad spectrum recycling company in the world.

As a theoretician, Paul Palmer constantly strove to extract the social and economic meaning of his recycling work. As an innovator, he was the subject of many treatments in the media as well as several governmentally funded studies of the recycling business.

In the eighties, as regulations made it all but impossible for a small company to work hands-on with chemicals in California, he turned to designing byproduct reutilization programs by long distance. As the President of Chemsearch Corp., he assists companies to market or rework excesses from their chemical processes or inventories of raw materials.

He currently makes his home among fruit trees, grapevines and horses in Sebastopol California. His web page offers hundreds of excess chemicals to potential buyers around the world.

Hazardous Waste Management, Issue Analysis from Discussions at Four Public Meetings, December 1975, published by EPA in 1976. No Document number assigned. Original full meeting transcript includes the submission of Zero Waste Systems Inc., which is also available from the author.

This doctrine of "inherent wasteness" was literally believed and acted on by many people of this era, striving to be "correct". Environmental laws made no provision for legal recycling. Officials, struggling to make sense out of nonsense, discussed whether it was enough for a person to regard a chemical as a waste for one microsecond to render it unreusable or whether the length of time had to be longer.

I attended many such "informational" meetings and made a collection of dozens of the announcements. I recollect approximately one every week in the Bay Area. Small industry people were unquestionably terrified and their fears were widely discussed.

His name was Rolf Hill. I spoke with him frequently.

Federal Register, Vol. 45, No. 98, page 33119

http://www.sonic.net/chemsurf/rtheory.html includes many treatments of related subjects.

Switchable Reflections Make Electronic Ink, report by Meher Antia, Science 285, p. 658 (1999)

Added values are tracked more carefully outside the US where business taxes are often based on them.

The Ecology of Commerce by Paul Hawken, p. 70, Publ. Harper Business (1993)

Environmental Engineering: Energy Value of Replacing Waste Disposal with Resource Recovery, R. Iranpour et.al, Science 285, p. 706 (1999). This article reveals its civil engineering roots by uncritically accepting the assumptions of the GI, for example, that recycling of bottles means finding a use for broken glass, that recycling should be paid for by reuse of materials and that anything which happens to not be wanted is a waste and can be buried.

Chapter 11

Innovation Towards Environmental Sustainability In Industry

Paulo J. Partidário
National Institute for Engineering and Industrial Technology (Portugal)

1.   INTRODUCTION

INETI – The National Institute for Engineering and Industrial Technology is a public agency dedicated to research, development and demonstration (R,D&D), and to technical, technological and laboratory assistance. Within the Portuguese Ministry of Economic Affairs, but having its own legal identity, scientific, administrative, patrimonial and financial autonomy, INETI envisages contributing to the modernization of industrial companies, services and the technological systems, which support them.

INETI's scientific and technological activities are integrated in coherent programmes and projects in the following priority areas: Information Technologies, Materials and Production Technologies, Energy Technologies, Environmental Technologies, Biotechnology, Fine Chemistry and Food Technology. Important activities are also undertaken in the following areas: Engineering and Training Management, Engineering and Information Management, as well as Technology and Innovation Management.

INETI's link with industry is its main distinguishing feature – engaging it in a technological partnership with companies – especially in applied research and technological development carried out within a framework in which the industrial application of results is always present and governs the preparation and drawing up of projects and activity plans. This partnership also takes on other forms, for instance: consortia with industry, collaboration in business projects, participation in business consortia with a view to the adjudication of works and projects, openness to new business opportunities in industry's "traditional" sectors, and support to companies in the optimized use of the new technological processes. INETI has been responsible for a number of R&TD projects, which include company contracts and EC contracts (BRITE/ EURAM, ESPRIT, JOULE, ECLAIR, FLAIR, SPRINT, STEP etc.) in addition to contracts with the Portuguese State in the area of Space Technology and the modernization of Armed Forces.

Industrial economy, resulting from technological progress over the last two hundred years, still is deep focused on the optimization of the production process, looking forward the reduction of costs and constant improvement in the quality of goods. Nevertheless, modern industry faces innovation as a key driver for change, where systems are becoming more complex and integrated. To promote innovation through a dynamic efficiency (Klein, 1977), a constant shifting set of alternatives, namely technological, has to be taken into account. That enables a dynamic company, industry or economy to be flexible, answering effectively to its constant changing external environment.

Innovation has shown, within interplay between continuity and change, to bring progress as well as unintended side effects. Present concerns about environmental sustainability derived precisely from imperfections, namely from industrial production and subsequent externalized costs. Main emphasis on environmental debate is currently becoming more integrated, focusing particularly on resources consumption and by-production of pollution, on a multimedia approach. It is thus increasingly focusing on how to create conditions for a more sustainable production, shifting from end-of-pipe technologies to more complex issues, such as resource depletion, ecoefficiency, waste prevention and reduction, bio-diversity, or global climate change.

Despite a huge increase in material productivity, since the beginning of the Industrial Revolution, material consumption has increased due to a faster increase in demand. However, in a general trend that has began in early 1970s, economical systems supported by technological and structural change effects have been increasingly producing with a lower intensity of materials, energy and pollution per unit of output (Considine, 1991; Hettige et al, 1992; Casler & Afrasiabi, 1993; Klodt, 1994). This trend is in fact supporting the emerging new paradigm of sustainable production (see Ulhöi, 1995), which can be defined as designing, producing, distributing and disposing or recycling products in such a way that, within all stages of a product lifecycle, the associated environmental impacts and resource use levels are at least in line with Earth's estimated carrying capacity.

Industrial companies started changing attitudes assuming more innovative and proactive environmental strategies due to market opportunities and new competitive advantages. This change in attitude, answering to environmental problems, has been called the 'greening of industry' (Fisher and Schot, 1993). Moreover, OECD countries are already endorsing sustainable development as a key policy objective, which calls upon preventive strategies such as cleaner production. Cleaner production may be defined as the continuous application of an integrated preventive environmental strategy applied to: processes (conserving raw materials, energy, eliminating toxic materials, reducing emissions and wastes), products (reducing negative impacts along the life cycle, from raw material extraction to the ultimate disposal of the product) and services (incorporating environmental concerns into designing and delivering services), to increase ecoefficiency and reduce risks to humans and the environment (UNEP, 1994). In a selection process, within a cleaner production progressive approach, UNEP uses the following groups of options: good housekeeping, on-site reuse and recycling, inputs substitution, technology retrofitting or new design and modification of product specifications. In each case, once options have been identified, they should be evaluated, like other investment or technical innovation options.

However, in spite of such a formal commitment, and though cleaner technologies enable the use of resources in a much more effective way, the overall material consumption is likely to increase. In industrialized countries, demand is growing and, in fact, there is still a trend concerning the use of short-life material intensive products. In many developing countries, the increasing population and standard of living will also lead to an important increase in material consumption.

From Ehrlich and Holdren's (1971) contribution concerning the environmental load created by human activity, and the strategic thinking that resulted from the research conducted by the Dutch Council for Nature and Environmental Research (Weterings and Opschoor, 1992), it is expected that the overall environmental burden on Earth will require an increase in ecoefficiency by a factor 10 in the next fifteen years, reducing pollution (emissions, wastes) and the consumption of non-renewable resources, to achieve the right balance between production and population growth in the one hand, and the carrying capacity of the environment on the other. Taking this assumption, it follows that, in that period, we have to learn to cause just one-tenth of the current environmental impact, and to use energy and raw materials ten times more efficiently. A revolutionary approach to these needs is supported by the Dutch governmental programme for Sustainable Technology Development (Vergragt and Jansen, 1993) and consists in the fulfillment of social needs in 2050 with a reduction of environmental impacts twenty times more efficiently, i.e. by a factor 20.

In companies, environmental issues are emerging as important issues within strategic management (Roome, 1994), once they show a relevant impact on business, influencing structurally industries and their subsequent products, processes, operations, and the organization.

Corporate environmental strategy is thus currently progressing from compliance, reacting to environmental regulation, to innovative but also proactive forms, i.e. doing better than the regulative authority requires. Within that scope, as N. Ashford points out (1993), there are conditions to make (in)tangible profits from reducing environmental impacts. An important number of case studies already exist following this trend (see European Commission, 1994; Ebeafi, 1998).

In fact, proactive companies are exhibiting to need less end-of-pipe investments as well as productivity increases, shorter production cycle times, enhanced employee morale and enrolment in incremental innovation processes, lower energy consumption, reduced material handling and storage costs, safer working conditions, and also income from the sale of non-interesting waste.

Moreover, as environmental policy and stakeholders pressure will go on increasing, companies awareness regarding the advantages of pollution prevention and ecoefficiency will increase. Those companies taking a longer-term view, being flexible and sensible to environmental issues, would be able also to proactively consider and take those issues as potential key competitive factors. To do this, the organization must be able to look carefully at how all its functions and divisions may have interactions with environmental issues.

2.   CONTINUITY OR CHANGE - THE ENVIRONMENTAL SUSTAINABILITY AS A DRIVER FOR INNOVATION

2.1.   The role of technology

The evolution of mankind showed that it takes many years before actions can effectively percolate society (Volland, 1987). Besides a resource availability condition, another key issue is that many actors generally play a specific role in bringing about the actual transformation of a technological regime.

Technology is an artifact with a dual nature, once it integrates technical configurations (a physical object with physical properties) and functions to which it is addressed. On the other hand, the concern and search process about how technology development can drive to leapfrog jumps, requires long-term oriented research and innovation planning, and above all an adequate interpretation of the innovation process, i.e. the overall activities by which new products and processes are developed and used.

Galbraith (1967) provided a clear definition of technology as 'the systematic application of scientific or other organized knowledge to practical tasks'. Later, Freeman (1982) provided a fuller definition distinguishing between innovation and technological innovation: 'Strictly speaking, technology is simply a body of knowledge about techniques, as the word itself implies. But it is frequently used to encompass both the knowledge itself and the tangible embodiment of that knowledge in an operating system using physical production equipment'. '...Technical innovation' or simply 'innovation' is used to describe the introduction and spread of new and improved products and processes in the economy and 'technological innovation' to describe advances in knowledge'.

Technology plays a crucial role within the decision-making process on continuity or change. Technology development depends not only on the single company but also on a complex process (Kemp & Soete, 1992). Besides the market, at least three other selection systems can be distinguished: a selection environment, a techno-economic system or network (which also involves socio-institutional elements), and a national regulatory system.

Technological developments within most sectors are based on so-called 'technological paradigms'. These are relatively ordered paths of technological development shaped by the technological properties, the problem solving heuristics and the cumulative expertise involved (Dosi & Orsenigo, 1988).

Technological paradigm is a concept, which was developed in analogy with Kuhn's concept of scientific paradigm. Defines the technological opportunities, like directions for further innovations, and some basic procedures and believes on how to exploit them (Dosi, 1988). Within a given technological paradigm, certain basic designs can be found. These are concrete technological systems or products that largely remain unchanged over extended periods of time, e.g. personal computer of the past 15-20 years, (which contrasts with the main frame computer). As far as PC is concerned, its basic design has not changed, although capacities, speeds, functions, and the like have enhanced tremendously. Another interesting example is computer microchips: the present paradigm prescribes certain semiconductor materials, micro-electronic multi-transistor design rules, and electricity as the driving force behind the system.

The concept of technological regime (Nelson & Winter, 1977) is comparable to the notion of technological paradigm developed by Dosi (1988). It is derived from the observation that problem-solving activities of engineers were not fine-tuned to changes in cost and demand conditions, but were relatively stable, focusing on particular problems and supported by certain strong notions, their potential and how those problems could be dealt with.

Technological regime is defined as "the whole complex of scientific knowledge, engineering practices, production process technologies, product characteristics, skills and procedures, institutions and infrastructures that make up the totally of a technology". It represents, therefore, a complex context where there is the pre-structuring of the kind of problem-solving activities that will condition the different elements, particularly socio-cultural behavior, and the structure that will enable (or constrain) certain changes. Within this complex frame, however, the accommodation between its elements is never perfect, once there will be always tensions and the needs for further improvement.

Referring to a regime rather than to paradigm results from the need to socially embed technological breakthroughs, to include rules in an explicit way. Rules that mean either a set of commands and requirements, and also roles and practices that are being established or that are not easily dissolved. Analyzing the existing drivers, and required conditions, on the frame of technology dynamics for a company's decision to follow an evolutionary or more revolutionary innovation path, it is important to call upon macro, meso and micro-level work approaches.

In every change of a techno-economic paradigm which has so far occurred, the new paradigm had already emerged and developed within the previous one (based on technology already well established) (Freeman, 1982). Radical technologies are relatively crude at the time of their invention and need to be improved and better-adapted to user needs (incremental innovations). They are only able to compete in specialized markets. These early market niches are important for the further development of the new technology. Besides providing necessary financing means, the experiences of the users are an important source of information in helping firms further to improve the product. Radical technologies may also benefit from accumulated experience in other sectors, and from the presence of a network in which it can be easily introduced (Schot, 1992; Schot et. al, 1994).

At a macro level of the production system, it is important to call upon a system approach to technology selection. Large technical systems, according to Hughes (1983), can be seen as a special kind of regime, where material connections and the building up of an infrastructure are crucial to its diffusion. This creates special effects, and leads to what Hughes called "momentum".

On a meso/ micro level of the production system, different contributions should be referred since Schumpeter' landmark (1939). Abernathy and Utterback (1978) and Utterback (1994) for instance, provided a new understanding of the complex interaction of product and process technologies. In this latter work, the focus is not just about technological innovation but also about the important interconnections with core competencies and organizational learning. Utterback portrays the changing rates of major product and process innovation as an industry moves from its initial 'Fluid phase', through the 'Transitional phase', to its final 'Specific phase'. Firms that do not have the core competencies, which cannot learn and that cannot generate or adopt radical innovations will fail to survive as an industry moves from one generation of technology to the next other.

In a study on process control technologies in various sectors of industry, Hagedoorn & al (1988) have shown a practical application of Nelson & Winter (1977) concept. It was shown that over more or less extended periods of time, certain technological 'regimes' prevail, where a technological regime can be defined as a body of rules applied by a technological community. Each regime may yield a number of actual technological systems (hardware, software, organization), each representing a certain 'technological trajectory' within the paradigm.

Thus, the continuous process of innovation can be described on the basis of consecutive technological regimes, each being characterized by certain key elements. This means that technological developments may be defined at the level of regimes and basic designs.

A technological trajectory is thus the activity of technological development along the economic and technological trade-offs defined by a basic design. For instance, still within the PC example as a basic design, the successive generations of microprocessors form a technological trajectory that started in early 1970's, which lends itself to further qualitative and quantitative description (characteristics, numbers sold, etc). The monitor screen and operating software systems are other key-elements having specific technological trajectories.

At a micro level a real test of acceptance occurs for a new technology, which will greatly influence its rate of diffusion. The way firms choose their technology depend very much on what their critical mass (education and training of specialized employees) believe it is feasible. In a given industrial sector, these technicians tend to have a set of basic design principles, which guide technological choices, to form a technological regime.

A technological regime, and its subsequent body of knowledge, can thus lead to a kind of 'blindness' to alternative technological solutions (path-dependency) making resistance to alternative solutions uptake, being very difficult for quite new alternatives to enter the existing technological portfolio. It shouldn't mean however that innovation doesn't take place, but the occurring incremental innovations are mostly add-on innovations that don't change the basic processes, thus not imposing more far-reaching implications, which could lead to the redesign of processes.

Apart from technological regimes, specific characteristics of the industrial sector influence the adoption of new technologies. Generally, knowledge of how a new technology will pervade mature sectors is largely absent, however for the uptake of a certain technology, factors can be defined and tested, and their expected positive or negative influences emphasized.

The economics of the sector is also an influencing factor. Environmental innovation based on pollution prevention is often assumed to provide simultaneous environmental and economic benefits, by savings on energy and materials for instance. However, this is not automatically the case in sectors with a strong price-based competition, having a low added value on products and a relevant effort on economies of scale. This leads to a situation where additional necessary investments in environmental technology can simply entail extra costs. Thus, it is extremely important whether there are any economic benefits to be expected from new technologies.

Besides technological regimes and sector characteristics, empirical studies on the relationship between regulation and technological change show that regulation, within an interaction process (Irwin & Vergragt, 1989), is one of the many external stimuli that affect a company's technological strategy and that reactions to regulation are often consistent with the historical patterns of technological change in a given industrial context. There are evidences that under a higher pressure on environmental regulation, certain countries promoted not just the necessary environmental measures but created also competitive advantages and new business opportunities as well (see for instance Ashford et al, 1979; Porter, 1990; Ashford, 1993; Porter and van der Linde, 1995a; Nehrt, 1998).

Technological innovation and diffusion, on the other hand, are essential for regulation to be well succeeded. Therefore attention must be given to encourage specific kinds of technological change in the appropriate industrial sectors. As far as pollution prevention is concerned, it has been argued (Ashford et al, 1979; Schot, 1992) that the key to success in pollution prevention is to influence the managerial decision taking process towards both technological change for production purposes, environmental concerns, and compliance purposes within interrelated rather than separable activities.

2.2.   The environmental challenge

In order to approach ecoefficiency gains in a consistent way, i.e. factor 10-20, having environmental sustainability as a driver, there is a latent demand for innovation progressing towards function and system-innovations, on an effective long-term perspective, thinking in leapfrog jumps and acting by steps, like it is represented in figure X.1. This calls for proactive behavior, and proactive strategies, that entail anticipative innovative approaches to problems enabling to take the initiative, performing planned systematic actions to achieve more than it is strictly necessary on a short-term perspective.

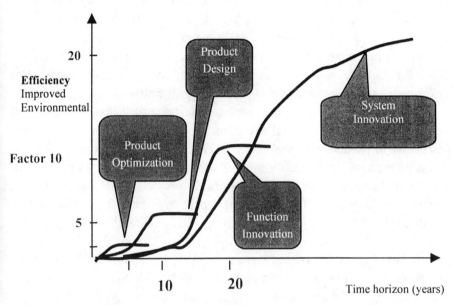

Figure 1– Transitions required within the innovation process, to achieve environmental sustainability. (Source: Brezet, 1997; Weterings *et al*, 1997)

Depending on the innovation process, at a certain moment in time, different generations of technology may exist next to each other. Though many forms of outcomes can be referred, we will consider a main distinction between two types of innovation: a normal innovation pattern, and a fundamental transformation pattern, resulting into incremental and radical innovations respectively.

Enabling different ecoefficiency gains, distinction can also be made between: optimization, redesign, function-oriented innovation, and finally system innovation. A system innovation can be divided into function innovations. A function innovation can also be subdivided into redesign, adjustment and optimization.

When considering manufacturing industry in particular, a special attention should be given to SMEs. Besides having a limited environmental awareness, they have also limited human and financing resources, as well as a limited access to research and technological support. Moreover, industry's environmental performance varies greatly comparing industrial sectors between different regions, existing also substantial differences between industrial sectors within the same region. Therefore, more sustainable production will remain a distant target, if industrial SMEs will not be adequately included in the overall effort. Within that scope, a fairly recent EC analysis (EC-DG III, 1995) showed that, excluding a relatively small number of industrial SMEs installations, general scenario includes objective but diversified existing difficulties, together with a limited environmental awareness, which reduces effective environmental compliance ability. Those limited resources are endogenous drawbacks, which make it difficult for SMEs to identify and understand, by themselves, their environmental performance in order to make ecoefficient improvements.

In research funded by the European Commission for instance, the overall participation of SMEs has exhibited an important growth in the past few years, due in particular to the effect of research policy measures, like co-operative research, exploratory awards and take-up actions. Twelve and a half thousand SMEs have taken part in R&D projects under the EU Fourth Framework Programme (FP4) and, of those, two thirds had never before participated either in transnational co-operative research, or in publicly-funded research of any kind (EU, 1998). They represented 24% of the overall participation in projects launched in 1997, compared to 18% in 1996. The FP4, which has been running since 1994, with the aim to promote scientific and technical basis of European industry, is currently being replaced by the Fifth Framework Programme (1998 – 2002). It is interesting to look back in order to analyze how far did it support cleaner production development. There were two main programmes in FP4 that addressed to cleaner production: programmes 'Environment and Climate' and 'Industrial and Materials Technologies'. In a recent analysis, Schramm and Hackstock (1998) concluded that cleaner production was more successfully stimulated under the programme 'Industrial and Materials Technologies'. In fact, despite the explicit call for tenders in the former programme within the issue 'cleaner technologies and cleaner products', it resulted in a relative low share (15%) of projects submitted (addressing mainly to cleaner technologies), when compared to projects submitted on end-of-pipe technology (69%). Those findings may suggest that cleaner production is more 'easily' addressed from an industrially integrated point of view (which seems to be obvious), but that technology entrenchment effects on R&D community, during the period of analysis, might also have been a barrier to shifts focusing particularly on pollution prevention approaches. As far as the innovation and diffusion processes on cleaner production are concerned, these are key aspects that must be understood in order to promote a higher rate of adoption.

Besides the examples referred above, other interesting case studies exist around the world, demonstrating the effects of industrial pollution prevention and ecoefficiency gains. In the United States, for instance, Project XL (eXcellence and Leadership) is a very recent national initiative that allows for the testing of innovative ways of achieving more cost-effective public health and environmental protection. A key objective is to trial cleaner, cheaper and smarter ways to achieve environmental results in a superior way to those achieved under current legislation, in conjunction with greater accountability to stakeholders The project is in its early stages, but looks forward to learn from the different pilot-scale projects to assist Government in redesigning current regulatory and policy setting approaches.

3.   THE INNOVATION PROCESS

The complex nature of the innovation process was already addressed on R. Rothwell's (1977) research: "Technological innovation is a complex technico/socio/economic process which involves a very intricate web of interaction, both intra-firm and between the firm and its economic, technical, competitive and social environment". As such, more complex models describe the relationship between many discrete decisions by specific actors and influence of external factors, which can be identified, and implications appreciated through close examination of precise situations.

The progress towards more environmental sustainability then requires that the innovation process should go beyond optimization, retrofitting and redesign based on incremental innovation, to focus particularly on breakthrough innovations.

An attempt to begin an understanding of the innovation process could be perhaps first to call upon J. Schumpeter (1939), who offered what could be considered one of the best working definitions. He suggested that innovation was not only the driving force behind economic growth, but also the determinant of cycles in business activity. His definition of innovation comprised already not just new products and processes but also new forms of organization, new markets and new sources of raw materials.

According to Schumpeter, technological change was a three stage linear process, comprising different categories of activities: invention (discovery of new ideas and methods), innovation (turning these ideas into commercial applications), and diffusion (the spread of these marketable products and processes to all available markets). Discrete activities that can contribute to innovation may include e.g. R&D, prototypes or basic research, but may also be not research based, being then frequently a trial-and-error process that contains many false starts. Line drawing between innovation and diffusion is in fact a complicated issue as innovations can rarely be adopted by new users without modification. When these adaptations become extensive enough, another innovation may have in fact occurred. The linear model, thus focus too little on the complex set of factors that can influence the process, namely the external environment in which the innovation takes place. Over time, this linear model of innovation process has given way to more complex models to emerge. But due to the complexity that is inherent to the process, most have their particular focus and fail to provide a general model of innovation.

After the invention stage, where a first working prototype is conceived, technological innovation may also have a more market-oriented definition, being considered the first commercially successful application of the new invention (Ashford, 1983). In fact, an innovation is often considered successful if it reaches the point of being sold. However, in many cases success in use is not guaranteed. It requires successful implementation for the innovation to be truly valuable. Indeed, for many innovations, one person's product is another person's process. The process of diffusion starts with the first adoption of an innovation. For the adopting company, the successful implementation of innovation can be critical to obtaining commercial benefits.

There are different types of outcomes from innovation. A convenient distinction is often made between product and process innovation, the former being a saleable new end product, and the latter a change in the production process. In addition, innovation could also be considered as well within new materials, new forms of working organization, new markets, new organizations and institutions.

Distinction is also made concerning the degree how an innovation occurs, being radical if it would bring major revolutionary changes (assuming a fundamental transformation pattern), or being incremental (normal pattern) whereas smaller evolutionary developments (adaptations) were attained, which are actually the bulk of innovative activity. Incremental innovation is thus a step-by-step co-evolutionary process, which retrofits and upgrades existing basic systems' design, where each company and networking structures are also modified incrementally. Radical innovation, on the other hand, focus on new fundamentally different technology, at a system level, where the new system is gradually co-existing/ replacing the existing ones along a learning curve, through changes in the systems' concept, and also with completely new forms of networking on a regional/ global basis.

Progression between patterns can also be stressed, in fact as transition patterns, which are related to a sort of major innovations. Although not carrying fundamental shifts, major innovations could be identified as a common route to major advances in performance at a sub-system level, that require new core competencies and carry over some problem solving routines and heuristics (Rycroft and Kash, 1999). While normal patterns of innovation do not require changes in public policy, transition and transformation patterns do have impacts at a public policy level, therefore requiring its change.

Radical innovations are discontinuities in the pattern of technological development. If an organization has matching competencies, it can build upon the new radical innovation. If it does not, then it must get them or face competition from companies that can manage the new radical innovation. These technological discontinuities can be threats and opportunities. They demand that organizations renew their products, processes and core technologies. Companies that successfully manage crossing one technological discontinuity to the next generation of new technology, generally follow a strategy of incremental product and process innovation. This can be profitable but there is the danger of becoming locked into that generation of technology. Once that happens, they will be overtaken by the next technological generation. Successful companies must be stable with regard to the current generation of technology, and flexible enough to cross future discontinuities to the next generation. Even so, as stressed by Christensen (1997) in a recent study about the way innovation occurred along four generations of hard disks for computers, the innovation networks and the key actors of a new hard disc generation always changed from one generation to the next one. It is important, therefore, to stimuli the structuring of a macro economy that contains strong incentives (e.g. economic incentives, regulation) for companies to change, adapt, and redefine the alternatives facing them (Klein, 1977).

The sources of innovation vary greatly (von Hippel, 1988)), thus basic assumption that product innovations are typically developed by product manufacturers themselves is seldom questioned. As von Hippel points out, the identification of who the innovator is, must have a major impact on innovation-related research, on firms' management of R&D, and on government innovation policy: "Whilst in some cases it is found that product manufacturers are indeed the innovators, through the utilization of an additional variable called 'the functional source of innovation' the manufacturer–as-innovator assumption may be replaced with a view of the innovation process as distributed across users, manufacturers, suppliers and others". The use of the functional source of innovation variable involves categorizing firms and individuals in terms of the functional relationship through which they derive benefit from a given product, process, or service innovation. For example, if they benefit from using it then they are considered users. If they benefit from manufacture, then they are manufacturers. However, the functional role of an individual or firm is not fixed. It depends on the specific innovation under examination.

A wide range of factors, at project or corporate level, is associated with innovation (Rothwell, 1992), which appears to be common, though with variable weightings, to the different manufacturing sectors, once (un)success depends on the qualities and abilities of players involved. Moreover, at different periods, during about the past fifty years, different perceptions of the innovation process and its performance in innovation have evolved up to now through five generations: 1stG - Technology push (1950s – mid 1960s); 2ndG - Demand pull (mid 1960s – early 1970s); 3rdG - Coupling model (mid 1970s – early 1980s); 4thG - Integrated model (early 1980s – early 1990s); and 5thG - Systems integration and networking (since the early 1990s). These types of innovation processes co-exist today, in one form or another. In fact, some firms face difficulties once they are entrenched within one process type and have difficulties to shift to a more appropriate model. Science-based firms or assembly-type firms, for instance, behave differently within the innovation process. In science-based firms, the process is internalized to a greater extent with limited input variety and will be largely sequential, at least in its early stages (e.g. 3rdG with limited functional overlap). In assembly-type firms, the best practice process will lie somewhere between 4thG and the forward-looking 5thG, and will involve a great variety of inputs and of players (i.e. networking process), and pressures to enhance speed and/or efficiency of innovation.

Abernathy & Utterback (1978) structured an important model, within the dynamics of innovation, in different industrial segments and throughout the economy. They examined the changing rates of major product and process innovation as industry moves through its various stages of maturity, while Utterback (1994) through an industry case study identifies the emergence of a 'dominant design' as an important step in the life of an industry.

Innovation is therefore seen as taking place within the company in response to 'recognition of need' and the technical means to do it as well. This simple linear model is, however more appropriate to certain industries that are engaged in mass production or continuous volume production. For Utterback "A dominant design usually takes the form of a new product synthesized from technological innovations introduced independently in prior product variants. A dominant design has the effect of enforcing or encouraging standardization so that production or other complementary economies can be sought. Then effective competition begins to take place, on the basis of costs and scale as well as of product performance.

Before a dominant design emerges, there is a wave of outsiders, new companies entering the business but, with the appearance of the dominant design, there is the beginning of a wave of exits. The emergence of a dominant design becomes something of a landmark and results in shifts at several different levels:

(i)   In the products: from radical to incremental, and from high variety to standardized;

(ii)   In the processes: from general purpose equipment with skilled manpower, to special purpose equipment requiring only low skilled manpower;

(iii)   In organizations: from organic forms with entrepreneurs to mechanistic forms with Tayloristic employees;

(iv)   In the markets: from unstable and fragmented to well defined with commodity like products;

(v)   In competition: from many small companies each with a slightly different product to an oligopoly with very much standardized products.

There are evidences that outsiders play an important role in bringing about radical innovations that may help to transform technological regimes (see e.g. Christensen, 1997). More generally, the dynamics of technological change is subject to developments outside technological regimes. On "The Refinement of Production" Mol (1995) shows how general social transformations, as highlighted by the ecological modernization theory, are reflected in three sub-sectors of the chemical industry (paints, plastics and pesticides). Such transformations, like the increasing importance of environmental criteria in the design of production processes, the growing participation of environmental NGOs in direct negotiations with economic agents and state representatives, and the transformation from bureaucratic top-down dirigism by government towards 'negotiated rule-making', are key factors in the dynamics of the technological change in the chemical industry.

Research carried out in other industrial sectors (e.g. aerospace, transportation), which are characterized by being low-volume producers or further upstream with industrial markets, made emerge on the other hand a totally different (non-Utterback pattern) model of innovation. These sectors have 'complex' systems of suppliers, producers, users and regulators and produce various generations of technology that are based on robust designs. They tend to involve governments and/or regulators in the innovation process for a variety of reasons, including safety (e.g. in large scale human transportation systems and nuclear power plants), the need for international standards (e.g. communication systems), the monopolistic nature of several of these sectors (e.g. power generation equipment), and the importance of their functioning for major parts of the economy (Miller *et al*, 1994).

In fact, the innovation process is a social driven activity, which depends from multiple conditions (technical, cultural and structural). It is always the result of choices as a result of social processes inside and around the company (Vergragt et al,1992) by individuals and organizations. Those choices are determined by the present state of the art, expectations regarding the future, internal and external constraints, creativity and social networks. Cultural factors are strongly determinant and difficult to influence. They concern perceptions and attitudes, lifestyles and beliefs. Structural factors concern economic (e.g. taxes) and organizational (e.g. establishment) aspects.

Ashford (1993) suggests four key organizational barriers to innovation in order to explain a lack, at a micro-economic level, to improve environmental performance even where it may lead to economic benefit: 1) Lack of information on the costs and benefits of environmental management; 2) Lack of confidence in the performance of new technologies and techniques; 3) Lack of managerial capacity and financial capital to deal with the transition costs of reorganizing the production process; and 4) Lack of awareness of the long run benefits of environmental management resulting in low priority being assigned to environmental issues. As far as technical barriers are concerned, there is a range of technical barriers, which limit the potential of environmental management initiatives over time. In fact, emphasis of environmental management initiatives that have been implemented in industry mainly addressed incrementally to process management. Thus, in this changing process, gradual adaptations on the existing production systems have prevailed.

Cleaner production projects in SMEs rarely promote revolutionary changes in the eco-efficiency of companies involved. Most cleaner production options reported by cooperative projects are incremental innovations that can be classified as optimizations or redesign. Optimization focuses on improving existing processes and does not essentially modify the system of production, but rather increases the efficiency of the system by making slight modifications (e.g. reducing heat losses, by better insulation). Redesign partly changes actual design of the existing processes, e.g. choosing to use materials that can be made suitable for reuse in the disposal stage, or that are less toxic. The system concept, thus remains basically unchanged. As far as new product specification and design is concerned, particularly in SMEs, it is a highly limited innovation process, depending on its degree of integration within the entire production chain, once companies with such size control only a small part of the product life cycle. A closer look at the motivations that are leading companies to currently implement cleaner technologies reveals several interesting factors. Costs are of primer importance, and cleaner technologies can be cost saving once they can promote reductions e.g. on consumption of resources, reductions on costs for safety, cleaning and remediation. Besides costs, companies concerned and committed to environment (and some already to sustainable development) emphasize their social responsibility, as well as the management of their public image.

Finally, also regulation particularly on product liability and safety, as well as on service liability, will enhance the closing of product responsibility loop (e.g. voluntary or mandatory take-back of consumer goods) leading to better products in a more efficient economy. At this stage, distinction should be made regarding recycling, once mandatory recycling leads to better recycling technologies, not to better products.

4. DIFFUSION AND ADOPTION OF TECHNOLOGY

Important issues on firms' decision-taking evaluation are therefore the relationship between the main internal characteristics of the firm that are determining its attitude to technical change and existing interactions between the firm and its regulatory environment, as well as with suppliers and customers within the product chain (i.e. the network). J. Bessant (1991) identifies networking as being powerful in technological diffusion particularly within industrial sectors consisting of sub-sectors made up of many small or medium sized firms. Case studies (e.g. Mulder & Bras, 1997) have revealed a number of different permutations of buyer-supplier relationships in manufacturing chains with patterns, which change over time.

Technology is generally not a free good, and that condition prevails in particular for competitive technologies and subsequent know-how to effectively operate them. After a specific innovation as occurred, it will be legally protected under the patent system rules, before being submitted to a diffusion process.

The process of diffusion starts with the first adoption of an innovation, but its spreading is a complex process, carrying successes and failures, whose activities and conditions depend on the nature of the innovation. Taking process innovations, for instance, depend a great deal on the adopting organization, rather than on the innovating organization.

The evolutionary model of technological development describes innovation as a process of variation and selection (Nelson & Winter, 1977). However, in evolutionary theory, there is a gap corresponding to an important intermediary process between variation process (in which technology is developed) and the selection process. In fact, in the variation process, new technological variations are developed within an uncertain process involving a heterogeneous group of actors, and then released into the selection environment. Thus, an actor network approach is needed once there is no independent variation and selection but a co-evolution of both the technology and the selected environment. Assuming that insight, focusing how links are made, the quasi-evolutionary model states that the variation-selection process is particularly influenced by the actors involved in the innovation process, whom try to anticipate and influence this process. The creation of technological niches enhances that approach (Verheul & Vergragt, 1995).

As R. Kemp (1994) points out, technical change is not a haphazard process once it proceeds in certain directions. Powerful mechanisms exist that reinforce the entrenchment of technology into the socio-economic system, however the transition to new technological trajectories, or even to a new paradigm, also benefit from accumulated experiences and from the presence of networks in which the new technology can be 'protected' and more easily introduced. The route to acceptance into a network, with subsequent diffusion, is highly complex but to reach a higher rate of diffusion, the new technology would benefit from the influence of that network if framed already at an early stage of development. Many technologies, at the time of their initial introduction, are often poorly developed in terms of performance characteristics and offer only few advantages over existing technologies.

Full and successful diffusion and adoption requires technical upgrading and price improvements. The nature of all these improvements may be slow as they need to be adapted to the requirements of the ultimate user at later stages in the innovation process. Kemp & Soete (1992) identified three key factors that affect the decision to adopt (or not) a cleaner technology: (i) the price and quality (i.e. technical characteristics) of the innovation that determine the costs and benefits for the potential user; (ii) the degree of information and knowledge available; (iii) the risk and uncertainty with respect to the economic consequences of the adoption of a specific innovation.

In an attempt to analyze what and how transfer takes place namely when environmental technologies are concerned, data was found to be scarce. From literature survey, about the kind of technology that is transferred, for instance from Europe to Asia, transfer fields concern industrial capital goods (German, Swiss, Swedish, Dutch and French equipment) in a wide range of technologies. Technology transfer, concerning hard and software *latu sensu* as well, take place e.g. in automotive, compact disk production, chips manufacturing, electronics and telecommunications, energy, food processing, or just packaging. To a very large extent, these technology transfers take place at a company level. Also joint ventures can be identified, and less frequently technology transfer assumes the form of licenses.

As far as cleaner technologies are concerned, there is not a clear evidence of that transfer once they are (or should be) integrated on the packages transferred.

Being the successful spread of marketable products and processes to all available markets, diffusion may occur within more than one sector of activity, being also the adoption not limited just to one country. At this stage, distinction should be made between international diffusion and technology transfer, once the latter besides being an imprecise term, has assumed very frequently just the form of acquisitions of technology packages without subsequent adaptation and contribution to the local innovation processes. Besides these aspects of imitability and learning curve attributes of an investment, also a different perspective for transfer could be added, when considering a flow of knowledge from the lab to industry.

Innovation and diffusion of environmental technologies, over the 1970s and 1980s, has experienced a substantial growth particularly in the three major industrialized countries (United States, Japan, and Germany), enhanced by the increasing awareness and concern about environmental protection (Lanjouw & Mody, 1996). Environmental innovations included end-of-pipe and also new technologies addressing to reduce pollution at the source. Trends in environmental technology innovation, as represented by patenting, reflected corresponding regulatory pressure in the three main leading countries, and spending on pollution control. Countries, such as Brazil, exhibited a relatively significant number of patents, most of which appear geared towards adapting imported technologies to local conditions.

World production of goods and services for pollution abatement and environmental protection reached US$ 200 billion in 1990 and there was an expected growth up to US$ 300 billion in 2000, at rates of 5-6% per year (OECD, 1992).

Together with the development of environmental markets, one of the consequences of this 'silent' main overall strategy is the change on corporate policy concerning add-on technologies, being progressively more comprehensive and using emission and waste prevention at product and/or process level while end-of-pipe solutions are used only when there are no better alternatives. Thus, the need to control and the correction of process outputs, according to the increasingly stringent specifications, the requirements of the permitting process or the public pressure, is becoming structurally simplified, within the production facility, and integrated in the corporate policy.

In recent decades, the European environmental equipment industry developed rapidly, experiencing continuous growth for years, largely due to high standards and regulations set by governments, which forced manufacturing industry to invest and develop more sustainable environmental technologies. This subject was analyzed by Porter (1990) who stated that companies, from countries with more stringent environmental regulations, should enjoy a first mover advantage from cost-reducing, pollution reducing technologies when environmental regulations get also more stringent in the countries of their competitors. The building and maintenance of competitive advantages for companies under such circumstances, and particularly when addressing to cleaner production, has in fact been demonstrated to be possible also under different regulatory environments (see f. instance: Porter and van der Linde, 1995a, b; Nehrt, 1998).

There are different factors that influence the international diffusion process of cleaner technologies *per se*. Some examples:

1. Not being a free good, competitive technology in particular (hard & software included), and its license of operation, are under commercial interests;
2. Special elements of technology (e.g. cleanness) can not be individualized from the main technology system;
3. Information about what is available at what price, and in particular the relative competitiveness of technology, is defective and distorted once the technology transfer markets are not yet very transparent;
4. Development of special feature adaptations to local situations need careful planning and personal contact, require high skill staff and participation, are expensive and time consuming;
5. Stimuli to enhance the transfer process, including learning curve aspects in a broader context, are being equated by international organizations and governments who are not the technology owners.

Adopting sustainability-driven development patterns, as a main goal, is partly depending on an effective technology diffusion process within sector and country adopters. That highlights the importance of careful planning and of the potential for leapfrogging according to local resources. A number of broad components have been identified by P. Durana (1995), as been critical to reduce risk and well succeed in technology diffusion efforts:

1. Technology fit. The technology should fit the biophysical and socio-economic environment of the site, be demonstrated successful under similar conditions, and demonstrated benefits to offer;
2. Personal contact. Technology is transferred best by people-to-people actions;
3. Adequate staffing. The right number and the right type of transfer agents, middlemen/ facilitators, and adopters are needed;
4. Participation. Stakeholders enrolment is critical from initial goal identification, choosing, planning, implementing, and evaluating and adapting;
5. Commitment and funding. Determination of costs should consider the full range from project initiation to self-support, and sufficient funding must be available.

In a more general framework, focusing on innovation and implementation within processes, and in order to help companies how to manage new process technologies, C. Voss (1994) proposed the following steps:

1. Pre-installation. Innovation of processes should not just be technically led, but to be most effective should support both the organization's strategic direction and the characteristics of the products to be produced. Complexity and uncertainty of the technology should match the knowledge and capability for the company to handle it, as well as the business needs. Evaluation should be based on the full system to be developed, not on parts of it.
2. Installation and Commissioning. Includes the effective interaction with suppliers, the use of appropriately composed cross-functional teams, and appropriate labor skills and availability.
3. Consolidation. The effective management of the post-installation consolidation phase can be crucial in obtaining success. Technical and user environment adaptations should be actively sought out. The implementation team should stay with the innovation until the main adaptations and learning effects have taken place. Appropriate organizational change should be actively sought out. Change should reflect the impact of the innovation on roles, communications flows and tasks. Performance measurements for those implementing and managing processes should match the objectives of the innovation.

New routes of research and improvement should begin giving main emphasis, not to the present products and processes, but to path-breaking solutions to meet needs and conditions required for a transition. Society will not penalize the use of unsustainable technologies unless more sustainable alternatives exist and have conditions to be viable. If yes, stimulus will also exist for business within technology development issues as well. If not, there will be no market conditions for profits, and that will penalize also the process of sustainable technology development.

Currently technological change, addressed in particular to solve negative impacts on environment, often proceeds incrementally along fixed paths ('path dependency') and thus, due also to the embededness of existing technologies in society ('entrenchment of technology'), the more radical changes required within the long-term ecoefficiency rates will not be easily brought about.

In fact, those incremental effects stand within the described path-dependence in technological and development trajectories, affecting negatively the identification, production and implementation of path-breaking solutions.

Current innovation practices are too compartmentalized to enable path-breaking approaches, causing a large mismatch between the societal and technological challenge and the magnitude of the necessary (factor 10 to 20 improvements in overall ecoefficiency) contribution to attaining sustainability. Thus, there is the need for a co-evolution of technological, structural and cultural innovations to realize the technical potential of such opportunities.

The exchange of ideas and information between research teams and societal stakeholder groups, and the mutual learning chances deriving from this approach, enables the integration of social needs within the three interrelated key dimensions for transitions: structure, technology and culture. It is therefore important to realize how to involve companies, and to make the process to be effective.

5.   HOW TO GET COMPANIES INVOLVED ?

5.1.   Proactiveness and environmental management systems

Being proactive (Charter, 1992) touches upon a broad range of aspects most strategy related of running a business. Proactive behavior means organizing one's business so as to be able to use the company's potential to benefit from opportunities and to avert threats, which may be anticipated in the environmental field.

Facing regulation, and assuming corporate compliance effort, four main reasons can be given to support such a strategy (Ryding, 1992): costs reduction, competitive advantage, improved public image and avoidance of inflexible regulations.

The ultimate test for the company lies in the market place, where revenues have to compensate for the costs and also secure a profit. Regarding market approach, different environmental strategies can be formulated accordingly (Steger, 1988; 1993):

1. Indifferent - with few environmental risks and few market opportunities;
2. Defensive - with major environmental risks and few market opportunities;
3. Offensive - with few environmental risks and major market opportunities;

4. Innovative - with major environmental risks and major market opportunities.

Though there is an obvious interest regarding the use of Steger's typology, it cannot be created the expectation that a company is easily classified according to a single strategy.

A company may adopt simultaneously, for different problems, two different strategies. Moreover, environmental management is described by various authors as an evolutionary process in which companies gradually evolve from a beginner to a pro-activist stage that in practice exhibit many feedback loops in this changing process. Within the second strategy, company responses to environmental regulation in an interactive process with the regulator actor. Environmental pressure is exerted upon the company also by other actors, and company answer pattern is one of the factors that influences how issues evolve. When regulation and company response are negotiated, three typical answer patterns have been observed: resist, comply/ accept, and compromise.

Companies need to overcome important constraints in order to improve their environmental performance and to develop and implement an environmental strategy.

In order to change its technology, a company must have the willingness, capacity and opportunity to change. Factors that determine a choice of one of those answers include the company's own interests and the stakes of other actors. Some key factors are:

(a) The company's negotiating power regarding that negotiating power of the regulator;
(b) Company's perception of pressure ('urgency');
(c) Company's ability to conform to pressure ('feasibility')

Market opportunities and/ or costs reduction, on the other hand, stimulate proactive strategies. Moreover, some environmental problems require more strategic approaches than others.

Strategic environmental approaches contrast, by opposition, with short-, medium-term technical and operational approaches. Walley & Whitehead (1994) concluded that environmental issues have operational and technical implications when either the manager's freedom of choice to answer, or the value involved, is 'medium to low'.

Environmental issues have strategic repercussions on company if their impact is high enough either to put core elements of company's business at risk or to change the company's cost structure, and if they allow managers considerable discretion in how to respond even through organizational learning and the development of new capabilities.

Illustrative examples of these are the following:

1. Emission control is an issue that requires an operational answer, because the tighter the government control is, the smaller is the room for managerial discretion on problem solving, while the cost impact may range from low to high;
2. Pollution prevention and the implementation of environmental management systems (EMS) are issues requiring a technical answer, because each distinct measure has generally a relatively small impact on the company's cost structure (main exceptions are equipment substitution and product changes) while managerial discretion is generally high.

Corporate environmental strategy is progressing from compliance (reacting to environmental regulation) to innovative but also proactive forms (i.e. doing better than the regulative authority requires), once there are conditions to make (in)tangible profits from reducing environmental impacts.

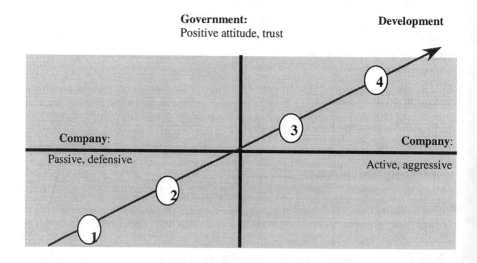

Figure 2- From Passive To Proactive Phases –
Phases of company's environmental management (incl. government stance):
1. Defensive (compulsion and sanctions); 2. Following (encouraging);
3. Active (facilitating); 4. Proactive (trust). (Source: VROM, 1997)

Companies within the same industry, struggling with the same type of environmental problems, may choose different trajectories and develop distinct diverging environmental strategies as a function of its specific external and internal conditions. The company must also match external pressures and opportunities with its own internal characteristics (Rothenberg et al., 1992), therefore there is not a single best model to apply and solve environmental problems.

Facing the longer-term, what the majority of companies can do is limited by what they know and what they already do. Companies can learn, but it is far easier to built upon their current knowledge, experience and skills, leading to incremental improvements, than to develop radically different solutions, although the very notion of SD would require them to do so. In this respect, often a change in corporate culture is called for, as well as a new managerial vision, new technologies, and a redefinition of the company's business as a service business rather than a unit of production, to address consumer demand.

A key point within this discussion regards the balance between material productivity and technological development in the one hand, and environmental performance, on the other. In a company, the adoption of such an attitude is supported in particular by the promotion of an adequate EMS, within the corporate overall management system.

5.2. Technology and strategy interaction

Innovation on products (e.g. design, lifecycle), processes (e.g. mass, energy and pollution intensity) and on the organization, is perccived as playing an important role in solving environmental problems and in progressing towards a more sustainable society. If the role of the individual innovative firm is very important in the development of more ecoefficient technology, where proactive behavior is crucial, front runners, followers and lagers regarding environmental issues will always exist, due to multiple (external, internal) conditions. A major contribution factor to progress with increasing ecoefficiency consists on the existing interactions, linkages and networks between suppliers, users, producers, regulators and other supporting actors creating mechanisms through which less innovative firms are able to contribute to the diffusion of upgraded more environmentally sustainable technology options.

There are different factors that shape the interaction between technology and strategy (Coombs, 1994). Broadly, innovation in the company is the expression of two market opportunities, based on technology and economy, consisting on transforming ideas into new applications of science and technology (new products, processes and services, or new attributes on existing ones) in a continuous changing process with a commercial purpose.

According to company's behavior, market demand, and user's expectations and needs are interpreted and converted into products, services and technologies by companies, whom organize the subsequent production.

Taken all those aspects, the description of the change that takes place in a company, under the influence of the environmental issue, has been mainly process-driven. The greater the environmental challenge accepted by a company, the more far-reaching is the change process. Roome (1994) established a link between the level of ambition on the area of the environment and the associated organizational changes. For that purpose, a distinction was made between first, second and third order changes:

1st order change -	Adds new techniques and technologies, but leaves the structure and values within the company unchanged;
2nd order change -	Consists of the gradual modification of existing organizational structures, systems, objectives and values within the company (may be supported e.g. by training of personnel at all levels);
3rd order change -	Consists of achieving excellence in the area of the environment (causing a break in the continuity of company's ideas and actions).

Moreover, J. Cramer (1997) pointed out three important variables, when a company is looking forward increasing its ecoefficiency rate:

1. The coincidence of increased ecoefficiency and market opportunities;
2. The internal structure and culture of the company (including the influence of a number of important actors in it);
3. The pressure from the immediate and wider social environment to take environmental measures.

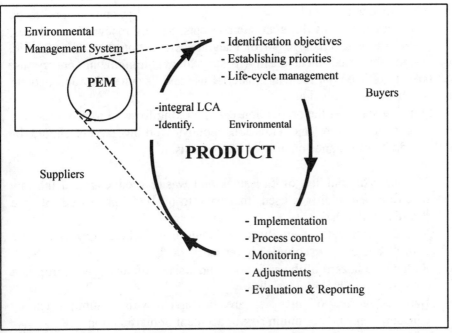

Figure X.3 – Product oriented environmental management (PEM). (Source: VROM, 1997)

That can be illustrated by an on-going shift from conventional EMS (process, site-oriented) to product-oriented environmental management (Fig. 3). It differs from process management, in that fundamental choices between products and raw materials affect the core of a company. Philips Sound & Vision perceived that trend and is implementing a very interesting product oriented strategy (Cramer & Stevels, 1997).

5.3.   Eco-Efficiency as a Way of Reference for Environmental Innovation

In order to manage ecoefficiency gains when performing input/ output analysis, two basic sorts of indicators may be used: (i) Environmental performance indicators like mass of emissions and wastes by production volume; material and energy consumed by production volume; or total wastes over time; and (ii) Eco-efficiency ratios like decreasing negative environmental impacts, through the minimization of the ratio : (Wastes/ Product Outputs)$_{min}$; or increasing efficiency and production yields, through the minimization of the ratio : (Inputs/ Outputs)$_{min}$

of the causes of emission or waste generation, and subsequent impacts over the environment as a whole, i.e. without generating pollution transfer effects. Ecoefficiency embraces different sub-concepts, namely renewable resources, integrated life cycle (cradle to grave), integrated production chain management, ecodesign, industrial metabolism (closing loops), minimizing energy use (improving efficiency), reducing material use and de-materialization (extend product life, reuse, recycling).

In a first stage, at the production level, looking forward to progressively perform a zero waste and emissions policy, working targets should be defined and performance quantified on the basis of:

- Minimize the amount of emissions and wastes produced, and the raw materials and additives used (that have to be 100% processed into the desired product);
- Reduce the levels of auxiliary materials used, where they cannot be entirely avoided, are completely reused;
- Reduce the hazard of the emissions and wastes produced by appropriate substitution of input materials;
- Both categories of material are processed with minimum energy consumption, and minimum risk of accident or malfunction.

From the main overall ecoefficiency concept "create more value with fewer resources, and less pollution", different elements can be defined as main objectives for an 'ecoefficient' company:

- Reduce the material intensity of goods and services;
- Minimize the energy intensity of goods and services;
- Minimize toxic dispersion;
- Enhance material recyclability;
- Maximize the renewable share of resources use;
- Extend product durability;
- Increase the service intensity of goods and services.

According to the rate of ecoefficiency that can be obtained in technological systems, different categories of innovations are considered: optimization, redesign, function-oriented and system innovation. System innovation can be subdivided into function innovations, and these in turn can also be subdivided into redesign, adjustment and optimization.

Optimization focuses on improving existing products, processes or infrastructures, increasing system efficiency by (slightly) modifying systems that already have a commercial use (no modification in the system concept) (1-5 years). Expected maximum improvements in ecoefficiency are of the order of factor 1,5.

Redesigning the actual design of existing products, processes or infrastructures will partly change specific features of the system e.g. by choosing to use materials that can be made suitable for reuse in the disposal stage (system concept remains largely unchanged) (1-10 years). In this case, the expected maximum improvement in ecoefficiency is of the order of factor 4.

Function oriented innovation, on the other hand, addresses to more far-reaching improvements (improved ecoefficiency by factor 10) that can be achieved by departing from the system concept and by developing (radically) new systems which perform the same function better (10-20 years).

Optimization, redesign and function innovation are indicative of ideal types within a continuum, requiring to be translated into company and even into individual actions. Transitions can only be implemented if the individual activities are focused toward a specific common goal, a human need in a new way with less environmental impact (but still economically successful). The number of degrees of freedom in this continuum becomes progressively higher: on the one hand, optimization refers to the system concept which is essentially a given factor, and on the other hand as far as function oriented innovation is concerned the system's function is the determinant issue impacting with the economic structure and also involving new (f)actors, namely concerning the establishing of new competitive market positions. The time scale required to perform these different types of innovations was roughly represented in fig. 1, where the environmental relevance of innovations can also be shown. Improving existing systems can lead to substantial improvements in efficiency, but at a given time the existing concept becomes fully developed and thus a plateau occurs which will have been reached in terms of environmental efficiency too. In a thermodynamic point of view, the maximum efficiency of a system was reached in that case. The next wave (there is always one) is always turning obsolete the latest technologies.

6. EXAMPLES TO BE FOLLOWED

Defining adequate benchmarks is very important to support the diffusion of ecoefficiency concept.

The United Nations Environment Programme (UNEP) began its cleaner production programme in 1989, involving more than one hundred organizations to implement cleaner production initiatives in over 60 countries. Together with the United Nations Industrial Development

Organization (UNIDO), it has established National Cleaner Production Centers in different countries. Moreover, UNEP's International Cleaner Production Information Clearinghouse provides data on about 600 technical cleaner production case studies all over the world.

Within the scope of pollution prevention and cleaner production technologies, Europe can already take useful conclusions from different initiatives (Christiansen & al, 1995). Among the most relevant European cleaner production cooperative projects, there are two initiatives that should be stressed. The PRISMA project (NL) and The Aire & Calder project (UK). Several other interesting examples about the impact of pollution prevention and cleaner production, conducted all over the world, could be stressed within: different countries (in very distinct market economy stages), and/ or very different industrial activity sectors (including distinct company's sizes, as well as contrasting technological and social conditions).

The Prisma project was launched in 1988 (Dieleman & de Hoo, 1993). Two main aims of the project were: (i) To show that in Dutch industry, the prevention of waste and emissions is possible in the short term and offers benefits both to companies and to the environment; and (ii) To formulate recommendations for an effective pollution prevention policy. The key findings were: (i) 10 SMEs participated; (ii) 164 opportunities for cleaner production identified; (iii) Many possibilities for cleaner production can be implemented within a short term: 34% of prevention/ minimization opportunities had a pay back of less than a year, and 49% had a pay back between 1 and 3 years; moreover considerable savings can also be achieved; (iv) Main opportunities identified were: 35% within process modifications, 25% within materials substitution, 20% within best practices, 10% within on-site recycling and 5% within product modifications; (v) A coherent package of preventive measures could be established for each company through the use of a Manual (adapted to the specific characteristics of each company) and carried out by limited person-power (including external supervision), showing after all that project approach and its methodology is useful and effective, though the Manual should be upgraded to include a feasibility study, environmental hygiene evaluation and the support to a (non)existing environmental management system.

The Aire & Calder waste minimization project was launched in 1992 (Johnston, 1996). The main aims of the project were: (i) To demonstrate the benefits of a systematic approach to emission reduction; (ii) To focus on procedural changes and cleaner technology; and (iii) To identify gaps in supply, technology and science. The key findings were: (i) 11 companies participated; (ii) 542 opportunities for waste minimization identified, but two types of options dominated: good housekeeping and technology modifications; (iii) cost savings to the companies in excess of 2 million pounds per year; (iv) 10% of minimization opportunities were cost neutral and 70% had a pay back of less than a year; (v) initial average savings in water consumption of 10% with potential to rise to 25%; (vi) initial savings in effluent discharge of 15% with potential rise to over 35%.

Considering 'dematerialization' initiatives, several multinational companies have already successfully implemented strategies like de-coupling turnover and profits from resource consumption and manufacturing volume (Schmidheiny *et al*, 1997). In most cases, the product used to perform the service remains the property of the service company. The product is taken back after use, and cleared or remanufactured prior to reuse. This creates a financing incentive for the company to increase the lifetime of the product delivering service. 'Schindler Elevators' is selling carefree vertical transport instead of elevators. Xerox is offering custom-made reproduction services instead of just selling photocopiers. Dow Europe sells services of chemicals instead of selling chemicals. Safechem and Dow Germany are renting solvents to dry-cleaners. Mobil Oil is selling engine oil quality monitoring instead of engine oil. Interface Inc. leases nylon carpets. A great number of companies already practice a voluntary buy-back or free take-back system, such as Eastman Kodak and Fuji, for their single-use cameras, or GE Medical Systems for medical equipment by any manufacturer.

The European Better Environmental Awards for Industry (EBEAFI), first presented in 1987, is a biannual award scheme that seeks to encourage, recognize and attract attention to the efforts of environmentally progressive companies in the European Union, and to inform the public of their achievements in that area. Within different categories like: cleaner technologies, ecodesign, managing for sustainable development, partnerships for sustainable development, and recovery of waste, the judging criteria for ebeafi'98 entries included: innovation, replicability, environmental benefits and market potential (Ebeafi, 1998).

7.   CONCLUSIONS

Environmental concerns are increasingly becoming a driving force in the innovation process, as they gradually percolate and get endogenised into business, and social structures.

Environmental gains towards sustainable development have to start with a considerable reduction on the consumption of resources. Such a 'de-materialization' however, is only feasible through innovation, which, in turn, needs to be driven by the economy. This issue, on the other hand, defines that a new management task is required to unlink economic success from resource consumption, being replaced then by resource utilization. Thus, companies must gradually pollute less, and consume fewer resources, per unit of product output.

While in the long run a great leap in efficiency is expected by changing the design, or by introducing completely new system concepts, in the short-, medium-term cleaner technologies are becoming an important contribution to reducing environmental impacts. Also more integrated technologies are an emerging trend, but it may take years before all those technologies enter the main stream. To enhance technology diffusion, case studies on cleaner production are important benchmarks to be used as working examples, particularly within companies in the same industrial sector. In order to understand the diffusion process, attention should be given not just to successful but also to unsuccessful case studies.

Concepts like cleaner production and ecoefficiency have a huge potential to become one of industry's contribution to sustainable development, either in industrialized or developing countries. However, different measures at a macro- and micro-level, are required to prepare the paradigm shift influencing the actual patterns of production and consumption. Moreover, industry cannot solve global problems alone, once it is one of the actors involved, together with consumers and public authorities. In fact, as demand increases faster than material productivity, leading to an overall increase of resources use, supply and demand need to be brought in line, in order to adjust resource consumption to a degree that is compatible with the carrying capacity of natural stocks (Weizsaeker *et al*, 1997).

REFERENCES

Abernathy, W.J. & Utterback, J.M. (1978). Patterns of Industrial Innovation, Technology Review 80, 7 (June/ July), 40 - 7.

Ashford, N.A.; Heaton, G.R. & Priest, W.C. (1979). Environmental, Health and Safety Regulation and Technological Innovation. In: Technological Innovation for a Dynamic Economy, Hill, C.T. & Hutterback, J.M. (eds), pp. 161-221, Pergamon Press, Inc. NY.

Ashford, N.A. (1993). Understanding Technological Responses of Industrial Firms to Environmental Problems: Implications for Government Policy. In: Fisher, K. and Schot, J. (Eds), Environmental Strategies for Industry, Island Press, Washington D.C., 277- 307.

Bessant, J. (1991). Managing Adv. Manufacturing Technology: The challenge of the fifth wave, Blackwell Ltd, Oxford.

Brezet, H. (1997). Dynamics in Ecodesign Practice, UNEP Industry & Environment 20, 1, Jan-June: 21-4.

Casler, S.D. & Afrasiabi, A. (1993). Input Composition and the Energy-Output Ratio, Structural Change and Economic Dynamics 4, 2.

Charter, M. (1992). Greener Marketing: Responsible business, Green Leaf Publishing, Sheffield.

Christensen, C. (1997). The Innovator's Dilemma, Harvard Business School Press.

Christiansen, K., Nielsen, B.B., Doelman, P. and Schelleman, F. (1995). Cleaner Technologies in Europe, J. Cleaner Prod. 3, 1-2: 67/70.

Considine, T. J. (1991). Economic and Technological Determinants of Material Intensity of Use, Land Economics 67, 1.

Coombs, R. (1994). Technology and Business Strategy, in: Dodgson, M. & Rothwell, R. (eds.), The Handbook of Industrial Innovation, Chapter 31, pp. 384 – 392, Edward Elgar Publishing Company, Vermont, USA.

Cramer, J. (1997). Environmental Management: From 'Fit' to 'Stretch', TNO Report STB/ 97/ 45, Apeldoorn, July, 31 pg.

Dieleman, H. and de Hoo, S. (1993). Toward a Tailor-made Process of Pollution Prevention and Cleaner Production: Results and Implications of the Prisma Project. In: Fisher, K. and Schot, J (eds), Environmental Strategies for Industry, Island Press, Washington D.C., pp 245-275.

Dosi, G. (1988). The Nature of The Innovative Process, In: Dosi, G., Freeman, C., Nelson, R., Silverberg, G., & Soete, L. , Technical Change and Economic Theory, Pinter Publishers, London, pg. 221-238.

Dosi, G. & Orsenigo, L. (1988). Co-ordination and Transformation: An overview of Structures, Behaviors and Change in Evolutionary Environments, In: Dosi, G., Freeman C., Nelson, R., Silverberg, G., & Soete, L., Technical Change and Economic Theory, Pinter Publishers, London, pg. 13-37.

Durana, P.J. (1995). Technology Transfer: Underpinning sustainable development, UNEP Industry and Environment April-September: 56-60.

Ebeafi (1998). European Better Environmental Awards for Industry, EC/UNEP, Van Rossum & Partners, Amsterdam.

Ehrlich, P. & Holdren, J. (1971). Impact of population growth, Science 171: 1212-1217.

European Commission (1994). Life Project 94/IRL/367 "Co-Operation for Environmental Improvement", Irish-Danish-French Co-Pilot Project on Environmental Administration and Audit, Cork RTC Press, Cork, Ireland.

European Commission (1995). Business Strategy and the Environment – Attitude and Strategy of Business Regarding Protection of the Environment, DG III/ EuroStrategy Consultants, December.

European Union (1998). Research and Technological Development Activities of the European Union: 1998 Annual Report, COM(98) 439.

Fisher, K. and Schot, J. (1993), Environmental Strategies for Industry, Island Press, Washington D.C.

Freeman, C. (1982). The Economics of Industrial Innovation, 2nd Edition, Frances Pinter, London.

Galbraith, J.K. (1967). The New Industrial State, Penguin, Harmondsworth, England.

Hagedoorn, J.; P.J.Kalff and J.Korpel (1988). Technological Development as an Evolutionary Process, Elsevier, Amsterdam.

Hettige, H.; Lucas, R.E.B. & Wheler, D. (1992). The Toxic Intensity of Industrial Production: Global patterns, trends, and trade policy, American Economic Review, Papers & Proceedings 82, 2.

Hippel, E. von (1988). The Sources of Innovation, Oxford Univ. Press.

Hughes, T.P. (1983). Networks of Power. Electrification in Western Society 1880-1930, pp. 1-17.

Irwin, A. & Vergragt, P. (1989). Re-thinking the Relationship between Environmental Regulation and Industrial Innovation: The Social Negotiation of Technical Change, Technology Analysis and Strategic Management 1, 1: 57-70.

Johnston, N.B. (1996). The Environmental Business - Corporate Case Studies: Case Study Two Waste Minimization/ Cleaner Technology, IPTS Technical Report Series EUR 16376 EN, 23 pg.

Kemp, R. & Soete, L. (1992). The Greening of Technological Progress: An evolutionary perspective, Futures 24, 5: 437-57.

Kemp, R. (1994). Technology and the Transition to Environmental Sustainability: The problem of technological regime shifts, Futures , 26, 10: 1023-46.

Klein, B. (1977). Dynamic Economics, Harvard University Press, Cambridge, Massachusetts.

Klodt, H. (1994). Energy Consumption and Structural Adjustment: The case of West Germany, Structural Change and Economic Dynamics 5, 1.

Lanjouw, J.O. & Mody, A. (1996). Innovation and the International Diffusion of Environmentally Responsive Technology, Research Policy 25: 549-571.

Miller, R.; Hobday, M.; Lerous-Demers, T. & Olleros, A. (1994). Innovation in Complex Systems Industries: The case of flight simulation, SPRU, October, University of Sussex, Brighton.

Mol, A. (1995). The Refinement of Production. Ecological Modernization Theory and the Chemical Industry, Van Arkel Publ., Utrecht.

Mulder, K. & Bras, R. (1997). Technology, Networks and the Management of Transformation Chains: Plastic Packaging and the Environment, Technology Studies 4, 2: 251-282.

Nehrt, C. (1998). Maintainability of First Mover Advantages when Environmental Regulations Differ Between Countries, Academy of Management Review 23, n.1: 77-97.

Nelson, R.R. and Winter, S.G. (1977). In Search of Useful Theory of Innovation, Research Policy 6: 36-76.

OECD (1992). The OECD Environment Industry: Situation, Prospects and Government Policy, OECD/ GD (21) 1, OECD, Paris.

Porter, M.E. (1990). The Competitive Advantage of Nations, MacMillan, London.

Porter, M.E. and van der Linde, C. (1995a). Green and Competitive. Harvard Business Review, September-October: 120-134.

Porter, M.E. and van der Linde, C. (1995b). Toward a New Conception of the Environment-Competitiveness Relationship, Journal of Economic Perspectives 9(4): 97-118.

Roome, N. (1994). Business Strategy, R&D Management and Environmental Imperatives, R&D Management 24, 1: 65 – 82.

Rothenberg, S.; Maxwell, J. & Marcus, M. (1992). Issues in the Implementation of Proactive Environmental Strategies, Business Strategy and the Environment 1, 4: 1-12.

Rothwell, R. (1977). The Characteristics of Successful Innovators and Technically Progressive Firms (with some comments on innovation research), R&D Management 7, 3:191-206.

Rothwell, R. (1992). Successful Industrial Innovation: Critical Factors for the 1990's, R&D Management 22 , 3: 221- 39.

Rycroft, R. and Kash, D. (1999). The Complexity Challenge: Technological Innovation for the 21st Century, Pinter, England.

Ryding, S.O. (1992). Environmental Management Handbook: The Holistic Approach – From Problems to Strategies, IOS Press, Amsterdam/ Lewis Publisher, Florida.

Schmidheiny, S.; Chase, R. and DeSimone, L. (1997). Signal of Change – Business progress towards sustainable development, WBCSD Publ., Geneva.

Schot, J.W. (1992). Constructive Technology Assessment and Technology Dynamics: The case of clean technologies, Science, Technology and Human Values 17 (1) Winter: 36-57.

Schot, J.; Hoogma, R. & Elzen, B. (1994). Strategies for Shifting Technological Systems – The case of the automobile system, Futures **26** (10): 1060-1076.

Schramm, W. & Hackstock, R. (1998). Cleaner Technologies in the Fourth Framework Programme of the EU, J. Cleaner Prod. **6**: 129-134.

Schumpeter, J.A. (1939). Business Cycles, I/II vol., McGraw Hill, New York.

Steger, U. (1988). Umweltmanagement – Erfahrungen und Instrumente einer umweltorientierten unternehmensstrategie, Frankfurter Allgemeine Zeitung, Frankfurt am Main/ Gabler, Wiesbaden.

Steger, U. (1993). The Greening of the Board Room: How German companies are dealing with environmental issues, in: Fisher, K. and Schot, J. (eds), Environmental Strategies for Industry, Island Press, Washington D.C.

Ulhöi, J.P. (1995). Corporate environmental and resource management: In search of a new managerial paradigm, European J. Operational Research 80: 2-15.

UNEP (1994). Cleaner Production - What is Cleaner Production and the Cleaner Production Programme ?, UNEP Industry and Environment October-December: 4-27.

Utterback, J.M. & Abernathy, W.J. (1975). A Dynamic Model of Product and Process Innovation, Omega **3**, 6: 639-56.

Utterback, J.M. (1994). Mastering the Dynamics of Innovation: How Companies Can Size Opportunities in Face of Technical Change, Harvard Business School Press, Boston.

Vergragt, Ph.J.; Groenewegen, P. & Mulder, K. (1992). Industrial Technological Innovation: Interrelationships between technological, economic and social analysis, In: Firm Strategy and Technical Change: Microeconomics or Microsociology, R. Coombs, P. Saviotti & V. Walsh (eds).

Vergragt, Ph.J. & Jansen, L.(1993). Sustainable Technological Development; The making of a Dutch long-term oriented technology program, Project Appraisal **8**, 3: 134-140.

Verheul, H. & Vergragt, Ph. J. (1995). Social Experiments in the Development of Environmental Technology: A bottom-up perspective, Technology Analysis & Strategic Management, vol. 7, 3: 315-326.

Volland, C.S. (1987). A comprehensive theory of long wave cycles, Technology Forecasting and Social Change 32:123-145.

Von Weizsäcker, E.U.; Lovins, A. & Lovins, L. (1997). Factor Four: Doubling Wealth – Halving Resource. The new report to the Club of Rome. Earthscan Publications Ltd, London.

VROM (1997). Environmental Management – A general view, Ref 99170/h/4-97.

Weterings, R. & Opschoor (1992). "The ecocapacity as a challenge to technological development", Rijswijk: Advisory council for Research on Nature and the Environment. Publication RMNO N° 74a.

Weterings, R.; Kuijper, J.; Smeets, E.; Annokkee, G.J. & Minne, B. (1997). 81 Options: Technology for Sustainable Development. Ministry of Housing, Spatial Planning and the Environment.

Voss, C.A. (1994). Implementation of Manufacturing Innovations. In: Dodgson, M. & Rothwell, R. (eds), The Handbook of Industrial Innovation, Edward Elgar Publ. Comp., Vermont, USA.

AUTHOR

Paulo J. Partidário is a Research Chemical Engineer at INETI (Portugal), in the field of environmental technology and industrial ecoefficiency. He holds an MSc degree in technology management and his current research interests focus in particular on environmental technology dynamics and on conditions for systems innovation.

Chapter 12

A Systematic Framework for Environmentally Conscious Design
Using Fuzzy House of Quality and Analytical Hierarchical Process Techniques

Michael H. Wang
University of Windsor

In this chapter, a systematic framework for environmentally conscious design purposes is outlined. It focuses on ensuring better environmental performance by means of selecting environmentally friendly design alternatives. Since the environmental performance of a product relies on varied issues including nature resources consumption, manufacturing and distribution efficiency, and end-of-life management, designers have to face complicated trade-offs during the processes of product design. Our framework contains two parts for dealing with the complicated trade-offs. These two parts are (1) the individual assessment for each life cycle stage, and (2) the overall assessment for entire product life cycle. The individual assessment tool incorporates House of Quality (HOQ) with fuzzy set theory and Analytic Hierarchy Proces (AHP), and is used in analyzing the interrelations between environmentally conscious requirements and product design criteria. The overall assessment method is based upon the concept of product life cycle design, which involves a four step analysis so that the comprehensive environmental effects can be captured at the up-front design stage.

1.   INTRODUCTION

Environmentally Conscious Design (ECD) is a practice by which environmental considerations are integrated into product development. In addition to the traditional design considerations, ECD considers functional requirements, producibility, assembly, serviceability, recyclability as well as environmental effects. It implies that ECD needs to consider the product's performance and environmental impacts throughout its entire life cycle. On the basis of ECD concerns, product design will consider the product's interaction with its physical environment throughout the product's life-cycle including raw material extraction, energy consumption to fabricate the product, transportation, use, disposal and the associated by-products generated from all processes involved in its life-cycle.

One of the most important parts in ECD deals with the product end-of-life management. There are two types of recycling plans for product end-of-life management: closed-loop and open-loop recycling. A closed-loop product recycling system as shown in Figure 1 involves reuse and remanufacturing of the components, or recycling of the materials to make the same product over again. A typical example is to recycle aluminum cans to produce the same or similar type of aluminum cans. The various design issues involved in the closed-loop recycling activity are comprised of Design for Disassembly, Design for Reuse, Design for Remanufacturing and Design for Recycling which are all associated with Environmentally Conscious Design, and will be concerned as criteria for the green product design framework.

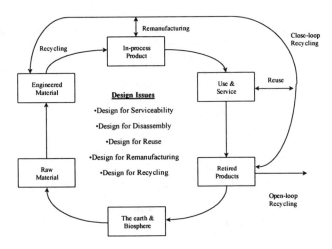

Figure 1. **The Closed-Loop Product Life-Cycle System**

Another type of recycling plan for product end-of-life management, open-loop recycling, reuses materials to produce a different product. Figure 2 shows the profile of open-loop recycling. A typical example of open-loop recycling is that of long fiber office paper which may be recycled to produce short fiber brown paper bags. The application of open-loop recycling will depend on the characteristics of the material. In addition, the process of open-loop recycling usually involves material degradation and waste generation. Thus, it is usually preferable to consider closed-loop recycling first when conducting the product end-of-life management.

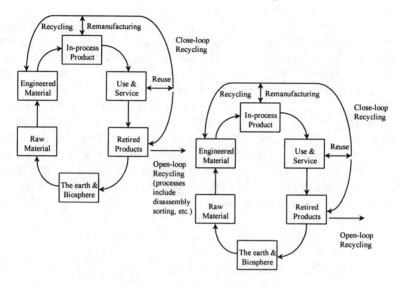

Figure 2. The Open-Loop Product Life-Cycle System

In this chapter, the proposed framework of environmentally conscious design can be divided into two parts. The first part is the individual analysis, in which fuzzy set theory and Analytic Hierarchy Process (AHP) are integrated into the House of Quality (HOQ). The second part of the framework is the systematic approach, in which a two-phase analysis is developed for illustrating the four stages of the product life cycle. The four product life-cycle stages are: (1) raw material consumption; (2) manufacture and assembly; (3) product distribution and use; and (4) management of end-of-life products. In the first phase of analysis, every product life-cycle stage will be examined individually. The overall assessment is addressed in the second phase of analysis.

HOQ is a fundamental part of Quality Function Deployment (QFD), which is used to streamline design activities and show a clear understanding of design tasks. One of the important functions of HOQ is to translate customer needs into physical engineering design solutions. In this Environmentally Conscious Design (ECD) framework, the element of customer needs is substituted by ECD criteria and the physical engineering design solutions are substituted by alternatives of design features. In other words, the HOQ is mainly utilized to map environmental constraints and concerns to the associated candidates of design features. Moreover, HOQ also provides a powerful evaluation mechanism for measuring the importance of those design features. Therefore, HOQ applied in this framework may enhance the analysis of environmental issues and the prioritization of various design alternatives.

The HOQ approach, on the other hand, has been known to contain some weaknesses. One of them is the use of imprecise artificial variables that contain ambiguity and vagueness of meaning. To address this problem in the ECD framework, fuzzy set theory is used to deal with the ambiguity and uncertainty in the data involved with relationship definitions in HOQ.

This ECD framework focuses on seeking better environmental performance by means of selecting environmentally friendly design alternatives. The environmental considerations used for guiding the selection include environmental impact reduction throughout the product life cycle, efficiency in the use of nature resources, waste control through production processes, and end-of-life management. As a result, the designed parts or products will meet environmental regulations, standards and ecological requirements.

2.   LITERATURE REVIEW

2.1   Environmentally Conscious Design (Green Design)

In addition to the traditional product design considerations that lead the product design to satisfy functional requirements and specifications, the goal of ECD is to develop more environmentally friendly products without compromising cost, quality, or manufacturing time. ECD also has been known as a systematic consideration of environmental impacts, resource depletion and human health over the whole product life cycle (Rhodes 1996). Therefore, for attempting to achieve the goal of ECD, attention to ECD should be paid throughout all the product design stages, which consist of conceptual design, detail design, manufacturing process determination and product end-of-life management. The framework proposed in this research provides a systematic approach to incorporate ECD into the product design process and directs the design to corresponding environmental requirements and regulations.

In general, ECD can be categorized into waste prevention and materials management (Watkins 1992). Waste prevention refers to reduction or elimination of wastes generation, which is performed during the design, or redesign of products and the associated production processes. To implement the behavior of waste prevention, the following guidelines should be considered (Fiskel 1993, Jovane 1993, US Congress Office of Technology Assessment 1992):

- Material substitution
- Waste source reduction
- Energy use reduction
- Life extension

The other category associated with ECD is materials management. Under this category, considerations of how to facilitate product recyclability are made. In other words, the application of product retirement is managed within this category. Material management should be arranged early in the product design stage. Accordingly, it is necessary to consider the following management alternatives that lead the product toward a better waste management (EPA 1993).

- Reuse – reuse is the additional use of an item after it is retired from a clearly defined duty. However, repair, cleaning, or refurbishing to maintain integrity may be done in transition from one use to the next.
- Remanufacturing – remanufacturing is a process that restores worn products to like-new condition. In a factory, a retired product is completely disassembled. Its reusable parts are then cleaned, refurbished, and stored. A new product is then reassembled from both old and new parts, creating a unit equal in performance and expected life to the original or a currently available alternative.
- Recycling – recycling is the reformation or reprocessing of a recovered material. Recycling may be defined as "the series of activities, including collection, separation, and processing, by which products or other materials are recovered from or otherwise diverted from the solid waste stream for use in the form of raw materials in the manufacture of new products other than fuel.
- Energy recovery – energy recovery is extracting energy from waste materials through incineration or other processes.

In the application of implementing the management alternatives of reuse and remanufacturing, a typical five-step processing stage has been identified which is also the product end-of-life processing stages (Bor 1994, VDI 1993, Johnson & Wang 1995):

- Disassembly;
- Cleaning components;
- Inspecting, testing and sorting components;
- Upgrading component or component renewal; and
- Reassembly.

With respect to the assessment tool of ECD, Life Cycle Assessment (LCA) is the most frequently used evaluation methodology. According to the Society of Environmental Toxicology and Chemistry (SETAC), the definition of the LCA process is "an objective process to evaluate the environmental burdens associated with a product, process, or activity by identifying and quantifying energy and material usage and environmental releases on the environment, and to evaluate and implement opportunities to effect environmental improvements." The assessment includes the entire life cycle of the product, process or activity, encompassing extracting and processing raw materials; manufacturing, transportation, and distribution; use/re-use/ maintenance; recycling; and final disposal. LCA is, therefore, used as an overall assessment tool, which evaluates the potential environmental effects of the design process, while ECD directs the design to minimize environmental impacts.

Auto manufacturers have made efforts worldwide in order to comply with government regulations such as take-back policy. For instance, in Europe, BMW's 1991 Z1 roadster model with plastic body panels was designed for disassembly and labeled as to resin type so they may be collected for recycling (Bylinsky 1995). In addition to BMW, Volkswagen has set up disassembly and recycling plants in order to comply with the upcoming recycling regulations. Consequently, they are trying to make an automobile out of 100 percent reusable/recyclable parts by the year 2000. In America, General Motors, Chrysler and Ford, formed the Vehicle Recycling Partnership (VRP) in 1991 to develop ways to recover and reuse as much of the fluff and metal scrap from motor vehicles as possible (Rosenberg 1992). Ford has also issued its own worldwide Recycling Guidelines, which suggest fewer and more dismantle-friendly fastener types, 'green' materials and component designs (Brooke 1993).

While the auto industry sees only potential benefits, some electric appliance manufacturers have already profited from launching their disassembly/recycling program (Penev 1994, Vergow 1994). For instance, Xerox has saved $200 million a year through reuse of parts; the focus on green design increased this amount by $50 million.

2.2   Design for Disassembly

Design for Disassembly (DFD) is a part of ECD and is strongly related to the goal of material management because DFD is the path toward better product post-life management. It will make disassembly practices easier so that post consumer products may be separated with less effort into parts and materials without contamination. For example, incorporating DFD into product design will eliminate the use of mixing copper and tin in steel. In short, DFD is a very important design concept for post consumer product reuse, remanufacturing and recycling (Philipps 1994), and may lead us to achieve the goal of ECD. The current DFD development may be found in the following perspectives:

- developing DFD guidelines;
- developing cost models for DFD evaluation;
- studying the possibility of using robots for automated disassembly;
- studying the feasibility of DFD related legislation for enhancing environmental preservation

The ultimate goal of DFD is to extend the life of used parts by means of reuse or remanufacturing, or to provide a new life of the used materials by means of recycling. Considering the definition presented previously, reuse is the most desired scenario of DFD, since the profit retrieved from "reuse" can be maximized. For achieving the success of DFD, much research has been done in this field that can be generally categorized as disassembly-oriented product design, economic analysis of disassembly efforts, and disassembly aided tools.

Within the realm of disassembly-oriented product design, one of the most valuable contributions of research is the development of guidelines (Boothroyd 1992, Dowie 1994, Jovane 1994, Simon 1991). Those guidelines can be classified as the areas of product structure, material use, and recycling principles and requirements.

Since complicated tasks are involved in ECD such as Design for Disassembly, Design for Reuse, Design for Recycling, etc., computer tools are becoming more and more important to support design practices. Som e available computer tools associated with ECD have been summarized in Table 1. These computer tools may enhance the analysis in certain product design aspects. Even though these computer tools will be applied in different areas, they are all undoubtedly aimed at supporting environmental improvement projects.

Table 1. List of ECD Software

Computer Tool	Applied Methodology	Application Area	Author
Design For Environment	Evaluate the environmental impacts by using a value assessment metric. Analyze the disassembly sequence by linking with the Design for Assembly software.	Financial return assessment of disassembly, disposal, reuse or recycling. Environmental impact assessment results from initial product manufacture and disposal, reuse or recycling.	Boothroyd & Dewhurst
Ecodesign expert system	Integrate a computer-aided design for the environment expert system onto a CAD platform.	The ecodesign model aims at encompassing the whole design process, i.e., from concept to detail	Poyner & Simon

		design.	
ReStar	Algorithm based approach on determining the optimal recovery plan based on tradeoffs between recovery costs and the value of secondary materials or parts.	It provides a design analysis tool for evaluating recovery operations, which may be used in product detail design stage.	Navin-Chandra
Disassembly Model Analyzer	Genetic Algorithms is applied for obtaining the profit-optimizing disassembly plan	It may be used for the modeling and analysis of disassembly for reuse and recycling.	Spicer & Wang

2.3   Life Cycle Design

Life cycle Design is a design pattern for understanding the interactions between product design requirements, economic systems and their environmental impacts (Keoleian 1998). It covers a broad spectrum, which comprises all of the various design-for-X (DFX) issues. The "X" in the term of DFX stands for all activities in a product life cycle, for example, Design for Manufacturing, Design for Assembly, Design for Service and Design for Disassembly are the parts of DFX (Agba 1996). Thus, life cycle design may be regarded as a more comprehensive design concept from the traditional design considerations such as DFA, etc. The life cycle design considers all stages of product life cycle. Within every product design process consisting of conceptual design, material selection, structure determination and process design, the environmental effects must be taken into account throughout all product life cycle stages.

Alting (1995) emphasized that life cycle design is an evolution of concurrent engineering, and exists in all cycle phases. More important, all cycle phases have to be taken into consideration simultaneously from the conceptual product design stage through the detailed design stage. Ishii et al. (1993, 1994) developed a framework that focused on post-manufacturing issues. The model started at the product structure analysis, followed by cost and compatibility analysis. Alting and Legarth (1995) have provided a good overview of life cycle design strategies. The overall and subsequent design strategies most often pursued in life cycle design are listed. Those design strategies applied throughout all design processes may be applied for the environmentally conscious selection of materials and components. It ends up with the state-of-the-art in life cycle design. A decision support tool dealing with the trade-off among strategies will be required for implementation of life cycle design.

3.   FRAMEWORK

This green product design framework has been organized as two parts. The first part is focused on the individual analysis, while the second part will be on the systematic approach.

The individual analysis is the fundamental element of this framework. It is a matrix analysis tool based on the House of Quality, called Modified House of Quality. In this modified HOQ tool, Analytic Hierarchy Process (AHP) and fuzzy set theory are incorporated into traditional HOQ so that it may be properly applied for Environmentally Conscious Design analysis.

The modified HOQ has a different matrix arrangement compared to the traditional HOQ. As mentioned, the matrix arrangement of traditional HOQ is intended particularly for translating the "voice of the customer" into product design processes. In this green product design framework, the modified HOQ is utilized to translate ECD requirements and constraints into product design considerations.

There are five steps involved in the process of implementing modified HOQ task. These five steps are:

- collecting the relevant ECD guidelines and constraints;
- consulting with the design team for the list of potentially feasible design alternatives;
- calculating the importance ratings in accordance with the results of pairwise comparison among ECD guidelines;
- determining the relationships between ECD guidelines and alternatives of design features in terms of fuzzy number representation;
- calculating the total importance score for each alternative of the design features

The second part of the framework is the systematic approach, in whi ch the principle of life-cycle design will be incorporated into the proposed green product design framework. The purpose of using this systematic approach on the basis of product life cycle is to capture the real environmental influence associated with adopting a particular design feature. In order to analyze the environmental impact properly, the systematic approach has been mapped as a two-phase analysis. The two-phase process can be outlined as follows.

Phase one: analysis of individual life-cycle stages, which includes:
- raw material extraction and processing;
- manufacture and assembly;
- product distribution and use; and
- management of end-of-life products

Phase two: overall life-cycle assessment.

As indicated above, a four-stage analysis of the product life cycle is conducted in phase one. The modified HOQ will be used in each life- cycle stage analysis. In other words, there are four modified HOQ matrices that represent four life-cycle stages respectively to fulfill the life-cycle assessment.

3.1   Modified House of Quality

HOQ is a powerful mechanism that is usually used to transfer customer requirements into product or service design. However, the weakness of HOQ includes a tedious implementation process and use of imprecise artificial expressions. In order to employ HOQ as an environmental evaluation tool as well as improve its implementation process, the configuration of HOQ has to be rearranged so that it can evaluate the various design alternatives in accordance with environmental design guidelines and regulations. The rearranged HOQ tool is called Modified House of Quality and that is shown in Figure 3.

With regard to the manipulation of imprecise artificial expressions in the traditional HOQ, the fuzzy set theory is used in modified HOQ for resolving the existing problem. In modified HOQ, a specific fuzzy number will be assigned to represent the strength of relationships between environmental requirements and design alternatives. As a result, the quantitative meaning contained in the relationships may be captured without vagueness.

3.1.1   Configuration of Modified HOQ

In order to incorporate the HOQ methodology into the application of green product design, a revision to the HOQ is required. The major change of configuration is to shift the correlation matrix from the top to the left- hand side. This modification of the configuration may allow the design team to measure the importance among the environmentally conscious requirements. Subsequently, the numerical importance ratings with respect to each environmentally conscious requirement may be calculated on the basis of the relationships identified in the correlation matrix.

The purpose of applying modified HOQ is to evaluate a set of feasible design alternatives and choose the most appropriate one. In order to achieve this goal, the design alternatives have been arranged to display in the topmost matrix so that the total absolute scores may be collected in the bottom matrix. Subsequently, the decision of selecting a preferable design alternative can be obtained on the basis of these absolute scores. Since the correlation matrix has been relocated, some of the matrices in the modified HOQ have different functions from the conventional HOQs.

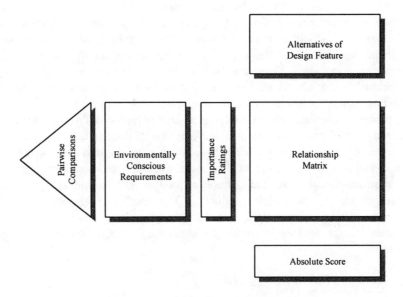

Figure 3. **Modified House of Quality**

3.1.2   Environmentally Conscious Requirements

Defining environmental requirements may be one of the most critical factors in the whole design process. A well- established set of requirements may enable the design to proceed more efficiently, since design alternatives will be evaluated based on how well they meet requirements. Generally, in terms of engineering considerations, design requirements contain design functions and design constraints.   In this study, the focus of design requirements is on environmental aspects, and the specific term, "environmentally conscious requirements", will be used in later sections.

Environmentally conscious requirements represent the environmental regulations and the ECD guidelines identified in the initial product design stage.   Since environmentally conscious requirements are product and process dependent, different industries will face a variety of environmental regulations and may want to consider various recycling strategies. Generally, the information in this matrix contains considerations of energy efficiency, material usage, pollution prevention, economic factors, design for disassembly, recyclability, and alternatives of end-of-life product utilization.

It is difficult to collect comprehensive data in this matrix because regulations are rapidly changing and recycling technology is continually improving. In other words, the design team sometimes has to forecast the future development of changing legislation and industrial requirements. Consequently, it will increase the degree of uncertainty and it will be more difficult for the design team to reach a consensus on the value of the importance ratings. There are two methods proposed to overcome this problem. One is developing a matrix-base worksheet to generate environmentally conscious requirements associated with the current product or process. The other one is applying Analytic Hierarchy Process (AHP) to obtain rational importance ratings that will be explained in later sections.

The matrix-base worksheets have been divided into four stages in accordance with the product life-cycle stages, including raw material extraction, manufacture and assembly, product distribution and use, and management of end-of-life products. The four-stage arrangement will comply with the systematic approach in the proposed green product design framework, as well as facilitate the generation of comprehensive design criteria.

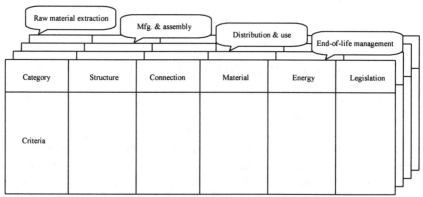

Figure 4. **Worksheets for Environmentally Conscious Requirements**

The worksheet for each life-cycle stage contains columns that represent considerations associated with environmentally conscious requirements. For the purposes of facilitating generation of environmentally conscious requirements and providing a rational weight for each requirement, the categorized considerations are further broken down into criteria. Figure 4 shows the configuration of the worksheets. The surface layer in Figure 4 represents the end-of-life management stage, while the other three hidden layers are raw material extraction, manufacture and assembly, and distribution and use, respectively.

In terms of ECD guidelines and associated environmental protection laws, there are five categories that can be identified. These categories are (1) product/component structure; (2) connection; (3) material; (4) energy consumption; and (5) legislation. These five categories are all or partially related to each of the product life-cycle stages.

In order to discover all the environmentally conscious requirements, the design team should look into the potential environmental impacts resulting from adopting a certain design alternative. The environmental impacts should be looked at with respect to depletion of nature resources, violation of environmental regulations, damage to ecological health and risk to human health. In other words, the generalized environmentally conscious requirements attempt to minimize the use of nature resources, comply with environmental laws, minimize the negative impacts outside of the production site, and eliminate harmful emissions to work/consumer exposure environment.

3.1.3   Design Alternatives

Design alternatives are the options that attempt to fulfill customer needs as well as design requirements/constraints. When the design team proceeds through the conceptual design stage, ideas will be generated in correspondence with customer requirements. The ideas are usually obtained from expertise, brainstorming, and/or experience. Since the generated ideas are not required for a preliminary evaluation, it may result in producing multiple design alternatives. Before moving on to further stages of product design, the created design alternatives have to be initially evaluated to screen out unfeasible ones. The proposed framework provides a means to sift out unfeasible alternatives by prioritizing the nominees of design features, for example, by regarding the design alternatives as objects to be analyzed. The generated design alternatives will be placed in the top matrix of the modified HOQ for succeeding analysis processes.

3.1.4   Importance Ratings

In this implementation step of the modified HOQ, an importance rating of each environmentally conscious requirement will be determined. As mentioned previously, imprecise and subjective weights will be assigned in the traditional HOQ application. Therefore, the Analytic Hierarchy Process is proposed as a method for obtaining rational weights in modified HOQ as an improvement over traditional HOQ.

While assigning importance ratings for the generated environmentally conscious requirements, the design team will face other challenges including seeking consensus, making tradeoffs and synthesizing judgment. Nevertheless, the analytic hierarchy process provides a flexible model that allows individuals or groups to organize ideas and derive the prioritized solution. In short, the expected advantages gained from utilizing AHP in modified HOQ include:

- generating comprehensive environmental requirements in a hierarchic structure;
- incorporating judgments and personal values in a logical way;
- enabling the establishment of priorities among environmental requirements by means of pairwise comparison;
- providing a method for tracking the logical consistency of judgments used in determining priorities

In the following sections, the basic steps of integrating AHP into modified HOQ are explored. The whole process will start with selecting relevant environmentally conscious requirements from the knowledge base. The pairwise comparison between components in each level of the hierarchic structure will be conducted next. Before moving forward to the overall synthesis step, a consistency check will be practiced in order to ensure that a proper pairwise comparison has been derived. Finally, the outputs will be the computed priorities that also represent the required importance ratings.

3.1.4.1   Hierarchic Structure

Structuring a hierarchy of environmentally conscious requirements may provide designers a clear view of design circumstances and a way to penetrate the problems. Basically, the structure is formulated by grouping environmentally conscious requirements into categories. This is the initial step of the Analytic Hierarchy Process. An well-organized hierarchic structure will enhance the succeeding prioritization operations. In essence, partitioning environmentally conscious requirements into more levels will provide a better description of the system. On the other hand, increased structure levels will result in a more complex computation for obtaining the priorities. These two factors are the major concerns while building the hierarchic structure.

A two-level hierarchy is used in this framework. These two hierarchic levels are category and criteria, respectively. This arrangement is also in correspondence with the worksheets developed for producing environmentally conscious requirements.

3.1.4.2 Pairwise Comparison Matrices

In applying the AHP, components of the hierarchy of environmentally conscious requirements are compared in pairs with respect to their relative impact ("weight" or "intensity") on a property or goal that they share in common. The results of pairwise comparison will be recorded in the correlation matrices in the modified HOQ template.

Compared to directly assigning relative weights to each component, pairwise comparison has advantages that give the analyzer a basis on which to reveal his or her preference by comparing two elements. The advantage of pairwise comparison can overcome the shortcomings of direct assignment of weights that is easy to misjudge for the analyzer and may result in inaccuracies.

Based on the hierarchy of environmentally conscious requirements, a matrix may be formed to compare the relative importance of categories in the first level with respect to the overall objective. Figure 5 shows the pairwise comparison matrix for the first level of the hierarchy with respect to the end-of-life management stage. The pairwise comparison matrix is constructed by providing the objective of comparison above and listing the elements to be compared in the heading row and the first column. Similar matrices are constructed for pairwise comparisons of each criterion in the second level with respect to those in the first level. That is, five more pairwise comparison matrices have to be constructed before obtaining the importance ratings along with environmentally conscious requirements. The goals for these five matrices are structure, connection, material, energy and legislation, which are the categories in the first analysis level.

Goal: End-of-life management	Structure	Connection	Material	Energy	Legislation
Structure	a_{11}	a_{12}	a_{13}	a_{14}	a_{15}
Connection	a_{21}	a_{22}	a_{23}	a_{24}	a_{25}
Material	a_{31}	a_{32}	a_{33}	a_{34}	a_{35}
Energy	a_{41}	a_{42}	a_{43}	a_{44}	a_{45}
Legislation	a_{51}	a_{52}	a_{53}	a_{54}	a_{55}

Figure 5. **Pairwise Comparison Matrix for the First Level**

In Figure 5, the variable a_{ij} are the relative importance of the components being compared with respect to the goal identified at the top. In order to present the relative importance, a scale of measurement is used that was introduced by Saaty (1980, 1985). Table 2 summarizes the scale used in this green product design framework.

There are two important properties of the pairwise comparison matrix. One is the reciprocal property, that is:

$$a_{ji} = \frac{1}{a_{ij}}$$

where the subscripts i and j refer to the row and column, respectively, where any entry is located.

Table 2. **Scale of Relative Importance**

Intensity of relative importance	Description of pairwise comparison
1	Both activities provide equal overall importance to the objective.
3	Moderate importance of one over another with respect to the objective.
5	Strong importance of one over another with respect to the objective.
7	Very Strong importance of one over another with respect to the objective.
9	Extreme importance of one over another with respect to the objective.
2, 4, 6, 8	A compromise is needed between two adjacent importance values.
Reciprocals	If activity i has one of the preceding numbers assigned to it when compared with activity j, then j has the reciprocal value when compared with i.

When comparing activity X with Y, a numerical value from Table 2 will be estimated for the importance ratio. The reciprocal value is then used for the comparison of activity Y with X. The other property is that the diagonal of the matrix is unity, since any component, which compares with itself will always give an equal importance. Based on these two properties, pairwise comparison matrices will have the form of numerical judgments as:

$$A = \begin{bmatrix} 1 & a_{12} & \cdots & a_{1n} \\ \dfrac{1}{a_{12}} & 1 & \cdots & a_{2n} \\ \vdots & \vdots & \ddots & \vdots \\ \dfrac{1}{a_{1n}} & \dfrac{1}{a_{2n}} & \cdots & 1 \end{bmatrix}$$

As long as the two properties hold, pairwise comparison matrices can be further simplified into triangular shaped matrices. The triangular shaped matrices have $n(n-1)/2$ variables reduced from n^2 variables and will comply with the modified HOQ configuration format.

3.1.4.3 Consistency Check

In AHP methodology, the consistency check is one of the most important issues. An analyzer reporting that activity A_1 is twice as important as activity A_2 and that A_2 is three times as important as activity A_3 is providing consistent judgment if the analyzer reports that A_1 is six times as important as A_3. If the analyzer reports any other value for the comparison of A_1 with A_3, the judgment is considered to be inconsistent.

Each pairwise comparison represents an estimate of the ratio of the priorities or weights of the compared activities. Saaty (1985) introduced the eigenvector method to calculate the overall weights in each pairwise comparison matrix for each level of the hierarchy. In order to measure the level of inconsistency, the consistency index (C.I.) and inconsistency ratio (I.R.) are utilized. The consistency index may be computed by using the formula as:

$$C.I. = \frac{\lambda_{max} - n}{n-1}$$

The formula for calculating the inconsistency ratio is as:

$$I.R. = \frac{C.I.}{R.I.} \times 100\%$$

where λ_{max} is the largest eigenvalue of the pairwise comparison;
          n is the number of elements being compared; and
          $R.I.$ is the random consistency index.

Saaty proposed a rule of thumb that the inconsistency ratio should be less or equal to 10 percent for acceptable results. Otherwise, it is recommended that a certain revision is required in pairwise comparisons.

3.1.4.4   Synthesis

Applying the environmentally conscious requirement worksheet in this proposed framework, a judgment and pairwise comparison in the two- level hierarchy (e.g. category and criteria) will be conducted individually. Since a set of priorities will be generated along with each comparison matrix, a proper weighting process will have to be performed before obtaining the final priorities (e.g. the importance ratings in modified HOQ). How these priorities are related to each other and how to synthesize these priorities into the importance ratings are described in this section.

Since the environmentally conscious requirements worksheet has been designed and constructed such that the components of environmental categories or criteria have no interaction between them, the assumption can be made that each level of the hierarchy is functionally independent. Based on the assumption and the AHP process, the importance ratings will be the weighted priorities from the bottom level assessments of the hierarchy structure.   Using the hierarchy structure described previously, the comparison matrices can be built as shown in Figure 6:

End-of-life management	Structure	Connection	Material	Energy	Legislation	Priorities
Structure						$_{\text{End-of-life}}P_{\text{Structu}}$
Connection						$_{\text{End-of-life}}P_{\text{Connec}}$
Material						$_{\text{End-of-life}}P_{\text{Materia}}$
Energy						$_{\text{End-of-life}}P_{\text{Energy}}$
Legislation						$_{\text{End-of-life}}P_{\text{Legisla}}$

Figure 6. **Comparison Matrix**

Where $_xP_y$ represent priorities (weights) in which subscript y are the comparison components with respect to the goal of subscript x.

Next, five comparison matrices may be built with respect to the five categories. Finally, the overall importance ratings with respect to the criteria may be calculated by multiplying the criterion's weight with the corresponding category's weight, which is:

$$_xP_z = {}_xP_y \times {}_yP_z$$

Where $_xP_z$ represent the importance ratings for all criteria z's;

$_xP_y$ represent the category weights with respect to the overall goal;

$_yP_z$ represent the criterion weights with respect to the corresponding category

3.2 Systematic Analysis

The second part of the proposed green product design framework is systematic analysis. The systematic approach will synthesize the individual analyses made by the modified HOQ. Essentially, this process of analysis is using the product life cycle approach that takes into consideration the entire life cycle of a product starting from raw material extraction and ending with end-of-life product management. In other words, the tasks incorporated in a product life cycle include mining, material processing, manufacture and assembly, use, and disposal. In addition, the transportation of materials between each pair of the life-cycle stages should also be considered.

The systematic analysis encompasses two phases: (1) life cycle design approach, and (2) overall assessment. In phase one, the product life cycle is divided into four stages. The modified HOQ will be performed in every product life-cycle stage in order to obtain the relative preference rankings with respect to alternatives of design features. In phase two, the results from each modified HOQ analysis will be synthesized by applying the AHP. Figure 7 shows the arrangement of the two phases.

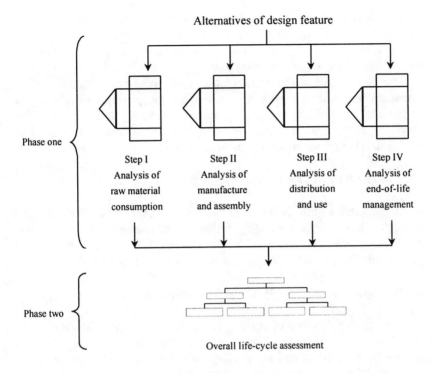

Figure 7. **Two-Phase Arrangement**

3.2.1   Life Cycle Design Approach

A product life cycle is organized into four stages in this proposed framework, which include:
- raw material extraction and processing;
- manufacture and assembly;
- use and distribution; and
- product end-of-life management

Raw material extraction means mining nonrenewable material from the earth. These materials usually need to be further processed into base materials by means of separation and purification. In addition to extraction from the earth, designers should count on other sources of raw materials as an input of this stage, which are recycled materials.

The second product life-cycle stage in the proposed framework is product manufacturing and assembling. In this stage, parts will be produced through a number of fabrication processes. Afterwards, the parts will be assembled into the final product and then released to the consumer market.

Distribution and use is considered as the third product life-cycle stage. The logistic design and packaging design are two major aspects with respect to product distribution. For product usage, designers should always keep in mind energy consumption and durability.

The last stage in the product life cycle is the product end-of-life management. After a certain time period of use, products will finally be discarded. In addition to dumping post-consumer products into landfills, there are at least three alternative strategies for dealing with end-of-life products. These three strategies are reuse, remanufacture and recycle.

3.2.2   The Role of Modified HOQ

Using a series of modified HOQ matrices, we could perform the concept of product life cycle design. In the following sections, we will describe how a series of matrices can enable the designers to capture the overall performance of a specific design feature throughout its whole life span.

3.2.2.1   Analysis of Raw Material Consumption

The focus of this step of analysis is on the first product life-cycle stage - raw material consumption. It attempts to capture the relationships between environmental aspects and the production requirements with respect to the raw material consumption and processing stage.

This step of analysis will begin with a survey of material requirements. In essence, it may be looked at in two ways, which are direct requirements and indirect requirements. Direct requirements consist of the mass of material consumption and relevant energy used for material processing, while indirect requirements will consider the by-product of the whole material acquisition process.

With respect to the mass of material consumption, the associated environmental considerations will at least include natural resources depletion and recyclability. During the process of material processing, some hazardous chemical substances will have to be added. Based on the principles of Environmentally Conscious Design, the criterion of how to minimize the risk to the environment resulting from use of hazardous materials should be addressed. Residue management, such as minimizing process wastes including air emissions, liquid effluents, and hazardous and non-hazardous solid wastes, is also within the system boundary of the product life-cycle stage of raw material consumption.

While the design team finishes this step of modified HOQ analysis, the results will indicate the preference of the alternatives of design feature under the constraints in the raw material consumption stage. These results will be used for the following overall assessment.

3.2.2.2 Analysis of Manufacture and Assembly

This step of analysis focuses on the product life-cycle stage of manufacture and assembly. The input elements include reused or remanufactured parts from the recycling of disposed products, and related material/energy consumption. The outputs are the finished goods.

Within the boundary of this step of analysis, issues to be considered for product design will encompass Design for Assembly (DFA), and Design for Manufacturing (DFM). With respect to DFA and DFM, cost concern and operation efficiency always have higher priority than other design factors. However, the complexity of decision-making is dramatically increased when environmental issues become part of the product design constraints.

3.2.2.3 Analysis of Use and Distribution

In this life-cycle stage, finished goods will be transported from manufacturing sites to distributors. Finally, products will be sold to customers for providing a service or for fulfilling a specific customer need.

During distribution and display on the shelf, packaging products will be made for providing protection and for aesthetic purposes. On the other hand, the packaging process will also generate a significant material waste. In this step of analysis, the design of the package is part of the considerations.

At the product use stage, energy consumption is the major concern, especially for electric appliances and automobiles. After a certain time of usage, the product might need some minor maintenance or repair. Therefore, design for serviceability is another subject that designers should take into consideration. In fact, there are some similarities between design for serviceability and design for disassembly, since partial disassembly might be unavoidable when maintaining or repairing a product.

Another modified HOQ analysis may be undertaken when the environmentally conscious requirements have been developed under the established boundary. The results will show the favored design alternative with respect to the life-cycle stage of use and distribution.

3.2.2.4 Analysis of End-of-life Management

All products will eventually get to the end-of-life stage when customers decide to retire the products. The strategies of a product end-of-life management include product reuse, part remanufacturing and material recycling. Hence, the inputs in this step of analysis are end-of-life products, energy required for disassembling these products, and energy required for sorting the disassembled materials and parts. The outputs are reusable or remanufacturable subassemblies and parts, recyclable materials, and associated residuals generated from the disassembly processes.

Regarding environmental concerns, the design for product retirement should be aimed at the product end-of-life management strategies of reuse, remanufacturing and recycling. For achieving any one of the goals of these strategies, the product disassembly must be accomplished. Therefore, design alternatives should always promote ease of disassembly.

The majority of environmental problems results from the residuals generated within the life-cycle stage of the end-of-life product. The rising level of municipal solid waste has caused a landfill site shortage. Many industries such as the auto and electronic appliance industries have started to develop new design concepts for adapting to the changing environmental requirements. This step of analysis evaluates design alternatives so that end-of-life product can be enhanced to secure the goals of reuse, remanufacture and recycle.

3.2.3   The Overall Assessment

The overall assessment is the second phase of the proposed framework. It involves bringing together the results of the previous four modified HOQ matrices and then evaluating the overall performance of the alternatives of design feature throughout the entire product life cycle.

When conducting the modified HOQ analysis, designers consider only the environmental criteria within a single life-cycle stage. The absolute scores obtained from each modified HOQ analysis represent the preference weightings with respect to the specific product life-cycle stage. The overall assessment will provide a means to synthesize the four steps of analysis and calculate the overall weightings. For example, the material used for automobile fenders may have two design alternatives, which are plastic and steel fenders. When considering the use stage, plastic fenders might gain higher weights because it would increase the ratio of miles-per-gallon due to mass reduction. On the other hand, steel fenders may be preferred for the end-of-life management stage due to the recyclability. The final decision will have to rely on the overall scores that may be obtained by using the method described in the following sections.

3.2.3.1   The Overall Scores and Defuzzification
The overall scores will provide designers an index that shows the weightings of design alternatives with respect to the overall goal – "green

product design." Since there are four stages in the life-cycle design approach, the overall score may be obtained by using the following equation that is the summation of the absolute score multiplied by the associated weight of life-cycle stage.

$$OS_j = \sum_{i=1}^{4} AS_{ij} \times W_i$$

Where

OS_j = Overall Score with respect to the jth design alternative;

AS = Absolute Score;

W_i = Weight of ith life-cycle stage.

Since the results obtained from the equation are fuzzy numbers, they have to be converted to so-called "crisp" numbers in order to make comparisons of the alternatives of design feature and to place the priority ranks. The process of converting a fuzzy number to a single real number is called defuzzification. Yager's Centroid method (1980) has been selected as the defuzzification method in this proposed framework. This method is to find the geometric center of a fuzzy number, corresponding to an x value on the horizontal axis.

By applying the defuzzification, the overall scores calculated could be transforming into crisp numbers. Accordingly, these crisp overall scores can be compared and ranked. The highest rank will be assigned to the alternative of design feature that has the largest "crisp" overall score.

4. CONCLUSION AND FUTURE WORK

The issues of ECD may be applied to a wide range including material selection, parts design and manufacturing process improvement. The ultimate goal of green product design is to conserve nature resources, minimize depletion of non-renewable resources and use sustainable practices for managing renewable resources. The varied considerations and constraints resulting from the practice of green product design will cause designers to face complicated trade-off. In other words, it is harder than ever for a designer to make a decision on green product design.

The proposed framework focuses on seeking better environmental performance by means of selecting environmentally friendly design alternatives. Two parts contained in the framework may facilitate the selection, so as to make the framework effective. These two parts are (1) the individual assessment tool, and (2) the overall assessment method.

The proposed framework and current research aim at the fulfillment of green product design requirements. Future work associated with further improvements of the proposed framework and the current green product design research has been identified as follows:

- Development of software and a comprehensive knowledge base for enhancing the operation and calculation of the proposed framework. The knowledge base should contain the current environmental regulations and standards and is linked to the interface of the software. The knowledge base should also be able to accommodate the rapidly changing environmental requirements.
- Efforts should be made to discover how to apply the design coordination approach that is a subdirectory of concurrent engineering for the proposed framework, since green product design is usually based on large-scale projects that involve negotiation between different professionals and the use of conflict resolution.
- A sensitivity analysis with respect to the measurement scale should be conducted. The sensitivity analysis will aim at finding how and how much is the influence by adopting different measurement scales.

REFERENCES

Agba E. I., Rigg, R. H. and Wahlquist, D. R., 1996, "Design requirements planning for Design-for-X implementation." Proceedings of the 1996 National Design Engineering Conference, March, 1-7.

Alting, L. and Legarth J. B., 1995, "Life cycle engineering and design." Annals of the CIRP, 42/2, 569-580.

Boothroyd, G., 1992, "Assembly automation and product design." Marcel Dekker, New York.

Bor, J. M., 1994, "The influence of waste strategies on product design." Materials & Design, Vol. 15, No. 4, 219-224.

Brooke L., 1993, "Recycling: what's next." Automotive Industries, May, 4344.

Bylinsky, G., 1995, "Manufacturing for reuse." Fortune, Feb., 102-112.

Dowie, T., 1994, "Green design." World Class Design to Manufacture, Vol. 1, No 4, 32-38.

Environmental Protection Agency (EPA), 1993, "Life-cycle design guidance manual: environmental requirements and the product system." Office of Research and Development, Washington, D.C.

Fiksel, J., 1993, "Design for environment: an integrated systems approach." Proceedings of the IEEE International Symposium on Electronics and the Environment, 126131.

Ishii, K., Eubanks, C. F. and Marco, P. D., 1994, "Design for product retirement and material life-cycle." Materials & Design, Vol. 15 No. 4, 225-233.

Johnson, M. and Wang, M., 1995, "Planning product disassembly for material recovery opportunities." International Journal of Production Research, Vol. 33, No 11, 31193142.

Jovane, F., Alting L., Armillotta, A., Eversheim, W., Feldmann, K., Seliger, G. and Roth, N., 1993, "A key issue in product life cycle: disassembly." Annals of the CIRP, 42/2, 651658.

Keoleian, G. A. and Menerey, D., 1994, "Sustainable development by design: review of life cycle design and related approaches." Air & Waste, Vol. 44, May, 645668.

Navin-Chandra, D. 1994, "The recovery problem in product design." Journal of Engineering Design, Vol. 5, No. 1, 67-87.

Penev, K. D. and Deron, A. J., 1994, "Development of disassembly line for refrigerators." Industrial Engineering, Vol. 26, No. 11, Nov., 50-53.

Philipps, K. K., Vadrevu, S., Olson, W. W. and Sutherland, J. W., 1994, "Concurrent engineering and the environment." S. M. Wu Symposium, Vol. 1, 349-354.

Poyner, J. R. and Simon, M., 1995, "Integration of ECD tools with product development." CONCEPT - Clean Electronics Products and Technology, 9-11 October, 54-59.

Rhodes, C. R., 1996, "ECD—more than just design for the environment." Printed Circuit Fabrication, Vol. 19, No. 11, November, 22-24.

Rosenberg, D., 1992, "Designing for disassembly." Trends, Nov., 17-18.

Ross, P. J., 1988, "The role of Taguchi methods and design of experiments in QFD." Quality Progress, Vol. 21, No. 6, June, 41-47.

Saaty, T. L., 1980, "The analytic hierarchy process." McGraw-Hill, Inc.

Saaty, T. L. and Kearns K. P., 1985, "Analytical planning." Pergamon Press.

Simon, M., 1991, "Design for dismantling." Professional Engineering, Nov., 20-22.

Spicer, A and Wang, M. H., 1996, "Disassembly modeling and analysis." Contract Report to Vehicle Recycling Partnership (VRP), Highland Park, MI, USA.

Tung, C. and Wang, M. H., 1998, "A systematic approach for product life-cycle design." Proceedings of Northeast Decision Sciences Institute 27th annual meeting, pp. 376-378.

U.S. Congress, Office of Technology Assessment, 1992, "Green products by design: choices for a cleaner environment." OTA-E-541.

VDI 2243 1993, "Konstruieren recyclinggerechter Produkte.." VDI-EKV, Beuth-Vertrieb.

Vergow, Z. and Bras, B., 1994, "Recycling oriented fasteners: a critical evaluation of VDI 2243's selection table." ASME Advances in Design Automation, DE-Vol. 692, 341-349.

Watkins, R. D. and Granoff, R., 1992, "Introduction to environmentally conscious manufacturing." International Journal of Environmentally Conscious Manufacturing, Vol. 1, No. 1, 5-11.

Yager, R. R., 1980, "On a general class of fuzzy connectives." Fuzzy Sets and Systems, Vol. 4, No. 3, 235-242.

AUTHOR

Dr. Michael H. Wang is currently a Professor in the Department of Industrial & Manufacturing Systems Engineering at the University of Windsor, CANADA. He received his Ph.D. degree in Industrial Engineering form the University of Iowa, an M.S.I.E. from SUNY- Buffalo, and a B.S.I.E. from Tsing-Hua University, Taiwan. His recent research interests include manufacturing system integration, concurrent engineering, and environmental conscious design and manufacturing. He has published more than 85 articles of these areas in international conference proceedings and professional journals such as International Journal of Production Research, International Journal of Environmentally Conscious Design and Manufacturing, and International Journal of Computer Integrated Manufacturing. He is a member of IIE, INFORMS, and SME.

Chapter 13

Environmental Attributes of Manufacturing Processes

John W. Sutherland & Kenneth L. Gunter

Graduate Research Fellow
Dept. of Mechanical Engineering – Engineering Mechanics
Michigan Technological University

This chapter examines the ISO 14000 environmental standard, the basic components of Environmentally Conscious Manufacturing, and gives some examples of how they may be integrated in support of an organization's environmental policy

1. INTRODUCTION

The standard of living in the U.S. is greatly dependent on the ability to competitively produce manufactured products. In the absence of the manufacturing enterprise, there is little need for technologists and engineers and those engaged in supporting business activities. Every design decision within such an enterprise carries with it manufacturing implications -- usually restricting the choice of materials and processes, and thus impacting competitiveness. In recent years, increased attention to the environment is presenting manufacturers with new challenges. The U.S. manufacturing industry produces approximately 5.5 billion tons of non-hazardous and 0.7 billion tons of hazardous waste each year (Zhang, et al., 1997). These wastes include: sand with additives produced by metal-casting operations, fluids from heat-treating, and welding gases. Ever more attention is being focused on reducing the environmental, health, and safety (EHS) consequences of process waste, as reflected by the tightening standards, increased fines, and growing litigation associated with the waste. It is clear that organizations that are to be competitive in the future must be able to avoid/minimize the costs concomitant with being (or not being) "green".

Manufacturing is but one step in the life cycle of a product (Fig. 1), each stage of which has associated EHS consequences. To be truly environmentally responsible, product designers and other decision-makers must adopt such a broad view of the product life cycle, and temper their decisions with the characteristics of the waste streams generated and energy used at each stage. For example, this means that a product designer should consider the environmental differences between the use of an aluminum or cast iron part not only in terms of cost and function, but also in light of the waste produced/energy consumed during mining, refining, manufacturing, use, and post-use life cycle stages. Of course, to supply decision-makers with information regarding the environmental consequences associated with their decisions means that knowledge must be available about the specific waste streams generated and energy consumed during each life cycle stage. For the manufacturing stage of the life cycle, this implies that EHS knowledge must be available on each unit operation. Researchers are only now beginning to develop this type of information, and given the vast number of manufacturing unit operations, much work remains to be performed in terms of describing the energy usage and waste stream character and quantity.

In addition to the decision-making problem described above that emphasizes process selection, the problem of process beneficiation also exists. With process beneficiation emphasis is placed on creating new processes and modifying existing processes to be more environmentally benign. In either case, some base knowledge about manufacturing process waste streams and energy is required. These waste streams may be in various states (solid, liquid, aerosol, and gas) and employ different transport mechanisms to reach a variety of receptors. The down-stream effects of the waste streams are quite varied. Some wastes may influence global warming, while others may pose health threats to factory workers. The effects of the waste on the environment and the health/safety of concerned life forms depend not only on the character of the waste but also on the concentration. Figure 2 illustrates this situation, and additionally suggests that the ultimate effect of a waste stream also depends on actions taken to interrupt the transport of the waste from the source to receptors (e.g., use of respirators by workers to prevent inhalation of paint fumes). Of course such actions have associated price tags.

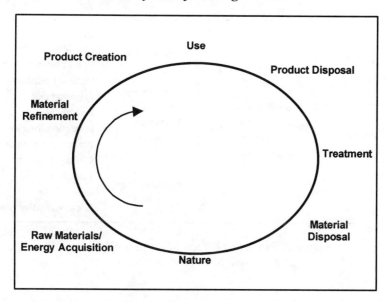

Figure 1: Product Life Cycle (Olson and Sutherland, 1993)

With this background, this chapter outlines a three-step methodology for manufacturing operation study that has as its goal an understanding of process waste streams and energy requirements. The first step of the proposed methodology, inventory analysis, is then applied in a general way to the following process classes:

- Casting Operations,
- Plastics Processing Operations,
- Machining Operations,
- Forming Operations,
- Surface Treatment Operations, and
- Joining Operations.

Several processes are examined in detail, and available knowledge concerning the waste streams is given. It may be noted that the chapter is focused only on assessing/ characterizing environment-related process outputs and does not consider their down-stream effects from an epidemiological/ecological standpoint. The chapter concludes with an examination of where environmental improvements can be achieved from a manufacturing viewpoint.

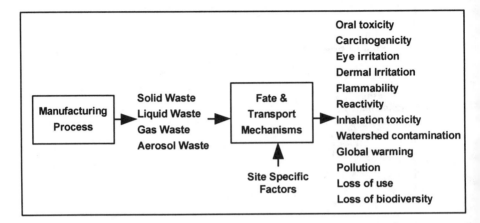

Figure 2: Forms and Effects of Process Waste Streams
(adapted from Sheng and Srinivasan, 1996)

2.   ENVIRONMENTAL PROCESS CHARACTERIZATION

2.1   From Waste Management to Pollution Prevention

Manufacturing operations are most often viewed as the set of processes employed to operate on discrete parts. Generally these processes utilize mechanical, thermal, and/ or chemical means to change some characteristic(s) of an input component to produce a final component -- an output. In addition to the processed part, operations have a number of other outputs, or by-products, that essentially represent manufacturing process waste streams. Choi, et al. (1997) suggest that such operations can be represented via an input-output form such as shown in Figure 3.

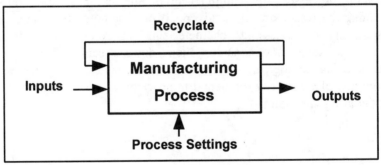

Figure 3: Input-Output Diagram (adapted from Choi, et al., 1997)

In considering the diagram of Figure 3, Inputs to a process could include: the part to be operated on, power, water, and air. It should be noted that these inputs could either enter the process deliberately or passively. Outputs could include: the finished part, scrap, gaseous emissions, and wastewater. The recyclate loop indicates that some process inputs are used over and over again, e.g., processing equipment and tooling. In addition to the substances and energy that enter and exit from the process "black box", another input class, Process Settings, is also present. These settings specify how the equipment/process is to be performed, e.g., spindle speed, temperature, and cycle time.

Given that manufacturing processes can be described with input-output representations, attention now turns to how this may be used to characterize the waste streams and energy consumption associated with processes. It is proposed that a three-step methodology may be performed, viz.,

1. Conduct a process inventory,
2. Quantify the input and output mass and energy flow rates, and
3. Describe the outputs as a function of the inputs.

The first step of the methodology (Process Inventory) requires that all the material and energy inputs and outputs be completely enumerated. The methodology at this point simply demands a listing of all the substance and energy types, not the amounts. The second step of the methodology (Quantification) requires that the amounts or rates of substance/energy input and output be identified for the complete listing developed in step 1. The degree of difficulty required to go from step 1 to step 2 in the methodology may be large. It might not be too taxing to identify the amount of aluminum entering/exiting from a turning process, with the input shape specified by the casting geometry, and the outputs being the finished shape and the volume of chips produced by the process. On the other hand, the cutting fluid input to a machining operation at a given flow rate, may be mostly recirculated (recyclate) and yet produce outputs such as fluid mist, gas, spillage, and chip carry off, the amounts of which are difficult to estimate. The third step of the methodology (Modeling) requires that the outputs be expressed as functions of the inputs. This modeling step may be informal (e.g., VOC production in the casting process is large because the pour rate is high) or more formally expressed as an empirical or mechanistic model (both of which may need information on the process settings). Again, there may be a tremendous leap in difficulty to go from step 2 to step 3, especially if it is desired to establish mathematical relationships between the outputs and inputs.

In considering the 3-step methodology described in the preceding paragraph it seems reasonable to discuss the merits of each step. For individuals just beginning efforts related to environmentally responsible manufacturing, step 1 (Process Inventory) is often a tremendous advancement and may reveal heretofore-unknown outputs that have environmental meaning. To make process comparisons, generally at least step 2 must be completed. If it is desired to improve a process, it is almost a necessity to have at least an informal model -- developed in step 3. The level of process specificity required increases from step 1 to step 2 to step 3. In addition, as noted, the complexity of the task increases as well. For these reasons, this chapter is largely focused on step 1.

During the execution of step 1 of the methodology (Process Inventory) a listing of the process inputs and outputs is created. It should be noted that all forms of a substance should be recorded. For example, a process may have water as an input and 3 water-related outputs: steam, wastewater, and water mist particles. In general, mass inputs/outputs from a process can be placed into one or more of the following categories: i) solid, ii) liquid, iii) gas, and iv) aerosol. Energy inputs/outputs should also be identified and also categorized where possible (e.g., electrical power, heat, radiation, and acoustical energy). The next section employs the Process Inventory concept in the analysis of a variety of manufacturing operations. In ending this section, it should be noted that while the methodology is largely focused on process characterization, and that such information is critical to the development of informed environmentally responsible manufacturing decisions, these decisions do not require the complete execution of the methodology.

3.   MANUFACTURING PROCESS INVENTORY

The preceding section outlined a 3-step methodology for manufacturing process analysis that has as its goal the development of environment-related process knowledge. The first step in the methodology is focused on identifying the inputs and outputs for a given process. This section is devoted to summarizing the recognized inputs/outputs for 6 process classes: i) casting, ii) plastics processing, iii) machining operations, iv) forming operations, v) surface treatment, and vi) joining. Specific processes within each class are discussed in greater detail.

3.1   Casting Operations

One of the shortest routes from raw material to finished part is that of melting a material and pouring it into a shape that approaches that of the finished component. The shape of the casting may be achieved through a wide variety of mould types (see Fig. 4) that differ in how they contain the molten material, their reusability, and how they are formed. Common casting processes include: sand casting, investment casting, permanent mold casting, and continuous casting. Solidification is achieved as the melt cools below the melting temperature of the material. Once the casting has cooled, it is removed from the mold and the resulting part may receive additional treatment.

The types of parts produced by casting processes vary widely. Some casting processes are used to produce bars, ingots, and sheets that may be further processed by other manufacturing operations. Many parts, however, are used in their "as-cast" form. Many shapes can be cast in net, or near-net shape, i.e., manufactured to, or nearly to, the desired dimensions. Final machining operations such as boring and honing holes for closely fitting shafts, or spot-facing of surfaces for mating with other parts, may be necessary to ensure final dimensional accuracy.

While there are numerous types of casting processes, the following basic steps apply to many of them (EPA, 1997):

- Pattern Making. Many casting operations utilize a pattern to form the mold. This casting step is associated with the production of the pattern.
- Mold and Core Preparation and Pouring. Mold and core preparation refers to the activity of creating a cavity into which molten metal will be poured.
- Furnace Charge Preparation and Metal Handling. This step refers to the task of heating the raw material and transporting the molten raw material to the mold.
- Shakeout, Cooling, and Sand Handling. This is the act of removing the cast part from the mold, and reclaiming the used sand for future molds.
- Quenching, Finishing, Cleaning, and Coating. A cast component often receives additional treatment to improve either its visual appearance or metallurgical properties.

Table 1 summarizes the waste streams associated with many of the sub-processes associated with the casting process.

Table 1: Casting Process Waste Streams (adapted from EPA, 1997)

Sub-Process	Material Inputs	Air Emissions	Wastewater	Residual Wastes
Pattern Making	Wood, metal, plastic, wax, polystyrene	VOCs from glues, epoxies, and paints	Little wastewater	Scrap pattern materials
Mold and Core Preparation and Pouring				
Green Sand Molding	Green sand and chemically-bonded sand	Particulates, metal oxide fumes, carbon monoxide, organic compounds, hydrogen sulfide, sulfur	Wastewater containing metals, elevated temperature, phenols, and other organics from	Waste green sand and core sand potentially containing metals

	cores	dioxide, and nitrous oxide. If chemically bonded cores are used: benzene, phenols, and other hazardous air pollutants (HAPs).	wet dust collection systems and mold cooling water	
Chemical binding systems	Sand and chemical binders	Particulates, metal oxide fumes, carbon monoxide, ammonia, hydrogen sulfide, hydrogen cyanide, sulfur dioxide, nitrogen oxides, and other HAPs.	Scrubber waste water with amines or high or low pH; & wastewater containing metals, elevated temperature, phenols, and other organics from wet dust collection systems and mold cooling water	Waste mold and core sand potentially containing metals and residual chemical binders
Lost foam	Refractory slurry, polystyrene	Particulates, metal oxide fumes, and HAPs	Little wastewater	Waste sand and refractory material potentially containing metals and styrene
Furnace Charge Preparation and Metal Melting	Ingots, scrap, returned castings, fluxing agents, ladles & other refractory materials	Products of combustion, oil vapors, particulates, metal oxide fumes, solvents, hydrochloric acid	Scrubber wastewater with high pH, slag cooling water with metals, non-contact cooling water and wastewater containing metals if slag quench utilized	Spent refractory material potentially containing metals and alloys, dross and slag potentially containing metals
Shakeout, Cooling, and Sand Handling	Water and caustic for wet scrubbers	Dust and metallic particulates, VOCs from thermal sand treatment systems	Wet scrubber wastewater with high/low pH or amines, contact cooling water with elevated temperatures, metals and mold coating	Waste foundry sand and dust from collection systems, metals
Quenching, Finishing, Cleaning, and Coating	Paint and rust inhibitor, raw castings, water, steel shot, solvents	VOCs, dust and metallic particulates	Waste cleaning and cooling water with elevated temperature, solvents, oil and grease, and suspended solids	Spent containers and solvents, steel shot, metallic filings, dust and wastewater treatment sludge

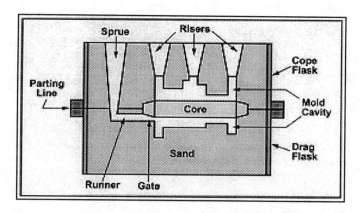

Figure 4: Mold for a Sand Casting Process

3.2 Plastics Processing Operations

A wide variety of manufacturing operations exist for the processing of polymers. The term polymer refers to any substance in which several (often thousands) molecules or building units (mers) are joined into large, more complex molecules. There are essentially three types of polymers: thermoplastics (large chain molecules held together by weak van der Waals forces), thermosets (chain molecules held together by extensive cross-links), and elastomers (highly kinked chain molecules with periodic cross-links). With a thermoplastic polymer (e.g., ABS, polyethylene, polypropylene) state. As heat (energy) is applied to a thermoplastic the material melts and the molecules can move freely relative to one another. Once in a liquid state, the viscous fluid is then delivered to a mold or die and allowed to cool and assume the desired shape. A key feature of thermoplastics from an environmental standpoint is that they may be reheated material initially in powder or pellet form is heated until it enters a liquid and re-formed over and over again. Thermosets (e.g., polyester, phenol-formaldehyde) and elastomers (e.g., natural rubber, styrene butadiene) have cross-links that connect the chain molecules. These cross-links are chemically introduced when the thermoset or elastomer has been placed into the desired shape. Because of the cross-linking, once a thermoset or elastomer product has been formed, the application of additional heat will not permit the part to be reshaped. If enough heat is supplied, these polymers will burn.

The production of plastic products often includes the following operations (EPA, 1995a):

- Plastic resin creation. Plastic materials often have additives mixed with them to realize some final product characteristic.
- Molding operation. Examples include: compression molding,

transfer molding, injection molding, encapsulation, and blow molding.

- Finishing. After a product is created, post-forming operations such as adhesive bonding and surface decorating may be undertaken to finish the product.

These operations produce wastes and utilize resources that may be placed into the following four categories: Chemicals, Wastewater, Pellet Release, and Fugitive Emissions. These categories are briefly described below.

3.2.1   Chemicals

The following list of chemicals often are used as additives in the preparation of plastic resin.

- Lubricants - stearic acid, waxes, fatty acid esters, and fatty acid amines
- Antioxidants - alkylated phenols, amines, organic phosphites and phosphates, and esters
- Antistats - quaternary ammonium compounds, anionics, and amines
- Blowing/foaming agents - azodicarbonamide, modified azos, OBSH, and HTBA
- Colorants - titanium dioxide, iron oxides, anthraquinones, and carbon black
- Flame-retardants - antimony trioxide, chlorinated paraffins, and bromophenols
- Heat stabilizers - lead, barium-cadmium, tin, and calcium-zinc
- Organic peroxides - MEK peroxide, benzoyl peroxide, alkyl peroxide, and peresters
- Plasticizers - adipates, azelates, and trimellitates
- UV Stabilizers - benzophenones, benzotriazole, and salicylates

3.2.2   Wastewater

Contaminated wastewater can arise from three sources: i) cooling/heating water, ii) cleaning water, and iii) finishing water. The EPA (1995a) reports that the use of water for temperature control only has produced treatable pollutants when water has directly contacted bis(2-ethylhexyl) phthalate (BEHP). Cleaning water has been found to contain oil and grease, organic carbon, phenols, and zinc (phenols and zinc are toxic pollutants), that result in high levels for biochemical oxygen demand (BOD), total suspended solids (TSS), chemical oxygen demand (COD), and total organic carbon (TOC). Finishing water has been observed to contain BEHP, di-n-butyl phthalate, and dimethyl phthalate, all toxic pollutants.

3.2.3   Pellet Release

Pellets are inert, but they are occasionally accidentally discharged into wastewater, where runoff may carry them into wetlands, oceans, etc. Pellets represent a concern because they may be ingested by birds and marine species. The EPA (1995a) has indicated that "the discovery of a single pellet in storm water runoff is subject to Federal regulatory action."

3.2.4   Fugitive Emissions

Under the high heat and pressure associated with the molding process, emissions may be produced. Because of the additives present within the plastic resin, these emissions may contain cadmium and lead.

3.3   Machining Operations

Machining is a general term that may be applied to all material removal operations. Conceptually, material removal operations should be avoided since they focus on eliminating material from a part with some inherent value. Technological advancements in casting and forming processes are constantly being sought so as to avoid unnecessary material removal operations. Still, material removal operations are widely used and are capable of creating geometries, surface finishes, and providing the precision not achievable by other operations. Traditional machining or cutting operations rely on a shearing mechanism in which the action of a sharp "cutting tool" is used to remove material. Non-traditional machining operations do not rely on a shearing mechanism to remove material; instead, they utilize thermal, chemical, and electro-chemical means to eliminate unwanted material.

3.3.1   Traditional Machining Operations

Figure 5 depicts an input-output relationship for a traditional machining operation. As is evident from the figure, there are a number of outputs from the process in addition to the desired product. Recently, the role of cutting fluids in machining operations has received increased attention because of environmental and industrial hygiene concerns. Fluid splashing, spillage, and chip carry-off can lead to inadvertent contamination of groundwater with the fluid as well as metal fines. Treatment of used fluids prior to introduction into wastewater streams is not totally effective and also represents a significant and increasing cost. The industrial hygiene community is also directing ever more attention to cutting fluid exposure (Hands et al., 1996) because of the negative health effects due to dermal contact and fluid mist inhalation. Recent studies (Yue et al., 1999) have shown that mist concentrations as well as the number of super and submicron particles can be manipulated by changing machining conditions.

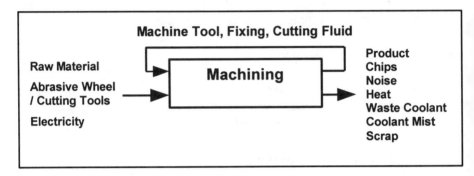

Figure 5: Input-Output Diagram of a Traditional Machining Process

3.3.2   Non-traditional Machining Operations

Non-traditional machining operations include electrical discharge machining, laser beam machining, electro-chemical machining, and photochemical machining. When compared to traditional machining processes, non-traditional operations are much less energy efficient. This fact has also been reported by Kalpakjian (1992). It should be noted, however, that non-traditional operations fill a niche in that they are often employed to work with hard or fragile materials that would be difficult to process using more traditional methods. In addition, these operations can produce features (shapes, textures, cavities) again not achievable by other processes. One common non-traditional machining operation is electrical discharge machining (EDM). In EDM, a negatively charged tool and a positively charged work piece are brought close to one another while submerged in a dielectric fluid (at low voltages, the fluid acts as an insulator). The voltage difference between the tool and work is increased until the dielectric fluid ionizes and a spark jumps across the tool - work gap; once the spark is produced, the dielectric fluid de-ionizes and is once again an effective insulator. The cycle repeats at a rate of 200 to 500,000 Hz, and the energy supplied by the sparks may produce local temperatures in the work of 12,000 °C -- a temperature high enough to vaporize the work material. In electro-chemical machining (ECM) the negatively charged tool (cathode) and positively charged work (anode) are brought close together. An electrolytic fluid is passed through the cathode (tool) and into contact with the anode (work). Across this fluid a reverse electro-plating process takes place; work material attempts to plate onto the tool but is washed away by the flushing action of the electrolytic fluid.

The EDM and ECM processes produce waste in solid, liquid, gas, and aerosol form. For example, the EDM process produces solid waste in the form of sludge and used filters. Tonshoff et al. (1996) report that if a mineral oil is used for EDM, airborne emissions may include polycyclic aromatic hydrocarbons, benzene, carbon monoxide, oil vapor/mist, and by-products associated with the breakup of oil and its additives. Additional airborne waste may include eroded metal particles from the work and electrode, e.g., particles of tungsten carbide, titanium carbide, chromium, nickel, molybdenum, and barium. Used dielectric fluid may contain measurable amounts of iron, lead, chromium, vanadium, carbon black, zinc, copper, cadmium, tungsten, nickel, cobalt, titanium, and molybdenum.

ECM also produces considerable amounts of pollutant waste including quantities of heavy metal sludge (metal oxides and hydroxides). These waste components can be toxic, hazardous, corrosive, or carcinogenic. The amount of residue generated is approximately 100 to 300 cubic centimeters for every cubic centimeter of removed material. Alloys typically machined through an ECM process contain 2 to 14% chromium and the sludge produced may contain a high concentration of Cr^{6+} ions.

3.4 Forming Operations

Many products are subjected, at some point during their processing, to plastic deformation. This means that the shape of the work is permanently changed without its volume being changed and without it being brought into the liquid state. Forming operations are those processes, which are used to produce this plastic deformation and may be categorized as either bulk deformation processes or sheet-working processes. Compared with other operations, forming operations produce relatively small amounts of waste. The raw material entering a forming operation may be a casting, whose waste streams have been described previously. Forming operations may require pre- or post-treatment to modify the surface character, metallurgy, or cleanliness of the workpiece. Of course, these treatments have associated environmental impacts. In terms of environmental issues directly relating to forming operations, attention centers on the energy required, lubricants employed to reduce friction, and flash (excess scrap material that must be removed from the part). Forming operations also typically require a die or some other deforming tool, the production of which again carries with it an environmental burden.

3.5   Surface Finishing and Treatment Operations

Products often receive a thermal or surface treatment at some point during their manufacture. Heat treating is used to control component metallurgy, painting is used to prevent corrosion or aesthetically enhance a product, cleaning operations are used to remove foreign material, and plating is used to protect or enhance the properties of the surface of the product. Many of these operations have negative environmental consequences. For instance, heat-treating requires energy to elevate the temperature of the part, and quenching oil may become contaminated when contacted with a high temperature part or give off hazardous air pollutants (HAPs). Table 2 lists some of the waste streams associated with treatment processes.

Table 2: Surface Finishing and Treatment Waste Streams (adapted from EPA, 1995b)

Sub-Process	Material Inputs	Air Emissions	Wastewater	Solid Wastes
Surface Preparation				
Solvent Degreasing and Emulsion, Alkaline Cleaning, and Acid Cleaning	Solvents, emulsifying agents, alkalis, and acids	Solvents associated with solvent degreasing and emulsion cleaning	Solvent, alkaline wastes, and acid wastes	Ignitable wastes, solvent wastes, and still bottoms
Surface Finishing				
Anodizing	Acids	Metal-ion-bearing mists and acid mists	Acid wastes	Spent solutions, wastewater treatment sludges, and base metals
Chemical Conversion Coating	Metals and acids	Metal-ion-bearing mists and acid mists	Metal salts, acid, and base wastes	Spent solutions, wastewater treatment sludges, and base metals
Electroplating	Acid/alkaline solutions, heavy metal bearing solutions, and cyanide bearing solutions	Metal-ion-bearing mists and acid mists	Acid/alkaline wastes, cyanide, and metal wastes	Metal and reactive wastes
Plating	Metals (e.g., salts), complexing agents, and alkalis	Metal-ion-bearing mists	Cyanide and metal wastes	Cyanide and metal wastes
Painting	Solvents and paints	Solvents	Solvent wastes	Still bottoms, sludges, paint solvents, and metals
Other Metal Finishing Techniques (e.g., Polishing, Hot Dip Coating, and Etching)	Metals and acids	Metal fumes and acid fumes	Metal and acid wastes	Polishing sludges, hot dip tank dross, and etching sludges

3.6   Joining Operations

Joining operations generally involve the placement of components in intimate contact, and then securing their relative positions through the addition of an adhesive/ filler material or through the application of heat to promote diffusive bonding. Joining operations produce a variety of wastes and consume considerable amounts of energy.

With adhesive bonding, epoxy resins may be used to join metallic or non-metallic materials to one another. Some of these adhesives are known to generate VOCs and produce other airborne contaminants. Components that are to be joined via an adhesive bond often require some surface preparation (e.g., cleaning and degreasing) to ensure adequate/intimate contact between the adhesive material and the components -- also having an environmental consequence.

Soldering and brazing operations add a molten filler material to the gap between components which when solidified holds components together. Much attention has recently been placed on the development/use of alternative solder materials because of the environmental consequences of lead within traditional (lead-tin) solders. Mechanical joining uses physical means to hold components together. Examples include: screws, snaps, rivets, and bolts. The environmental effects of such operations are largely associated with the energy required to perform the operation and the manufacture of the fasteners. It may be noted that unlike many other joining methods, mechanical joining operations can often be reversed, which may offer some environmental advantage in terms of product design characteristics.

Among the most common sets of joining operations are the welding processes. Welding operations can produce the following emissions: carbon monoxide, nitrogen oxides, ozone, dusts, and metallic fumes. Additional wastes that may be generated include used filler rods and electrodes, heat, and electromagnetic radiation (possibly resulting in retinal damage). The most frequently performed welding operation is arc welding. NIOSH (1988) has reported a number of health-related concerns associated with this process. For example, fumes and other airborne particulates produced as a result of condensation of vapors from base metals, coatings, and fluxes pose a health risk. It has been reported that this airborne particulate matter may include: lead, cadmium, cobalt, copper, manganese, and nickel. Other reported airborne contaminants include silica and fluoride compounds.

4.   ENVIRONMENTALLY RESPONSIBLE MANUFACTURING

In the previous section, the by-products of several different classes of manufacturing processes were examined. It was noted that many of these waste streams had environmental, health, and/or safety consequences. The energy efficiency (or lack thereof) of a few of the operations was noted. Attention now turns to identifying actions that may be employed to improve the EHS performance of these manufacturing operations.

As organizations examine their role and the impact of their actions on the environment, Figure 6 illustrates the trade-off between cost and environmental degrees of freedom. Generally, the earlier in the product and process design that the environment is considered, the more options that are available and the more environmentally benign will be the resulting decisions. Of course, changes made early in the design activity typically cost much less than changes made later. These facts have been reported elsewhere (DeVor et al., 1992) for characteristics such as quality and ease of manufacture.   One could conclude from Figure 6, therefore, that organizational efforts should be placed on factoring environmental considerations into the conceptual design of products. While this is the direction where organizations should be headed in the future, and concepts such as product stewardship and internalized environmental costs adopted (Graedel and Allenby, 1995), manufacturers must deal at present with environmental issues/problems directly resulting from their manufacturing operations. In addition, they must be attentive to the health and safety issues surrounding their processes.

This section provides both general actions as well as specific examples that are directed at the goal of environmentally responsible manufacturing. This goal may be achieved by adopting practices that reduce the consumption of hazardous materials, energy, and other resources, and reduce risks to worker health and safety.   Working towards the goal of environmentally responsible manufacturing not only has the benefit of improving EHS performance, but also cost effectiveness.

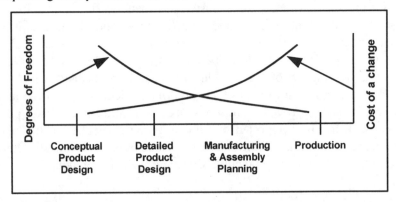

Figure 6: Environmental Degrees of Freedom and Change Costs at
Various Points in the Design of the Product and Process

From a manufacturing standpoint, there are at least 5 general actions that can be taken to be more environmentally responsible. These techniques are
1. Alternative Process Plan
2. Alternative Process Sequence
3. Process Change
4. By-product Utilization
5. Waste Handling

The actions listed above have been ordered in what would typically be the most effective (and probably most difficult) to the least effective (but perhaps easiest to achieve in the short term) alternatives. Additional detail about each action is described below. Furthermore, specific examples are given for each environmentally responsible manufacturing technique.

4.1 Alternative Process Plan

The set of operations that will be employed to produce a manufactured product are generally specified once sufficient detail is available about the dimensions and specifications of the product. There is generally significant flexibility in deciding which operations (and their order) to employ in creating a product. This process-planning task has historically been based primarily on the economics and cycle times associated with the various operations. Recently, process planners have been asked to place increased emphasis on quality during their decision-making. The goals of design for manufacturing and concurrent engineering have necessitated that product designers likewise consider how product-related characteristics will impact the downstream decisions of manufacturing planners. The objective of environmentally responsible manufacturing requires that all product and process decision-makers consider new metrics in addition to the ones traditionally considered.

As noted, process planners generally have some flexibility in deciding how a product is to be created. It is quite possible that one candidate process sequence could be preferable to another in terms of EHS characteristics. Figure 7 illustrates this point with two candidate process plans. Assuming that both plans are comparable in terms of cost, cycle time, quality, etc., the figure indicates that Plan 2 is preferable to Plan 1 because it is more environmentally responsible. This may mean that Plan 2 is more energy efficient, produces less hazardous waste, consumes fewer resources, and/or has a lower risk of endangering the health of the workforce. In general, process plans could involve completely different sets of operations, or they may be the same except for a single operation. Eliminating an operation from a process plan would be considered equivalent to creating a new process plan (this could involve changing the settings of the remaining operations).

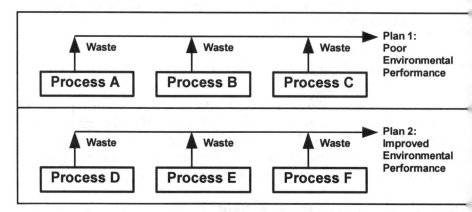

Figure 7: Alternative Process Plans Producing
Different Environmental Performance

Examples that illustrate this technique include: avoiding a painting operation by employing a grinding operation that produces an aesthetically pleasing product surface appearance; casting features into a product using cores so as to avoid machining operations; utilization of hard turning techniques to eliminate a grinding step; and net-shape forging followed by a finish machining pass as an alternative to multiple machining passes. It is worth noting that a more environmentally responsible process plan may also produce parts at lower cost or have reduced cycle times.

4.2   Alternative Process Sequence

In addition to process planners specifying the processes within a plan, they may also have some discretion in varying the order of the operations. In the context of a plan such as that of Plan 2 in Figure 7, this may mean that the process order is switched from D-E-F to D-F-E. While such a switch would mean that the same operations are performed, this does not necessarily mean that the EHS characteristics of the resulting plan would be the same. For example, consider a product that is presently laser welded following a machining operation. Because cutting fluid remains on the part after machining, it is vaporized during welding and subsequently condenses to form a mist that represents a risk to worker health. It may be possible to consider switching the cutting and welding operations to avoid this problem. Another relatively simple example could be the movement of an inspection operation earlier into a process sequence to avoid wasteful processing of defective products.

4.3   Process Change

The previous techniques have focused on the set of operations (and their sequence) that are used to manufacture a given product. With the process plan specified, decision-making degrees of freedom are reduced to the point that environmental improvement must be sought within an operation. The problem at hand is how to responsibly perform a given process given the degrees of freedom available only at the process. This means that a critical examination of such operation characteristics as process settings, inputs, and process procedure are required:

A. Process Method. Even though a given process has been specified, this does not mean that there is no latitude in how the operation is to be performed. Certainly, those methods and equipment types that provide for better EHS performance should be favored, assuming they still affect the same desired change to the product being operated upon. For example, Hands et al. (1996) report that machine tools with original equipment manufacture (OEM) installed enclosures are better than retro-fitted structures at containing fluid mist. So, assuming equal capability of two machines at performing a machining operation, it is preferable to use the one with the OEM enclosure. As another example, given the option of selecting from comparable machines, one should consider their energy efficiency. As a final example, some organizations are changing their traditional air spray painting method to an electrostatic finishing method in order to be less wasteful.

B. Process Inputs.   Manufacturing operations take some input component and modify it in a particular way to accomplish their objective. Often, other materials (apart from the product material itself) are used to achieve this goal. The mission of environmentally responsible manufacturing requires that these materials should be selected wisely. For example, in part cleaning operations, alkali-based systems may be preferable to solvent-based systems; changing the type of core sand used in a casting operation may allow the core sand to be recycled; and some stamping lubricants can remain on a part and be burned off during a subsequent heat treat operation as opposed to using a lubricant that requires a washing operation.

C. Processing Conditions. This refers to such characteristics as machine settings and the ambient conditions under which an operation is performed. Again, these conditions should be selected with consideration for the environment along with attention to other factors such as cost, cycle time, and quality. Examples include: increasing the drip time for a part coming out of a plating bath; reducing the temperature hold time for a compression molding operation to a minimum so as to reduce energy consumption; and reducing the amount of metalworking fluid applied during a cutting operation.

D. Material Recycling. Many operations employ materials that can, in principle, be used over and over again in the process, assuming that they are recaptured. Unlike the previous three points, this has less to do with the planning of an operation than it does with process operation. Obviously, if materials are recaptured and reused, they are not disposed of to the environment as quickly and less recharging will be required -- so, in-process recycling should be established where it does not exist. Furthermore, since many operations already use some form of in-process recycling (e.g., part washers), maintenance of existing systems is also critical to maximize the life of the recyclate. Capture and reuse of paint run-off is becoming more widespread as is the recycling of sand from casting processes. Evaporative losses of fluid recyclate should be minimized. Filtration technology may be one way to effectively prolong the life of a fluid recyclate.

E. Process Control. As a visit to any plant or job shop would reveal, there are often significant differences between process theory and practice. Rarely do processes run exactly as they were envisioned on the drawing board, either through lack of discipline or changes made in response to production needs. It is important to remember, therefore, that such changes may have a concomitant deleterious EHS effect. For example: workers may remove their respirators, eye protection, or protective clothing; operations may not be maintained in a state of statistical control and therefore produce a large number of defective products; and machine enclosures may be left open to permit easy access. For such occurrences, it is generally desirable to return the operation to that of the original plan.

Attention to the above referenced process characteristics often represent the first step in an organization's journey to environmentally responsible manufacturing.

4.4   By-product Utilization

Even if attention has been directed to the process plan and operation study in order to minimize waste, it is very likely that some by-products (i.e., materials not directly associated with the product) will still be generated. This may be because in-process recycling is not possible, or is not 100% efficient, or it may be due to the process technology being employed. Often, there is some inherent value in this waste (by-product) assuming that it can be collected and concentrated efficiently. One challenge is often identifying a suitable application for the by-product. Examples of this technique include: reuse of scrap or flash from plastics processing operations; burning used oil (cutting fluid, heat treat) as fuel; selling used waste pickling acids as feedstock for fertilizer manufacture; using non-hazardous waste foundry sand as construction material; and recovery of waste heat from operations to supplement a facility's heating requirements.

4.5   Waste Handling

Of course, there will always be situations where waste cannot be prevented through judicious process selection and optimal process operation, or utilized as a by-product elsewhere. For such situations, it is prudent to focus on those actions that can minimize the waste problem. This can be a serious issue in terms of costs -- a large drum of water contaminated with oil and metal fines may need to be handled as hazardous waste. Changes being contemplated by the Federal government with regard to regulations associated with wastewater treatment facilities may mean even higher costs. Anecdotes about mixed aluminum and cast iron chips in a moist environment resulting in a thermitic event are common. These examples all point to the need for care in waste segregation/separation. Some success can be achieved simply through care or sorting schemes to avoid contamination. Other cases may require mechanical separators (e.g., filters, cyclones, centrifuges, settling tanks) or other means (e.g., chemical, electro-magnetic) to segregate the waste into its constituents. It is possible that such approaches may produce materials of sufficient purity that they can be recycled or find utility elsewhere.

4.6   Education/Training

In addition to the five general techniques/approaches described in this section for achieving environmentally responsible manufacturing, an even more fundamental action is required: education/training. No plan or strategy for improving the environment can be successful without management commitment/involvement and buy-in from the workers-- all of this requires significant education. All personnel must come to at least accept the importance of EHS issues and the fact that long-term organizational success requires efforts to address these issues. It has been stated previously that perhaps the most effective way to bring about positive environmental change is early in the design of any system when the degrees of freedom are the largest and perhaps the least amount of information is available. It is at this point, where system concepts are defined and philosophical approaches are established, that education and training can avoid environmental problems.

It is envisioned that a successful environment education program will exist at three levels: i) system view, ii) technical issue focus, and iii) environmental practice. The "system view" would be an overview of EHS issues and would address how these issues affect the competitiveness of an organization. This overview should be suitable for all employees within an organization. For scientists, engineers, and decision-makers focused on product design, manufacturing planning, purchasing, plant layout, production, industrial hygiene, and waste treatment, the second level -- "technical issue focus" would be of interest. Here, emphasis would be placed on the definition of EHS metrics and how decisions affect them. The third level, "environmental practice" would be directed at production personnel and would address their role in realizing organizational environmental objectives.

5. SUMMARY AND CONCLUSIONS

This chapter has focused on several issues related to environmentally responsible manufacturing. It has outlined a three-step methodology for manufacturing operation study that has as its goal an understanding of process waste streams and energy requirements: i) process inventory, ii) quantification, and iii) modeling. Six general process categories were then discussed and the process inventory concept was applied to each one. Several processes were examined in detail, and available knowledge concerning their waste streams was presented. Attention then turned to identifying actions/techniques that promote environmentally responsible manufacturing. These actions called for attention to: i) alternative process plans, ii) alternative process sequences, iii) process changes, iv) by-product utilization, and v) waste handling. The importance of education and training on environment-related issues was identified as a critical overarching organizational principle.

In closing, a retrospective look at manufacturing reveals that throughout history, many organizations have failed because of their inability to respond to trends in the marketplace. In the early 1900's companies that did not implement the principles of scientific management developed by men such as F. W. Taylor went by the wayside. In the 1970's and early 1980's this phenomenon occurred again. Japanese companies, practicing the quality principles advocated by men such as W. E. Deming, entered the U.S. marketplace. The Japanese companies were playing by a different set of rules, and U.S. companies that failed to adopt these new ideas became uncompetitive and went out of business. Today, many international companies have begun to make significant progress in internalizing environmental considerations into their business practices. In today's global marketplace, U.S. manufacturers are competing head-to-head with these international companies that offer environmental advantages. It remains to be seen whether the U.S. manufacturers will adopt the principles of environmentally responsible manufacturing or end up as historical footnotes.

REFERENCES

Choi, A. C. K., H. Kaebernick, and W. H. Lai (1997). "Manufacturing Processes Modeling for Environmental Impact Assessment." *Journal of Materials Processing Technology.* Vol. 70, pp. 231-238.

DeVor, R. E., T. H. Chang, and J. W. Sutherland (1992). *Statistical Quality Design and Control: Contemporary Concepts and Methods.* Macmillan.

EPA Office of Compliance Sector Notebook Project: Profile of the Rubber and Plastics Industry (1995a). EPA/310-R-95-016.

EPA Office of Compliance Sector Notebook Project: Profile of the Fabricated Metal Products Industry (1995b). EPA/310-R-95-007.

EPA Office of Compliance Sector Notebook Project: Profile of the Metal Casting Industry (1997). EPA/310-R-97-004.

Graedel, T. E. and B. R. Allenby (1995). *Industrial Ecology.* Prentice Hall.

Hands, D., M. J. Sheehan, B. Wong, H. Lick. (1996). "Comparison of Metalworking Fluid Mist Exposures from Machining with Different Levels of Machine Enclosure." *American Industrial Hygiene Association Journal.* Vol. 57, No. 12, pp. 1173-1178.

Kalpakjian, S. (1992). *Manufacturing Engineering and Technology.* Addison-Wesley.

NIOSH Criteria for a Recommended Standard: Welding, Brazing, and Thermal Cutting. (1988). U.S. Dept. of Health and Human Services, NIOSH, Pub. No. 88-110.

Olson, W. W. and J. W. Sutherland. (1993). "Research Issues in Demanufacturing." *Transactions of NAMRI/ SME.* Vol. 21, pp. 443-450.

Sheng, P., and M. Srinivasan (1996). "Green Agents for Distributed Manufacturing." *Japan-USA Symposium on Flexible Automation.* July 7-10, Vol. 2, pp. 1111-1118.

Tonshoff, H. K., R. Egger, and F. Klocke. (1996). "Environmental and Safety Aspects of Electrophysical and Electrochemical Processes." *Annals of the CIRP*. Vol. 45/2, pp. 1-16.

Yue, Y., K. L. Gunter, D. J. Michalek, and J. W. Sutherland. (1999). "An Examination of Cutting Fluid Mist Formation in Turning." *Transactions of NAMRI/ SME*. Vol. 27, pp. 221-226.

Zhang, H. C., T. C. Kuo, and H. Lu. (1997). "Environmentally Conscious Design and Manufacturing: A State-of-the-Art Survey." *Journal of Manufacturing Systems*. Vol. 16/5, pp. 352-371.

AUTHORS

Dr. John W. Sutherland received his B.S., M.S., and Ph.D. degrees from the University of Illinois at Urbana-Champaign and is presently Professor and Associate Chair of the Department of Mechanical Engineering - Engineering Mechanics at Michigan Technological University. He has an active research program that focuses on environmentally responsible manufacturing. He has authored or co-authored over 100 technical papers in various journals and conference proceedings. He was the recipient of SME's Young Manufacturing Engineer Award, SAE's Ralph R. Teetor Educational Award, Michigan Technological University's Distinguished Teaching Award, and was selected for a Presidential Early Career Award for Scientists and Engineers.

Kenneth L. Gunter received his B.S. and M.S. degrees in Mechanical Engineering from Michigan Technological University, where he is currently a Doctoral Fellow. His research, supported by the U.S. Department of Education, is directed at air quality issues in manufacturing. He is a member of ASME and SME.

Chapter 14

Environmental Decision Support Systems
A Tool for Environmentally Conscious Management

Steven P. Frysinger, Ph.D.
James Madison University

1. INTRODUCTION

As the complexity of our environmental management problems has increased, so has the need to apply the information management potential of computing technology to help environmental decision-makers with the difficult choices facing them. Environmental information systems have already taken many forms, with most based upon a relational database foundation. Such systems have helped greatly with the day-to-day operations of environmental management, such as chemical and hazardous waste tracking and reporting, but they have two critical shortcomings, which have prevented them from significantly improving the lot of environmental scientists and planners tackling more strategic decisions.

Traditional environmental information systems ignore the crucial spatial context of virtually all environmental management problems, and they offer little or no support for the dynamics of environmental systems, both manufacturing and otherwise. Fortunately, a relatively new category of system, called an Environmental Decision Support System (EDSS), shows real promise in both of these areas.

1.1 What are Environmental Decision Support Systems?

Environmental Decision Support Systems are computer systems, which help humans make environmental management decisions. They facilitate "natural intelligence" by making information available to the human in a form which maximizes the effectiveness of their cognitive decision processes, and they can take a number of forms (Guariso and Werthner, 1989)

As defined here, EDSSs are focused on specific problems and decision-makers. This sharp contrast with the general-purpose character of such software systems as Geographic Information Systems (GIS) is essential if we are to put and keep EDSSs in the hands of real decision-makers who have neither the time nor inclination to master the operational complexities of general-purpose systems. Indeed, it can be argued that most environmental specialists are in need of computer support, which provides everything that they need, but *only* what they need. This point becomes more critical when it is understood that many important "environmental" decisions in design and manufacturing, for example, are not made by environmental specialists at all, but are instead made by professionals in other disciplines.

1.2 The Need for Environmental Decision Support Systems

The development of environmental policies and generation of environmental management decisions is currently, to a large extent, an "over the counter" operation. Technical specialists are consulted by decision-makers (who may or may not have a technical background), to assist in gathering information and exploring scenarios. Because of the inaccessibility of data and modeling tools, decision-makers must consult with their technical support personnel with each new question, a time-consuming and inefficient process.

If the data and analytical tools could be placed within reach of the decision-maker, they would be able to consult them more readily, and would therefore be more likely to base their decisions upon a technical foundation. In some instances, the availability of environmental decision support determines whether or not a product design or manufacturing process will indeed be "environmentally conscious". This is the premier reason why Environmental Decision Support Systems, of a sort described in part herein, are necessary if we are to achieve a higher quality in our environmental management decisions and obtain more protection with our finite resources.

1.3 Organization of this Chapter

In order to more fully understand the characteristics and structure of an effective EDSS, we must review and evaluate the constituents of this technology, which is by its nature multidisciplinary. This will be undertaken in Section 2, which will address the foundations of environmental decision making from both a policy and a cognitive point of view, the realities of both being essential to an understanding of practical environmental management. Section 3 will address the scientific and technological environment relevant to a modern consideration of EDSS, introducing the major disciplines which are frequently an integral part of an EDSS. Section 4 will then illustrate the application of EDSS technology to environmentally conscious manufacturing.

2. FOUNDATIONS

Environmental Decision Support Systems address a problem domain of remarkable breadth, ranging from selection of an appropriate light switch for an automobile to the determination of community risk associated with stored chemicals. The character of environmental decisions and their surrounding issues is central to the design of a successful EDSS .

2.1 The Nature of Environmental Management Decisions

To understand environmental management decisions, we must first identify the decision-makers. The stereotypical image of an environmental manager is an environmentally trained business manager given the responsibility for avoiding fines and other sanctions, and perhaps pursuing "beyond compliance" goals, all within the constraints of finite (and generally tight) budgets. Indeed, many environmental decision-makers fit this description.

However, these individuals also have their counterparts in the regulatory arena (such as agency compliance officers). Furthermore, critical environmental decisions are often made by market researchers, product designers, and manufacturing process developers. Naturally, the level of environmental expertise these individuals possess is highly variable. Nonetheless, all of them can and do make critical environmental decisions. It is therefore incumbent upon the toolbuilders - including EDSS architects - to craft systems and processes that will help to bridge the gap between technical expertise and the decision-maker, so that the benefits of this expertise may be realized.

2.1.1 Characteristics of the Problem

Environmental decision-makers are clearly a diverse group of people faced with a diverse group of problems. The breadth of their problem domain, in fact, defines the need for eclectic individuals with tools to match.

In general, environmental decision problems are

- *Spatial*, in that most human activities and their environmental impacts are associated with a place having its own characteristics which influence the decision;
- *Multidisciplinary*, requiring consideration of issues crossing such seemingly disparate fields of expertise as atmospheric physics, aquatic chemistry, civil engineering, ecology, economics, geology, hydrology, toxicology, manufacturing, materials science, microbiology, oceanography, radiation physics, and risk analysis;
- *Quantitative*, because the constituent disciplines themselves are highly quantitative, and because the costs and ramifications are generally so significant, that objective metrics are desired to help mitigate controversy;
- *Uncertain*, in that while the elements are quantitative, the scarcity of data and nascent state of the constituent disciplines leaves many unknowns;
- *Quasi-procedural*, since many environmental decisions are tied to a regulatory or corporate policy framework which specifies the steps by which a decision is to be reached, and because the threat of liability dictates a defensible audit trail for the decision process;
- *Political*, reflecting the fact that environmental management is driven by public policy, influenced by such considerations as economics, social impacts, and public opinion.

The diversity of these characteristics of the problem domain make effective environmental decision support extremely challenging.

2.1.2 Implications for Environmental Decision Support

Because of these factors, it is not practical to contemplate a generic decision framework for environmental management. Even if it were possible to capture all of the elements necessary to address the great variety of decisions to be undertaken, the system so built would be virtually unusable. The environmental manager is already confronted with a vastly complex problem space; one of the first jobs of the decision support system is to simplify this space, offering them everything that they need to make the decision at hand - but *only* those things.

Therefore, while our definition of EDSS includes the integration of multiple supporting technologies (such as simulation and GIS), we further restrict this definition to stipulate that EDSSs are focused on a particular decision problem and decision-maker. Thus, they are not general-purpose tools with which anything can be done - if only you knew how to do it.

Rather, they are particularly tailored to the problem facing the analyst, and offer a user interface, which is optimized for this problem.

The focused nature of such EDSSs improves the user's interaction with the computer system, allowing the user to concentrate on the probl em at hand and the information and tools needed to solve it. It also dictates a software architecture that facilitates the development of sibling systems embracing different decision problems with an essentially common user and data interface (Fry singer, 1995). Such a family of focused EDSS siblings offers user interface simplicity, in that the siblings share interaction style, organization, and fundamental approaches (where appropriate), while maintaining the focus each sibling has on its particular decision problem.

2.2 Task Analysis of Environmental Decision Making

The focused approach to EDSS design advocated here dictates the use of a human factors engineering technique, called *task analysis*, to support the specification of a particular EDSS for a particular problem.

As defined in the human factors community, "...task analysis breaks down and evaluates a human function in terms of the abilities, skills, knowledge and attitudes required for performance of the function" (Bailey, 1982). The EDSS designer must endeavor to understand the decision problem, and all of the factors that must be considered in solving it. In addition, the "social history" of the problem must be understood, since there will (in general) already be a number of different approaches to solving a given environmental management problem. For a system to support an analyst in arriving at a credible decision, the various competing approaches must be considered, and possibly accommodated.

A major stumbling block in task analysis is the fact that very few individuals can accurately explain the way in which they actually arrive at a particular decision. They can tell you how they think they *should* do it, and they can often develop a *post hoc* analytical rationale for their decision, but people are generally unaware of the actual process by which they make decisions. Thus, other instruments must be used to understand the decision process, ranging from observation and interview up through controlled experimentation to determine the influence of different variables on decisions.

In the environmental area, this is further complicated by the fact that there are often guidelines or regulations dictating the way in which decisions are *supposed* to be made about a particular problem. These do indeed dictate certain aspects of the process, but often leave a great deal unspecified. For example, the United States' Resource Conservation and Recovery Act (RCRA) requires that a waste facility be monitored by a network including at least one upgradient and three downgradient wells in order to assure that no hazard to the public health results from the facility. However, though the legislature was specific about this detail, they made little effort to assist the manager in deciding where or how many (above four) wells are to be installed. Furthermore, the language of the act would suggest that certainty is required with respect to the detection of leaks, though no reasonable person would argue that this is either theoretically or economically achievable. Implicit in this example is the issue of uncertainty, which, because of its importance in environmental management, deserves further attention.

2.3   Management of Uncertainty

Uncertainty is implicit in environmental decision-making. Complex technical decisions must be made regarding events - both in the past and the present - that depend upon many different variables. Solutions to such problems often depend on the use of various mathematical modeling techniques. These techniques, in the main, attempt to predict the future performance of a complex system on the basis of relatively sparse empirical data. The predictions drawn from these modeling studies form the basis for the entire industrial process to follow, including such expensive decisions as the design of a product and its associated manufacturing processes. Ultimately, the environmental effectiveness of the product throughout its lifecycle, in terms of protection of human health and reduction of environmental risk, depends upon these results.

However, these modeling studies are unavoidably visited by uncertainty of various types, ranging from conceptual model uncertainty - associated with the selection of assumptions necessary to choose the model(s) - to parameter uncertainty resulting from sparse empirical data, noisy measurements, and the general difficulty associated with measuring critical parameters.

2.3.1   Sources of Uncertainty

Uncertainty in such environmental management problems exists because of a lack of empirical data, errors in the data, incorrect models, and the general non-determinism of nature.

The first of these, a lack of empirical data, is easy to understand; we routinely live with imperfect knowledge of the current state of systems, owing to lack of data (in a usable form). This and the second (errors in the data) are the ones typically addressed in scientific and engineering studies when the goal is to reduce uncertainty. The usual approach is to collect more data, and to attempt to reduce the measurement error in the data collected.

The third reason, the use of incorrect models, is recently receiving more attention in environmental management. As environmental managers come to accept that model building (whether mental or mathematical) is an essential part of problem solving, the disagreements as to which m odels are correct become more apparent. Some would argue that a model is correct to the extent that it accurately predicts the future behavior of the system; the limiting factor for environmental problems is the complexity of the system in question. And here is where an interesting human factor emerges.

As mathematical models are expanded to attempt to account for more of the fine details of the natural system under study, the mental models of the analyst become inadequate. While humans are capable of recognizing and apprehending in a *gestalt* sense the breadth of complex systems, they are ill equipped to mentally manage the myriad simultaneous details attending such systems. It can be argued that we build mathematical models precisely because we cannot manage such details mentally. Yet, as we build these models, they too become more complex than we can fully grasp, resulting in a great deal of effort and controversy associated with the development of the mathematical models. Many environmental modelers spend more time studying their models than studying the natural systems they emulate.

This problem becomes especially acute when the decision-maker is not the developer of the mathematical model, because an opportunity exists for mismatch between the analyst's mental model and the quantitative mathematical model they're attempting to use. This results in uncertainty, both subjective (i.e. lack of confidence on the part of the analyst) and objective (i.e. a measurable variability in the decisions made by several analysts or by one analyst on several occasions). Ultimately, this uncertainty finds its way into public perception, causing the public at large to wonder how to interpret the products of science and engineering (the public's awareness of the modeling debate surrounding global warming is a good example of this).

Finally, the fourth cause of uncertainty in environmental problems arises out of the nondeterministic character of the natural environment, at least as it is currently understood. We should not expect to eliminate uncertainty entirely in solving environmental problems. Like the other three, this cause of uncertainty applies to both spatial and aspatial data, and some adaptive approaches have been proposed to help analysts arrive at accurate descriptions of the uncertain natural parameters (e.g. Heger et al, 1992).

Unfortunately, humans tend to have some difficulty in reliably making probabilistic judgments (Hogarth, 1987). There is a tendency toward a "fish-eye" view of uncertainty, in that perception of unfamiliar issues or events is related to familiar ones, resulting in distortion not unlike the familiar cartoon maps showing "The New Yorker's view of the World". This is evident in studies examining human perception of risk, and applies to probabilistic judgments more generally.

Quantification of uncertainty has been widely acknowledged as a critical issue in risk assessment (see, for example, National Research Council, 1993). A variety of methods for managing uncertainty have been studied (Morgan and Henrion, 1990), most of which are beyond the scope of the present chapter. One of these, which figures prominently in EDSS , involves the use of computer simulation methods to quantify the uncertainty associated with a model result, *conditioned* on the correctness and appropriateness of the model for the problem at hand.

2.3.2   Stochastic analysis

In considering the uncertainty of quantitative models, one considers the output of the model to be some function of one or more input coefficients. These coefficients become the parameters of a numerical representation of the model. The quantitative uncertainty in the modeling solution, then, results from the combined uncertainties of the input param eters.

Stochastic analysis of uncertainty is predicated on the ability to articulate the probability distributions of each uncertain parameter and then iteratively solve one or more model equations involving these parameters. To accomplish this, samples are drawn from the parameter distributions, most often employing Monte Carlo or Latin Hypercube sampling methods.

To generate N Monte Carlo samples from a given probability distribution, one first produces the corresponding cumulative distribution function (CDF). The ordinate of the CDF, which ranges from zero to one, is then sampled uniformly, and the corresponding abscissa values are taken as pseudorandom samples of the target distribution.

Latin Hypercube sampling, a variation on the Monte Carlo method, forces the uniform samples drawn on the ordinate to cover the entire range (zero to one) by dividing the axis into N equal-width bins. From each bin a sample is drawn, with uniform sampling within each bin. This modification helps to ensure that the tails of the target distribution are sampled, and therefore can result in convergence on the target distribution in fewer samples than the unmodified Monte Carlo method.

To solve environmental models using such stochastic methods, one solves the model equation iteratively, each time using parameter values drawn from the uncertain parameter distributions by the methods just described. The set of results of these calculations form, themselves, a distribution which aggregates the uncertainty of each of the parameters, and whose characteristics can be used to describe the model. The moments and upper and lower quartile bounds of such a calculated distribution can be directly employed in decision-making based upon the model. For example, if one calculates individual exposure to radionuclides using such an approach, the CDF of the distribution of results can be used to find the probability that exposure will exceed 25 mrem/year. It has been demonstrated (McKone and Bogen, 1991) that the use of such methods can help to avoid the "creeping conservatism" which often results from the use of upper bound parameter values alone to model risk.

3. CONTRIBUTING DISCIPLINES

There are several disciplines that interact with and are integrated by Environmental Decision Support Systems as defined in this work. This section will introduce the most prominent of these, with a special focus on the particular areas of intersection and contribution. This treatment cannot be construed as a fair representation of any of these disciplines as a whole; rather, it is intended to provide a sense of the interdisciplinary nature of EDSS, and to illuminate some of the opportunities for interdisciplinary research associated with EDSS.

3.1 Environmental Science

Environmental Science is itself an interdisciplinary field, integrating Biology, Chemistry, Mathematics, and Physics in the context of environmental protection and management. There is a distinctively applied, anthropocentric orientation to Environmental Science; it differs from such fields as Ecology in that it approaches the study of our environment with an eye toward human needs and use of the environment, and therefore addresses the science, engineering, and management practices which will help to conserve environmental resources for human benefit. This is not to imply that Environmental Scientists as a lot do not place value on nature in and of itself, but that their professional lives are more focused on natural *resource* protection, where the word resource refers to human needs and wants. This distinction is significant for the present EDSS discussion only because, as a practical matter, nearly all environmental decisions are anthropocentric. Even in the relatively rare cases where economic resources are available for "pure" ecological protection or remediation, the decisions made must necessarily consider cost/benefit as best they can in order to justify the use of the limited funds. Therefore, *worth* is an important element of virtually every practical environmental decision, and its analysis is most definitely in need of assistance from EDSS technology.

The contributions of Environmental Science to EDSS begin with the basics. In some instances, we are interested in the basic science involved, with no particular environmental twist, such as the solubilities of chemicals in water, the partitioning of a chemical between the vapor or aqueous phases, the chemical equilibrium of carbon dioxide and water, or the physics of radioactive decay. In others there is a distinctly environmental angle, such as the adsorption of chemicals on soil particles, or the avian toxicity of a pesticide. The line between these two cases is blurred, which is one of the reasons that the basic sciences are so readily integrated into Environmental Science pedagogically.

Of special interest to EDSS are Environmental Science's contributions in mathematical modeling of environmental processes. In this context, Environmental Science integrates such disciplines as Geography, Hydrogeology, and Meteorology, along with various associated engineering disciplines, notably Civil and Chemical Engineering. In some fields mathematical models are employed to help discover the truth about the phenomenon under study, with the (usually optimistic) goal of arriving at *the* model which describes the way the process works. In contrast, Environmental Scientists develop models primarily in order to accurately predict the future (or sometimes past) behavior of the system, without suffering the delusion that the model works the same way the system does. Model fidelity - the degree to which the model reflects the way the system actually works - is usually of secondary concern in Environmental Science. Model robustness - the degree to which the model predicts system behavior under varying conditions consistent with the stated assumptions - is of primary concern.

The focus of environmental modeling is prediction, useful because it can help us to understand what has happened, or what will happen. Such models are central to Environmental Decision Support Systems, and in fact to environmental decision making in general. Though some environmental managers would profess to distrust models, and prefer to make predictions through some other means, they fail to realize that these other means invariably include *mental models* of the system. Mental models may not be mathematical, but they are most certainly models, and bear all of the constraints that apply to models.

These constraints can nearly all be reduced to one axiom: a model is only as good as the assumptions that accompany it. In the case of environmental models, there are always significant assumptions needed in order to apply a particular model to a particular situation. Assumptions could arise in an attempt to cope with uncertainty in future events (such as the number of inches of rain that will fall next year), or in an attempt to simplify the problem to make it more tractable (such as modeling groundwater contaminant transport in two dimensions rather than three). Assumptions in environmental models aren't *bad*; indeed, they are necessary. However, they must be made and validated consciously during model building, and not forgotten when the model is applied. Part of the role of EDSS in the application of environmental models is to help the decision-maker to acknowledge, and to an appropriate extent participate in, the assumptions made and validated. In some systems, this is accomplished by requiring the analyst to explicitly state their assumptions respecting the models to be applied.

Another multidisciplinary grouping, drawn from the Health Sciences, can be included in Environmental Science in this context, although they are not traditionally grouped together in an academic environment. Health Science is here taken to include various branches of medicine, toxicology, and epidemiology. These disciplines provide crucial information regarding the ultimate human health ramifications of the systems or actions under study. For example, this would include the first phases of risk assessment, wherein the relationships between human exposure and human health effects are explored and described. Like other aspects of Environmental Science, this (collective) discipline also contributes models to environmental decision support. These models, both analytical and empirical, assist with such tasks as dose-response calculation and uptake prediction.

3.2   Information Systems Engineering

Information Systems Engineering (meant here to include Computer Science and its kin) is also a multidisciplinary field. Not surprisingly, Information Systems Engineering and several of its associated technologies plays a key role in Environmental Decision Support Systems. We will explore four of these, which are of particular importance to EDSS.

3.2.1   Geographic Information Systems

A central feature of virtually all environmental decisions is their spatial context. Geographic Information Systems (GIS) are computer software systems which directly target the management, analysis, and display of spatial information, and which are therefore crucial in an effective EDSS.

There are many GIS packages available, differing in the details of their design. However, some key design features are common to virtually all commercial or public-domain GIS offerings. (A more complete introduction to Geographic Information Systems may be found in Runoff, 1989.)

Current GISs represent spatial information as layers of two-dimensional data encoding different spatial data elements, analogous to (and in fact derived from) the traditional mapmaker's technique of drawing different map features on separate layers of transparent material. These layers can then be overlaid in whatever combination is desired to produce a map showing those features that are of interest. For example, one might overlay a property (lot/block) map onto a soils map in order to evaluate the soils present in individual lots for septic suitability analysis. These two-dimensional layers are typically managed as one of two data types, vector and raster. Early in the history of GIS packages would use either one or the other of these two data formats, but they are now both supported in common GIS products.

The *vector* data format, as its name implies, represents spatial objects (such as building lots or soil regions) as polygons formed by sequences of vectors, or line segments, each of which is in turn represented by its endpoints (in whatever reference system, such as latitude/longitude, is convenient). Some spatial objects (such as roads or rivers) are represented simply as vector sequences, which do not close into polygons. Finally, some objects (such as drinking water wells) may be represented as a single point. While the structures discussed above represent the location of the spatial objects, they do not describe the attributes of the objects. Such attributes are typically represented in a relational database, which is linked to the spatial description by an identifier field. Thus, if one selects the polygon representing a soil region, for example by clicking the mouse within that region, the GIS would first determine the identifier of the polygon which contained the mouse pointer, and then use this identifier to extract attribute information (in this case soil classification) from the relational database. In fact, when the spatial objects are drawn on the computer screen, one or more of the attribute fields can be used to determine such drawing options as line color or type, or polygon fill color or pattern. In this way a color-coded soils map can be displayed, at the same time that the information used to produce it is available to other computer software. Foremost among the virtues of the vector approach to spatial data representation is the fact that the points (which are the building blocks of all types of spatial objects) can be expressed with a level of precision limited only by the computer's number representation. (Of course, this has no bearing on the accuracy of the data so represented.)

The *raster* data format takes an entirely different approach to spatial data storage. Data layers are represented as regular matrices, with the (normally square) cell dimensions determining the resolution of the layer. The name "raster" is related to the raster display of modern cathode ray tube (CRT) displays, which are composed of rows and columns of pixels. However, there is no actual correspondence between a GIS raster layer and a CRT's pixels; the data in one cell of a GIS raster layer can be drawn using one or more CRT pixels. In a raster representation of a soils map layer, each cell of the raster contains a value corresponding to the soil category within that cell. If the cell dimension is, for example, 30 meters, then the soil category assigned to the cell is that of the soil that dominates the 30 by 30 meter area represented. It is obviously quite a simple matter to display a color-coded soils map by mapping a raster's cell values onto the video memory's pixel values through a color lookup table. This results in display operations that are somewhat faster than can be achieved with a vector (polygon) display. Alternatively, the raster layer's cells can contain key values providing connectivity to a relational database, similar to vector systems, although this approach is used less often. In any case, the precision of the spatial representation using raster data structures is limited by the data storage available for each raster. If one wanted 10 meter rather than 30 meter resolution (supposing one had corresponding information resolution), the space required to store the layer would increase by a factor of 9.

The chief advantage of a raster data structure is the ease with which one can perform calculations oriented toward the intersection of two or more layers. For example, if one defines septic-suitable areas as those that have a sandy loam soil *and* a slope of less than 10%, one can produce a new layer by performing a cell-by-cell comparison of the soils layer with a slope layer (which itself could be produced by analyzing an elevation layer).

Such calculations are common in natural resource management, which has resulted in raster-oriented GISs dominating these fields. On the other hand, in areas where precise locations are important (such as tax maps or pipeline location), vector-oriented GISs have dominated. Since most GIS packages have migrated into a hybrid orientation, supporting both data structures and conversions between them, one no longer has to make the choice when purchasing the software, and can choose the structure appropriate for the problem at hand.

As was hinted during the foregoing discussion, GIS technology includes more than the simple storage and display of map layers. A critical component of GIS is the analytical suite, which permits calculations, comparisons, and manipulation of data layers to produce either new derived layers or simple answers. For example, given a soils layer, any competitive GIS can very simply report the area represented by a particular soil type, either as a percentage of the whole or in such units as acres, hectares, or square meters. Likewise, most GIS packages permit more sophisticated spatial statistics, such as the generation of rasters by interpolation of contour maps, or conversely the generation of contours from rasters. Such analytical capabilities differentiate GIS packages from more simple mapping packages.

These analytical capabilities have increasingly permitted GIS technology to be the basis for decision making in many contexts, not the least of these being environmental management. GIS capability is now a standard in nearly all organizations undertaking environmental analyses, with the useful side effect that many sites of interest have already compiled significant repositories of GIS data pertinent to their problems. However, GIS largely remains an over-the-counter operation. Because of relatively complicated user interfaces, exacerbated by rather breathtaking secondary memory (disk) storage requirements for GIS data, most organizations maintain something along the lines of a GIS department or group. Decision-makers, if they recognize that they have a problem that can be addressed by GIS methods, approach this group with the problem description and enter the group's service queue. For some problems, this sort of specialist attention is necessary. GIS groups tend to be staffed by individuals with considerable knowledge of cartography and the tricks necessary to manipulate map data without corrupting it. On the other hand, a good deal of GIS capability is in principle within the grasp of workers from other fields, but the tools and/or data themselves are not available. In integrating GIS technology into Environmental Decision Support Systems, we attempt to address the latter problem, not the former. For the subset of GIS -tractable problems which can be approached by the non-GIS specialist, integration into a decision tool addressing their larger problem will solve the batch oriented, over-the-counter bottleneck which more often than not results in GIS methods not being used where they might otherwise be put to good effect. Another way to think of this is to consider that EDSS can bring some elements of GIS to decision makers in such a way that they needn't know it's GIS.

3.2.2   Computer Data Representation via Graphics and Sound

While the Geographic Information Systems technology just described goes a long way toward providing display capabilities for environmental management problems, it does not satisfy all such needs. First of all, GIS displays are overwhelmingly two-dimensional in nature, with a strong bias toward representing data in map format, or "plan view". Many GIS packages also provide a so-called "2.5-dimensional" representation capability wherein map layers containing elevation information in the raster cells are drawn as surfaces from a user-specified perspective. While often useful, such displays are not by themselves adequate.

For many environmental management problems, true three-dimensional displays are helpful. For example, when evaluating the behavior of a modeled airborne contaminant plume, the analyst should be able to navigate about (and through) the three-dimensional plume in order to get a better feel for its shape and character; contour plots fail to communicate this information. Computing and displaying such volumetric renderings rests squarely within the domain of Information Systems Engineering. The algorithms required to efficiently draw, shade, and cast virtual light upon three-dimensional objects drawn on a two-dimensional computer screen are the result of considerable research in the field of computer graphics. Many of these algorithms have been known for quite some time, but the ability to use them to generate very sophisticated volumetric displays in near-real-time is relatively new, especially on common desktop computing hardware. These tools have begun to play an important role in environmental decision support, and will be integrated into EDSS platforms with increasing frequency.

More recently, however, advances in personal computing have included the development and widespread dissemination of what's been called multimedia technology. This suite of computer capabilities has added photographic and motion picture display to the more conventional computer graphics world, and has also added high-fidelity sound generation capability to the platform. The ability to include photographs (such as site familiarization photos) and videos (such as a sequence capturing the removal of a well core) has the potential to greatly enhance the information delivery potential of Environmental Decision Support Systems. The audio capabilities have an obvious use in delivery of speech (such as in on-line help or cooperative work situations), but also support the use of sound to represent quantitative information, which cannot readily be accommodated by available visual display channels. (See, for example, Kramer, 1994, or Fry singer, 1990.)

3.2.3 Supercomputing and Networking

A third area of Information Systems Engineering which has the potential to significantly impact EDSS design relates to the execution of computationally intensive models. Historically, such computations have been relegated to a segment of computer technology called supercomputers. Supercomputers may be operationally defined as computers that are both fast and expensive enough that few of them are in existence. This rather awkward definition is necessary to account for the fact that current personal computers offer a level of computational throughput which would have been considered supercomputing twenty-five years ago. It is pointless to attempt to define supercomputers in absolute performance terms, because the technology advances so rapidly as to render such boundaries obsolete in very few years.

Nonetheless, it may be presumed that no matter how fast individual workstations become, there will be still faster computers which are few in number but which are made available to a wide population. In this work, such supercomputing technology is considered in combination with digital networks because high-bandwidth data networks have made it possible to consider linking supercomputers with personal workstations in such a way that the interactive user need not be aware that computations have been "contracted out".

In some sense, this sort of approach would represent a distribution of the EDSS architecture across multiple, remotely located machines. This view is especially appropriate if one distributes the data or code repository functions as well. For example, one might keep national meteorological data in a disk farm associated with a NOAA (National Oceanic and Atmospheric Administration) supercomputing facility, which might also store and maintain modeling codes that have been submitted to a quality assurance process. An individual EDSS being used to evaluate potential emissions from a factory might make use of these data and codes, as well as the supercomputer power, to solve a local air-modeling problem. Avoiding the need to distribute the data and codes saves a considerable amount of space (which would have been redundantly consumed on every similar EDSS platform), and also reduces the risk of data (or code) contamination.

In any case, the environmental models currently in use already stretch even high-end workstation capabilities to the point that analysts might wait several days for a single iteration of a model to execute. As computer throughput increases, more iterations of the Monte Carlo simulation will be executed, and more complicated models employed. Although individual workstations can satisfy many environmental management computation requirements, there will be a need for supercomputer access in environmental decision support for quite some time.

3.2.4   Expert systems

Finally, Information Systems Engineering offers the Environmental Decision Support System a technology to help capture and deliver the knowledge of experts in particular problem domains. EDSS is predicated on the notion that *human* intelligence is needed to make responsible environmental management decisions. *Artificial* Intelligence (AI) might therefore seem anachronistic in this work. However, although Expert Systems research has indeed grown out of AI research, the connection stops there. Expert Systems offer the possibility of providing advisors to environmental analysts, for example, to help them choose the assumptions and parameters of their conceptual model of the problem (Heger et al 1992). In this sense, the interactive user has the benefit of aggregated advice from many experts who would otherwise be unavailable, but still has the last word. This area of research in EDSS is the most prospective, and much work remains to be done before it can be claimed that Expert Systems technology has contributed substantially. Nonetheless, there is great potential for a productive collaboration.

3.3   Decision Science

The term Decision Science is used here to refer collectively to the various fields of investigation, which attempt to provide quantitative (or at least controlled qualitative) structure to the decision-making process. This includes sub disciplines ranging from Statistics and Geostatistics, through Operations Research and linear programming optimization, to classical and Bayesian probability theory. While such formal decision methods are only sparingly applied in current environmental decision frameworks, it can be expected that this will increase in the future, if for no other reason than they provide some accountability for the decision process and remove some of the air of subjectivity from it.

There is a formalism associated with Decision Science, the terms of which are fairly intuitive. To begin with, a *decision* itself is a choice between alternatives. These alternatives are compared according to some *criteria*, the measurable evidence upon which the decision is to be based. A criterion can be a *factor*, which enhances or detracts from the suitability of an alternative, or a *constraint*, which limits the alternatives under consideration. In order to combine criteria for evaluation and action, one employs *decision rules*. These include procedures for aggregating criteria into a single index, along with an algorithm for comparing alternatives according to this index. Decision rules can be *choice functions* (sometimes called objective functions) or *choice heuristics*. The former provide a mathematical method for alternative comparison, typically involving some form of optimization. The latter provide an algorithm or procedure to be followed, sometimes with a stopping rule to indicate when the procedure should terminate and the solution either taken or the search abandoned. For example, if one seeks to fit a linear equation to a set of data points, one can solve the conventional linear regression equation, which sets a derivative to zero to solve for the minimum cumulative squared error. This would be a choice function. Alternatively, one can solve the equation iteratively while varying the coefficients according to some prescription, stopping either when this same error metric is "small enough" (but not necessarily a minimum) or when the number of iterations has exceeded one's patience. This choice heuristic *might* result in the same solution as the choice function, but in examples such as this one it probably will not. On the other hand, there may not be unique analytical solutions to the problem at hand, leaving heuristic approaches the only game in town.

There is usually a specific *objective* of the decision at hand, and the decision rules are structured in the context of this objective. When there are multiple criteria, which must be considered in the decision, this is termed a *multi-criteria evaluation*, in which some method for combining the criteria must be selected. More complicated is the *multi-objective* case, in which there are multiple objectives, which may be complementary or may conflict.

While there are a great many techniques available from Decision Science, two are commonly employed in environmental decision making.

3.3.1   Linear Programming

Linear programming methods are usually associated with Operations Research. They are typically applied to optimization and resource allocation problems where there are linear relationships between problem parameters, both objectives and constraints. The linear equations describing the constraints associated with decision variables are solved simultaneously to define a solution space or feasible region (in as many dimensions as there are variables). The linear objective function is then evaluated to determine its minimal or maximal value (for cost functions or benefit functions, respectively). If this optimal value, plotted in the space of the decision variables, is contained within the feasible region defined by the constraints, then an optimal, feasible solution has been found. Given this structure, linear programming solutions strongly resemble conventional (multiple) linear regression methods, solved either graphically or iteratively. These methods are frequently used in optimization problems such as cost/benefit analysis for monitoring or remediation systems, or allocation of monitoring wells along a site perimeter.

Vogel (1991) cites an example of this form of systems analysis applied to a so-called conjunctive use problem in which the best balance of water supply sources (surface and groundwater) is sought, with the goal that the total system-yield under coordinated use exceeds the sum of the yields under uncoordinated use. Given an annual water demand of K, the decision maker seeks to find optimal values of groundwater withdrawal (G) and surface water withdrawal (S) such that $G + S >= K$. Naturally, there are various constraints on both ground and surface water usage (for example, there are maximum yields from each source), which taken together form the feasible region. If the objective function of this problem can be described linearly (i.e. there is some linear combination of G and S whose coefficients describe the normalized benefits of each source of water), then the family of curves (straight lines) representing this function under various coefficient values can be plotted on the decision axes overlying the feasible region. The selection of the optimal coefficients can then be made graphically.

For more realistic problems, there are many constraint equations and terms in the objective function, preventing the use of graphical methods. However, a variety of techniques have been developed to evaluate such systems mathematically. Even in cases where the exact optimal solution is intractable, linear programming has the potential to identify a range of solutions in the neighborhood of the optimal solution.

3.3.2 Decision Trees

Decision trees are associated with a decision analytic method, which accounts for both expected value and uncertain events. A hierarchical graphical structure is used to describe the structure of the decision problem. Nodes (vertices) in the tree are either decision nodes or chance nodes, depending upon whether the branches result from the decision-maker's choice or some uncertain event, respectively. Every decision tree has as its root a decision node, which is the first decision under consideration. Every branch in the tree eventually terminates in a "leaf" representing the outcome of that particular path through the tree, with its associated probabilities and expected value. Folding back the path probabilities and expected values of chance nodes (by multiplication), one can arrive at expected values for the decision nodes, and make an optimal decision based upon this value. However, to do this one requires some metrics for expected value of each outcome, and probabilities for each branch from each chance node. Furthermore, the decision and chance alternatives must be finite; one is selecting from a particular set of decision alternatives, rather than adjusting an operating point on a continuum.

4. APPLICATIONS OF EDSS IN INDUSTRY

The foregoing has provided a foundation for Environmental Decision Support Systems in general. Though the technology has seen most of its application in natural resource management and environmental remediation, there are many opportunities to bring the power of EDSS to bear on problems in industry. Three examples will serve to illustrate this point.

4.1 Integrated Factory Decision Support

Environmental compliance within a manufacturing environment is an information-intensive pursuit, and can be facilitated by the integration of information systems and repositories (Fry singer, 1997). The value of such integration can best be illustrated in the breach. When changes are made in the configuration of a manufacturing area, such as the movement of solvent baths from one location to another, the failure to include environmental compliance managers in the decision process can result in permit violations (for example, the movement of VOC sources from one vent stack to another can require air permit modifications). In general, layout of the factory floor can impact environmental performance, as well as compliance, so that one must consider environmental ramifications when attempting to develop and/or modify the manufacturing layout.

Geographic Information Systems (GIS), and their cousins the Computer Aided Design (CAD) systems, have been involved in facilities management for some time, but their use in support of environmental management is relatively new (Douglas, 1995). Combining the spatial plant design data with relational data describing such domains as materials inventory provides the basis for integrated decision making - integrating environmental management with overall plant management functions. Combining these with decision tools and simulation capabilities allows the manager to make superior decisions about plant layout, and improve their compliance record.

Beyond physical plant arrangements, this sort of information system integration can also go a long way toward reducing the cost of regulatory compliance. For example, in the United States the Emergency Preparedness and Community Right-to-know Act (EPCRA) requires annual reporting of the quantities and whereabouts of hazardous materials, with the intent of ensuring the safety of emergency responders in the event of fire or other disaster. This simple and arguably worthwhile requirement can result in a great deal of expense to a company whose information management and decision tools are not integrated. It is typical for such companies to issue annual inventory surveys to plant personnel, who then must physically locate and record such materials so that the regulatory reports can be completed. A far better alternative is to integrate the purchasing, storeroom, and environmental compliance software systems so that the flow of materials into and within the facility is generally known at any time. This not only permits the decision support system to easily produce the annual reports required for EPCRA, but also allows regular review of materials movements and usage, which in turn can facilitate such other tasks as tracking of air emissions calculated by mass balance.

This example illustrates an EDSS emphasizing GIS and relational database technology, and especially the integration of these technologies across organizational and functional boundaries within the operation.

4.2 Risk Management Planning

In the United States, provisions of the Clean Air Act require owners or operators of a stationary air pollution source with more than a threshold quantity of a regulated substance to submit a Risk Management Plan (RMP). Among other things, this plan must describe the accidental release prevention and emergency response policies at the source, the regulated substances handled, and the worst-case release scenarios and alternative release scenarios, including administrative controls and mitigation measures to limit the distances for each reported scenario.

This regulation provides a natural application for Environmental Decision Support Systems. To plan for a release of chlorine from tank cars on a siding, for example, one must evaluate the dynamics of material movement in the siding area, including variations in quantity of chlorine present at any one time. Then atmospheric transport models must be used to predict, for each of a variety of weather conditions, where the toxic gas is likely to go, and in what concentrations. Since weather prediction and atmospheric models are attended by a great deal of uncertainty, quantitative means must be employed to manage this uncertainty. Monte Carlo simulation, as described previously, can help to quantify the uncertainty given historical data upon which to base probability distributions.

This example illustrates an EDSS based upon modeling and simulation, with substantial support provided by GIS technology.

4.3 Design For Environment

A third example of the use of Environmental Decision Support Systems in the industrial context involves supporting the decision making process engaged in by product and process designers intending to minimize the environmental impact of their product. Designing for Environment (DFE) requires the availability of a great deal of information regarding alternative materials, components, and processes available for consideration by the designer. Such information is notoriously difficult to find, and when available its applicability to different situations is quite variable. To support the designer adequately, the system must make this information available for ready access, but it must also help the user to select only the information appropriate to the problem at hand, and perhaps also assist in the actual design decisions.

By integrating an expert system with a highly descriptive relational database, the EDSS can meet this need. Expert assistance (even if delivered by computer) is very appropriate in this situation, especially if one considers that the designer very likely has little training in environmental issues.

5.   CONCLUSION

Environmental Decision Support Systems, defined here as a class of information systems integrating several technologies in support of improved environmental decision quality, have served well in a variety of applications in natural resource management and environmental remediation. They offer similar benefits to the industrial environmental manager prepared to invest in their deployment. While not turn-key, off-the-shelf solutions, such systems, once developed, can earn their keep by helping to solve problems, which might otherwise be intractable.

REFERENCES

Runoff, S. (1989). Geographic Information Systems A Management Perspective. Ottawa: WDL Publications.

Bailey, R.W. (1982). *Human Performance Engineering.* Prentice Hall, London.

Douglas, W. J. (1995). *Environmental GIS: Applications to Industrial Facilities.* Lewis Publishers, Boca Raton, Florida.

Fry singer, S. P. (1990). Applied Research in Auditory Data Representation. In E. J. Farrell (Ed.), *Extracting Meaning From Complex Data - Proceedings of the SPIE/SPSE Symposium on Electronic Imaging.*

Fry singer, S. P. (1995). An Open Architecture for Environmental Decision Support. *International Journal of Microcomputers in Civil Engineering,* Vol. 10, No. 2, pp. 119-126.

Fry singer, S. P. (1997). New Approaches to Environmental Information and Decision Support Systems, in the National Association for Environmental Management's *Environmental Management Forum,* October 28-31, 1997, Dallas, Texas.

Guariso, G. and Werthner, H. (1989) *Environmental Decision Support Systems.* Ellis Horwood Books, Chichester, England.

Heger, A.S., Duran, F.A., Fry singer, S.P., and Cox, R.G. (1992) Treatment of Human-Computer Interface in a Decision Support System. In *IEEE International Conference on Systems, Man, and Cybernetics,* pp. 837-841.

Hogarth, R. M. (1987). *Judgement and Choice.* New York: Wiley.

Kramer, G. (1994). *Auditory Display: Sonification, Audification, and Auditory Interfaces.* Proceedings of the 1992 International Conference on Auditory Display, AddisonWesley.

McKone, T.E. & Bogen, K.T. (1991). Predicting the Uncertainties in Risk Assessment. *Environmental Science and Technology,* Vol. 25, No. 10, pp. 1674-1681.

Morgan, M. G., & Henrion, M. (1990). *Uncertainty.* Cambridge: Cambridge University Press.

National Research Council. (1993). *Issues in Risk Assessment.* Washington, DC: National Academy Press.

Vogel, R. M. (1991) Resource Allocation. In R. A. Chechile & S. Carlisle (Eds.) (1991) *Environmental Decision Making: a Multidisciplinary Perspective.* New York: Van Nostrand Reinhold, pp. 156-175.

AUTHOR

Steven P. Frysinger is Associate Professor and Environment Group Coordinator in the Integrated Science and Technology (ISAT) program at James Madison University, located in Virginia's Shenandoah River valley. Serving in this capacity since 1995, he teaches courses in environmental science, environmental management, and industrial hygiene, and leads a group of faculty dedicated to interdisciplinary education in the environmental sciences. He is also a Principal Technical Staff Member with AT&T's Environment, Health and Safety organization, having joined Bell Labs in 1982. He maintains his role in AT&T part-time, focusing on efforts to integrate environmental information systems and enhance environmental management through environmental decision support system technology.

Dr. Frysinger received his Bachelor's degree in Environmental Studies, concentrating in Physics, from William Paterson College. He completed Master's degrees in Computer Science and Applied Psychology/Human Factors at Stevens Institute of Technology, and earned his Ph.D. in Environmental Sciences from Rutgers University. He has had a varied career in environmental systems engineering and management, information systems engineering, and computer/human interface research, and has authored numerous papers and presentations in these areas.

Chapter 15

Decision Models For Reverse Production System Design

Jane C. Ammons[1], David Newton[2], Matthew J. Realff[3]

[1]*School of Industrial & Systems Engineering, Georgia Tech.*
[2]*School of Chemical Engineering, Georgia Tech.*
[3]*School of Industrial and Systems Engineering, Georgia Tech*

1.   THE EVOLUTION OF PRODUCTION SYSTEMS

Projections tell us that the growing levels of resource use in the industrialized world are not sustainable. In order for the current standards of material wealth to be maintained in the face of population growth, and the desire of many countries to enjoy similar standards, the resource usage levels must fall approximately 80-90% (Young and Sachs, 1994). This requires a radical rethinking of the industrial and social infrastructures that support this wealth.

Two ideas seem key. First, material products must be replaced by more complex patterns of interaction between products and services that lead to an overall reduction in the amount of material necessary to achieve the desired living standard. For example, this may result from a shift from individual ownership of a product, such as a car, to the purchase of transportation services that involves the use of the car by multiple parties. This will result in the intensification of use of the materials embodied in the car and an overall reduction in material needed to deliver the service of transportation.

Second, the materials embodied in products must re-circulate through the system rather than be discarded after use. Figure 1 shows the mode of material use employed for much of the 20th Century. This is a linear production system. First, virgin raw materials (oil, minerals, wood) are extracted, followed by their conversion by the manufacturing complex into products. After the products are used, they are discarded to landfill or at best converted into energy.

Figure 1. Linear Industrial and Consumer System

As an alternative, Figure 2 shows a closed-loop production system in which materials are recovered after use and put back into the industrial complex for the manufacture of new products. This second industrial paradigm is already practiced in some industries for some materials and products, such as cars and steel, beverage cans and aluminum, and nuclear power plants and uranium. A movement to a service-based economy and the closing of the production and production cycle complement one another. For example, a shift to leasing of products rather than outright ownership potentially makes it easier to manage the recovery of the material at the end of the product life. Similarly, designing products for easier maintenance, upgradability and eventual recycle may raise their material content and value, but this is no longer a competitive disadvantage since the material will only be "loaned" to the customer and its inherent value retained by the manufacturer.

Figure 2. Closed Loop Industrial and Consumer System

The transition of industrial and consumer infrastructure from the linear to the closed loop will be continuous. It needs to be managed to avoid massive displacement of people and wasted capital. Furthermore, it is not clear what the structure and scale of the closed loop system should be for maximum efficiency.

The current infrastructure favors the use of economies of scale in which large deposits of relatively concentrated materials are mined, converted into primary materials, and then further manufactured into products. In the closed loop system, however, the material reservoirs will be the places of use themselves, and hence will tend to correlate with areas of high density population. This suggests that a different topography of resource distribution will come to dominate the 21st Century and with it will come a new set of questions for industrial systems.

Figure 3 is an abstraction of forward and reverse production systems (RPS). The overall cycle shows that in the forward direction the manufactured value increases, but in the reverse direction the manufactured value is reduced as the value-added operations are "undone." The driving forces of recycling are the recovery of manufactured value, in a form in which re-use is possible, and the avoidance of waste disposal costs. These benefits must be balanced against the costs associated with sorting, various levels of demanufacturing and material recycling, and the new logistics arcs that created to connect these activities.

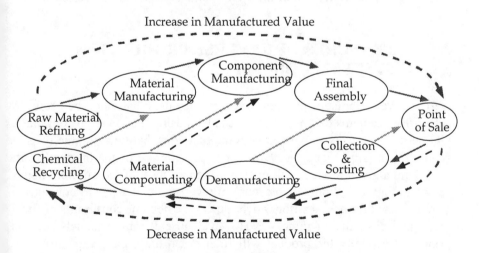

Figure 3. Material Flows in Forward and Reverse Production Systems

As the transition of industrial and consumer infrastructure from the linear to the closed loop evolves, key questions emerge. Will the reverse production system have to configure itself to match the existing forward component? Or will the long term economics drive the forward system to reconfigure itself to reflect the new sources of material?

These are profound system questions for the industry of the next century. In this chapter we will describe an quantitative approach for studying the questions of designing reverse production systems for the transition period from linear production systems to closed-loop ones. We assume that the products and their means of production will stay the same, but ask how to maximize the revenue of the reverse system as it connects to the forward production system. We will illustrate three issues that we consider to be important using industrial case studies of electronics and carpet reverse production systems design. These issues are

- *The level of aggregation of components of the infrastructure* – should sorting and preprocessing be carried out in a localized or centralized fashion, or at an intermediate level?
- *The uncertainty of supply and price* – how sensitive are our network designs to changes in the volume of material that can be recovered and the price of the material into which we can recycle it?
- *Cooperation of manufacturers* – given that materials occur together in products, much as they do in crude oil, what are the advantages for current secondary material companies collaborate in the collection of different products from the new reservoirs ?

2.   FEATURES OF REVERSE PRODUCTION SYSTEM PROBLEMS

The reverse logistics problem has many features that could be useful to categorize problems that might then be solved using the same type of model. One set of features is associated with the functions that lie along the reverse supply chain [*e.g.*, Flapper (1996)], including the use of the product itself. In general a given product will cycle through a subset of the different nodes shown in Figure 3. For example, currently very little, if any, carpet is reused or refurbished for use as a floor cover due to the obvious constraints of fixed shape and difficulty of removal. However, in electronic products such as printers, frequently the product will filter down the organization from the high-end users to those for whom the product function is less critical or regarded of lower value. In both of these products, each of the functions, reuse, refurbishment, component recycling, or material recycling involves one or both of the generic tasks of demanufacturing and remanufacturing.

Demanufacturing covers the set of specific tasks such as simple sorting, disassembly to various levels of subassemblies, or material separation to various levels of purity. *Remanufacturing* covers repackaging, repairing, reassembly and material compounding or synthesis. In addition, for these products, the reverse chain is prefixed by a collection task and suffixed by a redistribution one; only the nature of what can be redistributed changes.

We can identify two very distinct classes of reverse production problems. In the first, the functional chain of tasks is carried out without removal of the product from its current location. In this case the reverse logistics problem constitutes moving the functional chain to the location. Most examples of this are essentially repair, renovation, or refurbishment of large immobile or expensive to move structures. In the second, and more common case, the product is removed and enters the function chain where each function may be located in different places. The mathematical model given later in this paper will cover only the latter class of problems, although there could be products for which both types of logistics and manufacturing are involved.

The generic structure of the RPS system includes:

a) *the routes* for products and materials to take through the potential task network, *e.g.* reuse followed by refurbishment versus disassembly and material recycling;

b) *the allocation of functions* in the reverse chain to geographic locations, *i.e.* the physical length of the chain versus its functional length;

c) *the number and size of collection sites and processing sites*;

d) *the amount of material to allocate* to each potential end-use; and

e) *the modes of transportation* used to connect the physical sites.

Within this latter class of problems we can further refine our problem description by intersecting the features of the functional chain with the general properties of manufacturing and logistics problems which include characteristics corresponding to time, uncertainty, integrality, nonlinearity and complex dependencies.

There are two key features of the product that will help shape the structure of its reverse logistics system: 1) average and variability in frequency of product retirements, and 2) complexity both in materials and manufacture. The delineation of typical products according to these features illustrated in Figure 4.

The frequency of product retirements is dependent on two further features of the product, the length of use and the number in circulation. For example, food product packages like aluminum cans and PET (polyethylene teraphthalate) soda bottles have relatively short use cycles measured on the order of a few weeks, and newspapers with a daily use cycle. These occur in very high numbers, thus the frequency of product retirement is high. However, consumer durable goods, such as carpets can have use cycles from 5-10 years, and occur in medium numbers, leading to a lower frequency of product retirement. At the very low end of frequency, we have highly specialized types of products such as airplanes, ships and oil rigs.

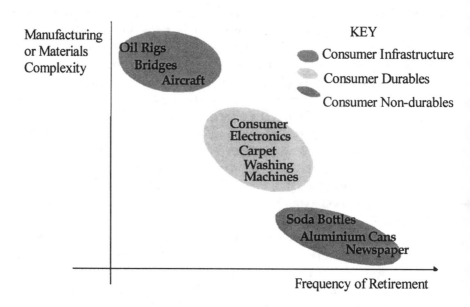

Figure 4. Product Features That Help Shape Reverse Logistics Systems

We hypothesize that the variability in product retirement decreases along the axes of frequency of retirement and number in circulation. Currently we support this with the following qualitative argument. If the frequency of retirement is high, then we have either a very short use time period or a very large number or both. If the number in circulation is high, then the law of large numbers acts to reduce the variation in the average return. In the case of short use cycles, aggregation of the items can be employed as an effective strategy to reduce overall average variability. This aggregation can be either in time, by holding inventory, or in space, by directing flows from many collection points to few reprocessing centers. For products with long use cycles that occur in low numbers, the exact converse is true. For these very low frequency retirement products, we hypothesize that although these numbers may be highly variable, we can track the low number of such items. Thus, collection effort will be minimal and that the infrastructure used to manufacture the items can be utilized to recycle them in many cases.

The complexity in manufacture and materials content of products will impact RPS structural issues (a) (b) and (c). Aluminum cans are essentially a single material, and beverage bottles have been redesigned to be only PET, but carpet is a close intermingling of at least four different materials, including a thermoset and a thermoplastic. If the manufacturing complexity is high, then the value of the product is implicit in its form and we will want to preserve that value in the RPS chain by as much reuse, refurbishment and sub-assembly and component recycling as possible. If the manufacturing complexity is low but the material value high, then we will want to ensure as efficient material recycling as possible without concern to preserve the form of the product. Although these look like divergent problem classes, we believe that by modeling the task structure of the RPS chain in a generic fashion they can be captured by the same model form and the final RPS structure driven by an appropriate objective function.

In this chapter, the reverse production system is envisioned as a network of recycling tasks that can be in different locations and interconnected by different modes of transportation. We are concerned with *strategic infrastructure decisions*, which we define to be the time evolution of the nature and capacity of the various recycling and transportation tasks and the material products of the system. The objective is to maximize the net revenue derived from the recycling system. We believe this to be a more useful objective than that of minimizing some measure of environmental impact. If environmental burdens, such as CO_2 emissions, are represented, they can be included alternatively as 1) terms in the objective, if subject to taxation, or 2) as constraints on quantities, if subject to regulation.

The recycling tasks are represented by simple transformations of material between states. A mass balance is assumed to hold across the task and the cost of performing the task is assumed to be linear in the quantity processed. In addition, there are fixed costs associated with the purchase of the equipment, the initial set up of the task, and the associated facility. A typical network of recycling tasks will start with collection tasks followed by sorting, disassembly, material recovery and final purification. The transportation tasks link locations and carry specific material states that are generated by the recycling tasks. They are assumed to be linear in the mass of material carried, a reasonable assumption at this strategic level, and the cost is dependent on the material type. The overall model has many similarities to the state-task network formulation used in chemical engineering scheduling problems (Pantelides, 1996).

An example of a network of recycling and transportation tasks in given in Figure 5. In this example, the sites are divided into two different types, collection and processing. The sorting task can be carried out in either type of location, and a solution may consist of a mixture of sorting done at remote collection points and centralized sorting at processing sites. There can be many collection sites and the location, number and when to open them is to be determined by the model. The transportation can be in two different modes, rail and truck. The processing sites themselves can be of two different types and can have several different processes that could be active at a site.

3. A MATHEMATICAL MODEL FOR STRATEGIC INFRASTRUCTURE DETERMINATION

The model consists of a superset of the possible recycling and transportation tasks and a set of locations at which each of the recycling tasks could be done. Thus, the discrete decisions to be made are the selection and allocation of the recycling tasks to the locations and the selection of the transportation modes to connect the locations over time.

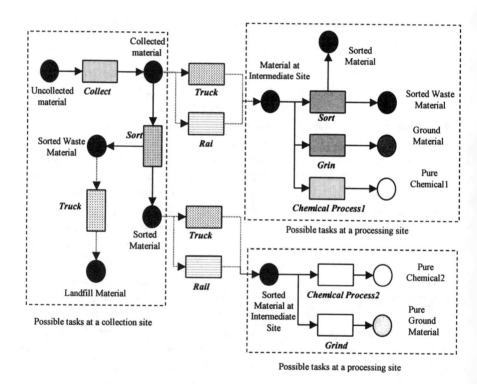

Figure 5. A Network of Recycling and Transportation Tasks

The continuous decisions are the volumes of material to be shipped and processed for delivery as final products. The objective consists of maximizing the net revenue over time, where net revenue is the recovered value minus the operating and capital expenditures. Verbally, the model can be stated as follows.

Maximize: *Net Profit* (Revenues – Operating and Fixed Costs)

Subject to: *Flow balances between sites*

(based on material consumed and produced by the tasks located at those sites).

Upper and lower bounds
on storage, transportation and processing of material at sites.
Logical constraints on sites,
such as the need to open a site before allowing tasks to be located there.

The form of the model is mixed integer linear program (MILP) and its solution can be approached through standard techniques and software. The MILP model uses the following notation for indices, decision variables, and parameters.   First, in Table 1 we define indices used to distinguish model elements and different parameters and variables.

Table 1. Model Indices and Superscripts

Symbol	Indices	Symbol	Superscripts
I	Sites	c	Collection
J	material type	r	Storage
M	transportation mode	s	Site
P	process type	h	Transportation
q	replication of recycling task	d	material sales
t	time period		

In Table 2, the notation for the decision variables used in the model is defined.

Table 2. Model Decision Variables

M_{ijt}	=	Amount of material collected of type j at site i at time period t
S_{ijt}	=	Amount of material stored of type j at site i at time period t
$H_{iji'mt}$	=	Amount of material shipped from site i to site i' of type j using transportation mode m at time period t

D_{ijt} = Amount of material sold at site i of type j at time period t

$_{ipt}$ = Extent of process p performed at site i during time period t

$y_i^{(c)}$ = 1 if collection is to be performed at site i, 0 otherwise

$y_{ii'm}^{(h)}$ = 1 if shipment is to be allowed between sites i and i' by transportation mode m, 0 otherwise

y_{ipq} = 1 if replication q of process p is to be allowed at site i, 0 otherwise

$y_i^{(r)}$ = 1 if storage is to be allowed at site i, 0 otherwise

$y_{ij}^{(d)}$ = 1 if material j is allowed to be sold at site i, 0 otherwise

$y_i^{(s)}$ = 1 if site i is to opened, 0 otherwise

Similarly, Table 3 presents the notation used to define parameters for the model. Using this notation, the mixed integer programming model for reverse production systems strategic infrastructure determination is stated in Table 4.

Table 3. Model Parameters

P_{ijt} = Price of selling material j at site i at time t

$K_i^{(r)}$ = Charge per unit per time period for storage at site i

$K_i^{(c)}$ = Charge per unit per time period for collection at site i

K_{ip} = Charge per unit flow per time period for process p at site i

$K_{ii'm}^{(h)}$ = Transportation cost in unit flow per distance from site i to i' for transportation mode m

$b_{ii'm}$ = Distance from site i to i' by transportation mode m

$f_i^{(s)}$ = Fixed cost of opening site I

$f_i^{(r)}$ = Fixed cost of storage capability at site I

$f_i^{(c)}$ = Fixed cost of collecting material at site I

f_{ip} = Fixed cost of process p at site I

$f_{ii'm}^{(h)}$ = Fixed cost for transportation from site i to site i' by transportation mode m

$e_{ijt}^{(c)}$ = Maximum capacity for collection of material type j at site i at time period t

$e_{ij}^{(d)}$ = Maximum amount of material type j that can be sold at site i in any time period

$e_i^{(r)}$ = Maximum amount of material that can be stored at site i in any time period

$e_{ii'm}^{(h)}$ = Maximum amount of material that can be shipped for site i to i' by transportation mode m

e_{ipt} = Maximum amount of material that process p can produce at site i at time period t

$\alpha_i^{(r)}$ = 1 if storage is allowed at site i, 0 otherwise

$\alpha_{ij}^{(d)}$ = 1 if material j can be sold at site i, 0 otherwise

$\alpha_{ii'm}^{(h)}$ = 1 if shipment by transportation mode m is allowed between sites i and i', 0 otherwise

α_{ip} = 1 if process p is allowed at site i, 0 otherwise

$\alpha_i^{(c)}$ = 1 if collection is allowed at site i, 0 otherwise

ρ_{jp} = proportion of material type j consumed by process p

$\rho^\bullet_{jp}$ = proportion of material type j produced by process p

4. INDUSTRIAL CASE STUDIES

Example uses of the mixed integer programming model to facilitate RPS infrastructure design are illustrated by studies for 1) an international producer and distributor high value network routing units, and 2) U.S. carpet recycling. Although the data used in the case studies are representative of general industry trends and qualitative differences, none of the numbers or networks can be interpreted as the operational reality of any existing company or system.

4.1 Case Study 1 – Network router recycling.

An electronics equipment manufacturer wants to explore the potential value of demanufacturing routers in order to recover and resale integrated circuits, gold, wiring, and certain other metals. Due to the location of its existing forward production and distribution facilities, and the location of markets where used units can be collected, the company is facing many potential locations as sites for RPS infrastructure worldwide. Thus the company is considering a global network of collection and processing for reuse and material recycling.

Table 4. Mixed Integer Programming Model for Reverse Production

Maximize   **Net Revenue**

$$\sum_t \sum_i \sum_j P_{ijt} D_{ijt} \qquad = \quad \text{Sales Revenue}$$

$$- \sum_i (f_i^{(c)} y_i^{(c)} + f_i^{(s)} y_i^{(s)} + f_i^{(r)} y_i^{(r)})$$

$$- \sum_i \sum_p \sum_q f_{ip} y_{ipq} - \sum_i \sum_{i' \neq i} \sum_m f_{ii'm}^{(h)} y_{ii'm}^{(h)} \qquad - \quad \text{Fixed Costs}$$

$$- \sum_t \sum_i \sum_j K_i^{(r)} S_{ijt} \qquad - \quad \text{Storage Costs}$$

$$- \sum_t \sum_i \sum_j K_i^{(c)} M_{ijt} - \sum_t \sum_i \sum_p K_{ip} \, \xi_{pit} \qquad - \quad \text{Collection and Processing Costs}$$

$$- \sum_t \sum_i \sum_j \sum_{i' \neq i} \sum_m K_{ii'm}^{(h)} b_{ii'm} H_{iji'mt} \qquad - \quad \text{Shipping Costs}$$

Subject to:

$$S_{ijt} = M_{ijt} + S_{ijt-1} - \sum_p \rho_{jp} \, \xi_{ipt} + \sum_p \rho'_{jp} \, \xi_{ipt}$$

$$+ \sum_{i' \neq i} \sum_m H_{i'jimt} - \sum_{i' \neq i} \sum_m H_{iji'mt} - D_{ijt} \qquad \forall \ i, j, t \qquad \text{Net conservation of flow}$$

$y_i^{(c)} \leq y_i^{(s)}$	$\forall \ i$	Logical variable requirements
$y_{ipq} \leq y_i^{(s)}$	$\forall \ i, p, q$	
$y_i^{(r)} \leq y_i^{(s)}$	$\forall \ i$	
$y_{ij}^{(d)} \leq y_i^{(s)}$	$\forall \ i, j$	
$y_i^{(c)} \leq \alpha_i^{(c)}$	$\forall \ i$	
$y_{ipq} \leq \alpha_{ip}$	$\forall \ i, p, q$	Capacity constraints
$y_i^{(r)} \leq \alpha_i^{(r)}$	$\forall \ i$	
$y_{ij}^{(d)} \leq \alpha_{ij}^{(d)}$	$\forall \ i, j$	
$y_{ii'm}^{(h)} \leq \alpha_{ii'm}^{(h)}$	$\forall \ i, i' \mid i \neq i', m$	
$y_{ipq+1} \leq y_{ipq}$	$\forall \ i, p, q$	
$M_{ijt} \leq e_{ijt}^{(c)} y_i^{(c)}$	$\forall \ i, j, t$	
$H_{iji'mt} \leq e_{ii'm}^{(h)} y_{ii'm}^{(h)}$	$\forall \ i, j, i' \mid i \neq i', m, t$	
$S_{ijt} \leq e_{ij}^{(r)} y_{ij}^{(r)}$	$\forall \ i, j, t$	
$D_{ijt} \leq e_{ij}^{(d)} y_{ij}^{(d)}$	$\forall \ i, j, t$	
$\xi_{ipt} \leq \sum_q e_{ipt} y_{ipq}$	$\forall \ i, p, t$	
$M_{ijt}, S_{ijt}, H_{iji'mt}, D_{ijt}, \xi_{ipt} \geq 0$	$\forall \ i, j, i' \neq i, p, m, t$	Variable restrictions
$y_i^{(c)}, y_{ii'm}^{(h)}, y_{ipq}, y_i^{(r)}, y_{ij}^{(d)}, y_i^{(s)} \in \{0,1\}$	$\forall \ i, j, i' \neq i, p, m, q$	

For illustration purposes, 20 potential collection sites are proposed: Atlanta, Austin, Boston, Caracas, Frankfort, Hyderabad, Istanbul, Johannesburg, Kobe, London, Los Angeles, Mexico City, New York, Paris, Rio de Janeiro, Rome, San Francisco, Singapore, Sydney, and Taipei. Seven potential sites for processing the used units are considered: Atlanta, Dublin, Hyderabad, Manila, San Francisco, Shanghai, and Singapore. Eight potential locations for end-product resale include Berlin, Kobe, London, Mexico City, New York, Rio de Janeiro, Seoul, and Singapore, although not every site is able to deal with every type of end-product. The location of these potential sites is found in Figure 6.

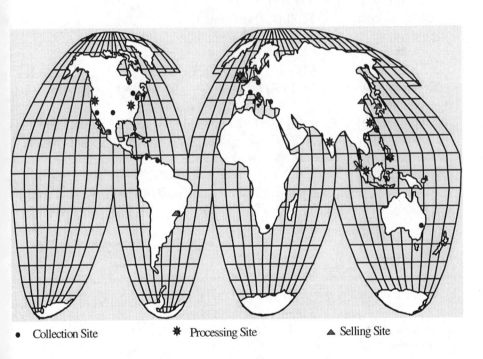

• Collection Site   ✳ Processing Site   ▲ Selling Site

Figure 6: Potential Sites For Electronics Case Study

Table 5 specifies many of the values used in the study. Prices for facilities and labor were determined partially by the location of the site. End product prices and collection volumes were a function of the specific location, whereas equipment, collection, disposal and transportation costs were assumed to be the same at every location. Since time is not considered to be a critical factor, there was no time penalty associated with longer transportation modes or with the delay associated with different processes occurring in different locations. For the purpose of this study, international tariffs, duties, and/or regulations were ignored.

Table 5. Parameter Values for Case Study 1

Parameter	Value *
Average unit size	545 kg
Average makeup:	
Gold -	0.0063% of weight
Other metals -	83.6% of weight
Wires -	10% of weight
IC chips -	12 to 13 chips
Waste -	6.0037% of weight
Waste disposal cost	$97 USD (per unit)
Collection cost	$500 USD (per unit)
Selling price:	
Gold -	$11,000 USD/Kg
Other metals -	$1.16(RD), $1.21(NY), $1.25(B), $1.32(K)USD/Kg
Wires -	$3.25(RD), $3.30(NY)(B), $3.42(K) USD/Kg
IC chips -	$0.5(S), $0.55(MC) USD/Chip
Fixed opening cost (thousand $US per year)	150(D)(H), 200(M)(Sh)(S), 250(C)(I)(MC)(Se)(T), 300(RJ), 400(At)(A)(Bo)(K)(LA)(NY)(R)(SF)(Sy), 350(F)(J), 450(L)(P)
Cost for sorting	$20,000 per year + $1.20(D)(H), $1.60(M)(Sh)(S), $3.60(L)(P) $2.00(C)(I)(MC)(Se)(T), $2.40 (J)(RJ), $2.80(F)(Sy), $3.20(At)(A)(Bo)(LA)(NY)(R)(SF) per ton
Cost for disassembly	$50,000 per year + $.033(D)(H), $.044(M)(Sh)(S), $.088(At)(SF) per Kg disassembled
Cost for process /separate metals	$3M (D)(H), $3.6(M)(Sh)(S), $6M(At)(SF) per year + $.022(D)(H), $.024(M)(Sh)(S), $.044(At)(SF) per Kg
Noxious waste cost	$102 USD (disposal/avoidance cost per unit)
Transportation costs:	
Truck -	$.00013 /Kg/mile
Ship -	$.000026 /Kg/mile

* City Abbreviations: Atlanta (At), Austin (A), Berlin (B), Boston (Bo), Caracas (C), Dublin (D), Frankfort (F), Hyderabad (H), Istanbul (I), Johannesburg (J), Kobe (K), London (L), Los Angeles (LA), Manila (M), Mexico City (MC), New York (NY), Paris (P), Rio de Janeiro (RJ), Rome (R), San Francisco (SF), Seoul (Se), Shanghai (Sh), Singapore (S), Sydney (Sy), Taipei (T)

The design of the company's RPS for the network router units is subject to uncertainty in the input data. We examine two particular sources of uncertainty, the take-back volumes that will be available, both initially and as the collection system expands, and the product design. We assume that the company is considering a product design change because, in certain required demanufacturing steps, the current design generates serious and significant amounts of noxious wastes.

This international RPS infrastructure design problem was formulated using the mixed integer programming model described above. The resulting formulation had 17,416 decision variables, or which were 2,578 binary variables. The model contained 18,199 constraints.

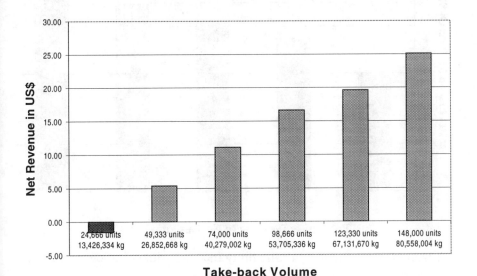

Figure 7.   Case Study 1: RPS Profitability as a Function of Take-Back Volume

The case study problem was solved using the commercially available solver AIMMS (Bisschop and Entriken, 1993). Solution times on a 200 Mhz desktop PC with 256 Mb RAM ranged from of 18 seconds to 40 minutes for various versions of the problem, with the typical times being around one minute.

To represent the uncertainty associated with initial and subsequent take-back quantities, the model was solved for different overall collection volumes. For each collection volume, an alternative infrastructure and corresponding reverse logistics solution was obtained. Figure 7 shows the sensitivity of the profitability to collection volume. The profitability breakeven point can be estimated to be approximately 25K returned network routing units per time period.

Similarly, the impact of the proposed product design change can be studied. Figure 8 shows the sensitivity of RPS profitability to relative reduction in noxious wastes generated by the recycling processes. The network solution did not change, and thus is purely a linear function based on the amount of waste reduced.

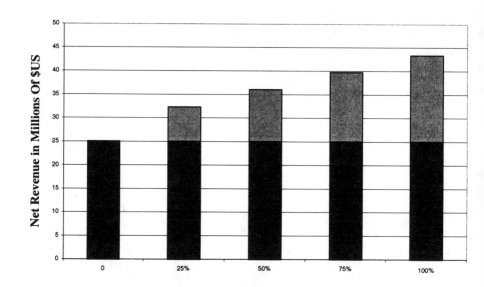

Percentage Decrease In Noxious Fumes and Waste

Figure 8. Case Study 1: Profitability Increase with Pollution Reduction

4.2   Case Study 2: Carpet Recycling

The following case study involves two hypothetical U.S. companies, A and B, that produce fiber for carpets. Carpet contains about 50% by weight of fiber and a significant portion of U.S. carpet has nylon, a polymer, as the face fiber. There are two different types of nylon fiber nylon 6 and nylon 6,6 which have different physical properties and which are composed of different monomer units. Nylon 6 is formed by polymerizing caprolactam, and nylon 6,6 by the condensation polymerization of adipic acid and hexamethylene diamene (HMD). The polymer can be converted back into its monomers by depolymerization.

Company A has developed a process for the depolymerization of nylon 6 to caprolactam, which we will refer to as DepolyA, and has a used carpet collection infrastructure partially in place. Company B has developed a different process of depolymerization of either nylon 6 or nylon 6,6 to HMD, referred to as DepolyB and has a different used carpet collection infrastructure partially in place. The decisions that remain to be taken are *what size* and *where* to establish the chemical recycling plants necessary to carry out the depolymerization processes and which collection sites to open to feed those plants. Company A is willing to build processing sites at Houston, TX, Birmingham, AL or Port Elizabeth NJ. Company B is willing to build processing sites at Bakersfield, CA, Chattanooga, TN or Kingston, ON. The potential processing and collection sites can be seen in Figure 9.

★ Potential processing sites for Company A

▲ Potential processing sites for Company B

● Potential collection sites of used carpet

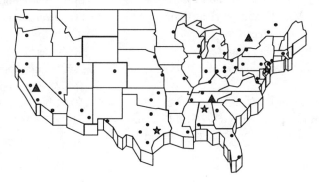

Figure 9. Geographical Distribution of Carpet Collection and Reprocessing Network

The different types of nylon carpet are sorted from each other and other carpet types by hand using relatively inexpensive equipment. The sorting can be carried out at a processing site or at the collection site. The fixed costs are relatively low at the processing sites, but at the collection site a workspace and storage facility must be used, and so $50,000 annually is added to the overhead cost of collection site sorting.

We are not aware of any depolymerization technique for either polyester or polypropylene carpeting, so carpet of this type is to be land filled, or ground into shoddy for resale. Shoddy can be used for furniture stuffing and the stream of dirt, calcium carbonate, and latex can be used as a soil enhancer. Shoddy is sold at a modest price at a processing site, while the soil enhancer is given to a local nursery or garden shop. Other waste from the processes and unwanted carpet after sorting is disposed of at a local landfill, with a corresponding tipping fee.

The cost of processing facility does not include processing costs, these can be incorporated by altering the price of the generated monomer material. Each site is limited to one depolymerization facility and Company B can also include grinding machinery at its sites. It is also assumed that either type of depolymerization process can triple its capacity by doubling the equipment cost. This is done to incorporate a notion of 'economy of scale' for the processing capability. Table 6 provides additional data used in the analysis.

Table 6. Prices Used in Scenarios.

Description	Value
Selling price of caprolactam (Schnell,1998)	$0.93 per pound
Selling price of HMD (Hexamethylene diamene)	$0.80 per pound
Selling price of soil enhancer	0
Selling price of shoddy per pound	$0.20 per pound
Cost to dispose of carpet (Lave, et al. , 1998)	-$0.025 per pound
Cost to collect	$0.02 per pound
Cost to sort used carpet	$0.01 per pound
Setup cost to sort at a processing site (60M lb. capacity)	$1,714 per annum[*]
Setup cost to sort at a non-processing site(60Mlb. capacity)	$51,714 per annum
Setup cost to grind (60M lb. capacity)	$1,534,000 per annum
Cost to grind	$0.01 per pound
Setup cost to depolymerize (100M lb. capacity)	$7,570,000 per annum
Setup cost to depolymerize (200M lb. capacity)	$11,430,000 per annum
Setup cost to depolymerize (400M lb. capacity)	$17,320,000 per annum
Cost to ship	$0.06 per ton per mile
Cost to open processing site	$1,000,000 per annum
Total amount of used carpet allowed to be collected by each company	~ 250,000,000 lbs.
Price charged by Company B to Company A for Nyon 6	$0.29 per pound
Price charged by Company A to Company B for Nyon 6,6	$0.15 per pound
Used carpet make-up	Nylon 6,6 - 43%
	Nylon 6 - 28.5%
	Polypropylene - 18%
	Polyester - 10.5%

[*]Annualized costs are based on taking the total capital costs and depreciating them over seven years to generate an annualized equivalent cost.

We will use the model to determine the *impact of cooperation* between the two companies by sharing collection resources. Scenarios 1A and 1B represent the two companies utilizing their existing infrastructure without any interaction with each other. In Scenario 2AB Company A sells sorted carpet to Company B, otherwise Company A is identical in setup to Scenario 1A, the situation is reversed in Scenario 2BA. Scenario 3 involves the exchange of material between the two companies: Company A selling Nylon 6,6 to Company B and Company B selling Nylon 6 to Company A. Scenario 4 maximizes the total net revenue, regardless of the impact on each individual company, and a processing site was required for both companies.

Table 7 reports the net revenues for the different scenarios to the nearest fifty thousand dollars. The overall conclusion is that the greater the cooperation, the more profitable the carpet recycling system becomes for both companies. In Scenario 2AB Company A has to pay a substantial premium to Company B to get its Nylon 6 since, due to the greater volumes Company B can use, it can employ better economies of scale. We used the marginal price from the optimization to determine the selling price of Nylon 6, $0.29. This explains the small increase in net revenue for Company B between scenarios 1B and 2AB.

In Scenario 2BA the price of Nylon 6,6 to Company B is substantially above the marginal value in the optimization, which is negative, since it is currently being disposed of by Company A. A price of $0.15 was chosen arbitrarily since this somewhat balanced out the increased profit of the system between A and B. The additional influx of nylon carpet enabled Company B to decrease costs by using a larger facility and taking advantage of the accompanying savings from economies of scale. The positive price and the increased efficiency accounts for the increase in profitability to both companies in scenario 2BA. Both companies share in the value of utilizing Nylon 6,6 rather than the previous disposal cost which Company A alone was paying.

Scenario 3 revenues are less than in scenario 2BA but more than 2AB since it is a mixing of the two scenarios and transportation costs are also more relevant. In scenario 4, since the overall network is optimized, transportation costs are reduced from the previous scenarios since it is no longer necessary for carpet collected from one company's network to go to it's processing plant before being resold to the other company. This is the reason for the jump in revenues between scenarios 3 and 4.

It must be stressed that the costs in the synergy case studies do not reflect actual costs, but instead illustrate ways in which our model can distinguish when and how cooperation can be mutually beneficial.

Table 7. Net Revenue of Optimal Solutions.

Scenario	Company A	Company B	A and B combined	Price of Nylon 6	Price of Nylon 6,6
1A	$11,700,000	-	-		
1B	-	$36,350,000	-		
1A + 1B	-	-	$48,100,000		
2AB	$14,250,000	$36,550,000	$50,750,000	$0.29	
2BA	$27,900,000	$46,950,000	$74,850,000		$0.15
3	$23,750,000	$45,200,000	$68,950,000	$0.29	$0.15
4	$25,700,000	$56,300,000	$82,000,000		

5.   SUMMARY AND CONCLUSIONS

Developing useful models for RPS design requires a fundamental understanding of complex system characteristics and driving forces. In this chapter, we have described key aspects of reverse production systems and used these insights to develop a mathematical programming model for strategic infrastructure design. To demonstrate the application of the model for decision making, two large scale case studies have been presented. In the first case study for the international recycling of network routers, questions were answered relative to the sensitivity of system net profit to recycling levels and to the impact of product design on net profit. The value of cooperation between two carpet recycling companies was explored in the second case study.

Currently, a transition to closed-loop production systems in underway across products, industries and geographical boundaries. Successfully accomplishing these transformations will require careful planning of reverse infrastructure and operations. The mixed integer programming decision model presented in this chapter has proved useful for exploring key questions and uncertainty surrounding RPS development.

6.   ACKNOWLEDGEMENTS

This research has been partially supported by United States National Science Foundation under grant #9800198 and the Consortium for Competitiveness in Apparel, Carpet, Apparel and Textile Industries (CCACTI) of the State of Georgia. The authors are grateful for the generous interaction and guidance provided from many industry experts, especially those from Mark Ryan of DuPont and Mike Costello of Allied Signal. We value the input from Georgia Tech Ph.D. student Selin Cerav and M.S. student Mohit Madnani. It should not be assumed that the results in the paper's case studies can be used to judge the economics of current carpet recycling efforts. Although the data is representative of general trends and qualitative differences, none of the numbers or networks described can be interpreted as the operational reality of any existing company or system.

REFERENCES

Barros, A.I, R. Dekker, and V. Scholten (1996), "Recycling Sand: A Case Study," *Proceedings of the First International Seminar on Reuse*, Eindhoven University of Technology, Eindhoven, The Netherlands, November 11-13, 23-33.

Benders, J. F. (1962), "Partitioning Procedures for Solving Mixed-Variables Programming Problems," *Numerische Mathematick 4,* 238 – 253.

Bisschop, Johannes, and R. Entriken (1993), *AIMMS, the Modeling System,* Paragon Decision Technology: Haarlem:B.V.

Carpet and Rug Institute (CRI) (1996), Carpet and Rug Institute Industry Review, CRI Dalton, Ga.

Daniels, Richard L. and Panagiotis Kouvelis (1995), "Robust Scheduling to Hedge against Processing Time Uncertainty in Single-Stage Production," *Management Science,* 41/2, 363-376.

Flapper, S. D. P. (1995), "On the Operational Aspects of Reuse," *Proceedings of the Second International Symposium on Logistics,* Nottingham, U.K., 11-12 July, 109-118.

Flapper, S. D. P. (1996), "Logistic Aspects of Reuse: An Overview," *Proceedings of the First International Working Seminar on Reuse,* Eindhoven, The Netherlands, 11-13 Nov, 109-118.

Gutierrez, Genaro J. and Panagiotis Kouvelis (1995), "A Robustness Approach to International Sourcing," *Annals of Operations Research,* 59, 165-193.

Gutierrez, Genaro J., Panagiotis Kouvelis and Abbas A. Kurawarwala (1996), "A Robustness Approach to Uncapacitated Network Design Problems," *European Journal of Operational Research,* 94, 362-376.

Huttunen, Anne (1996), "The Finnish Solution for Controlling the Recovered Paper Flows", *Proceedings of the First International Seminar on Reuse*, Eindhoven University of Technology, Eindhoven, The Netherlands, November 11-13, 177-187.

Kouvelis, Panagiotis, Abbas A. Kurawarwala, and Genaro J. Gutierrez (1992), "Algorithms for Robust Single and Multiple Period Layout Planning for Manufacturing Systems," *European Journal of Operational Research,* 63, 287-303.

Kouvelis, Panagiotis and Gang Yu (1997), *Robust Discrete Optimization and its Applications,* Dordrecht, Boston : Kluwer Academic Publishers.

Lave, Lester, N. Conway-Schempf, N., J. Harvey, D. Hart, T. Bee, C. MacCraken (1998), "Recycling Postconsumer Nylon Carpet," *Journal of Industrial Ecology,* 2/1, 117-126.

Mulvey, John M., Robert J. Vanderbei and Stavros A. Zenios (1995), "Robust Optimization of Large-Scale Systems," *Operations Research,* 43/2, 264-281.

Pantelides, C.C. (1996), "Unified Frameworks for Optimal Process Planning and Scheduling," *Proceedings Second Conference on Foundations of Computer Aided Operations (FOCAPO)*, CACHE Publications, 253-274.

Pohlen, T. L. and M. Farris II (1992), "Reverse Logistics in Plastic Recycling," *International Journal of Physical Distribution & Logistics Management*, 22/7, 35-47.

Realff, Matthew J., Jane C. Ammons, and David Newton (1999), "Carpet Recycling: Determining the Reverse Production System Design," *The Journal of Polymer-Plastics Technology and Engineering*, Polymer-Plastics Technology and Engineering, Volume 38, No. 3, 1999, pp. 547-567.

Schnell Publishing Company (1998), *Chemical Market Reporter*, 253/24, June 15.

Spengler, Th., H. Püchert, T. Penkuhn, O. Rentz (1997), "Environmental Integrated Production and Recycling Management," *European Journal of Operational Research*, 97/2, 308-326.

Young, John E., and Aaron Sachs (1994), "The Next Efficiency Revolution: Creating a Sustainable Materials Economy," *Worldwatch* Paper 121, Sept. 9-10.

Yu, Gang (1996), "On the Max-Min 0-1 Knapsack Problem with Robust Optimization Applications," *Operations Research*, 44/2, 407-415.

AUTHORS

Dr. Jane Ammons is an Associate Professor in the School of Industrial & Systems Engineering at Georgia Tech. Her areas of expertise include production systems design and analysis, including manufacturing systems and supply chains, reverse logistics, and continuous quality improvement. Dr. Ammons has worked as a plant engineer for an industrial manufacturer, is a registered Professional Engineer, and currently works with numerous companies and businesses on effective system design and operation. She is a former chair of the United States National Science Foundation Engineering Advisory Committee and is currently serving as a consultant on the Army Science Board.

Dr. Matthew J. Realff is an Assistant Professor in the School of Chemical Engineering at Georgia Tech. He received a bachelor's degree from the University of London, England, in 1986 and a Ph.D. from the Massachusetts Institute of Technology in 1992, both in chemical engineering. His field of research is process systems engineering focusing on design and operation decision modeling.

David Newton is a Ph.D. candidate in the School of Industrial and Systems Engineering at Georgia Tech. He received a bachelor's degree from Southeastern Louisiana University in Math Education in 1988 and taught high school for four and a half years. He received a Master's in Applied Mathematics from Nichols State University in 1995 and a Master's in Operations Research from Georgia Institute of Technology in 1997. His field of research is reverse production systems.

Chapter 16

Environmentally Sound Supply Chain Management
Implementation in the computer industry

Sara Beckman[1], Janet Bercovitz[2], Christine Rosen[3]
[1,3]Haas School of Business
University of California, Berkeley
[2]The Fuqua School of Business
Duke University

1.　INTRODUCTION

1.1　Industrial Ecology and Environmental Supply Chain Management

　　Traditional approaches to environmental management have focused on individual production units within the firm, and have attempted to minimize energy consumed and waste produced by each unit as a separate entity.[1] Increasingly, however, companies are adhering to the philosophy expounded by industrial ecologists and implementing plans to minimize energy consumption and waste throughout the entire life cycle of a manufactured good – from extraction of virgin materials, through processing of raw materials and energy, to parts fabrication and product assembly, and finally through the use and ultimate disposal of the product. (See, for example, O'Rourke, Connelly, and Koshland (1996), Tibbs (1991), Tibbs (1992), Lowe (1993), Richards et. al. (1994), Ayres (1993), Ehrenfeld (1994).)

---

[1] Traditional approaches to environmental management closely mirror traditional approaches to quality, cost and inventory management. Each individual unit of a manufacturing firm sought to optimize its own production activities according to the metrics it was assigned. Little if any coordination of effort across units was undertaken in the interests of optimizing for the whole.

Most industrial ecology research has concentrated on modeling the physical materials and energy flows at the heart of the industrial ecology system and on developing organizational strategies and engineering systems for optimizing those physical flows. The goal has been to find models that "connect different waste-producing processes, plants, or industries into an operating web that minimizes the total amount of industrial materials that goes into disposal sinks or is lost in intermediate processes" (Richards et. al., 1994, pg. 3). A simple industrial ecology model of the computer industry is depicted in Figure 1. Such models have made important contributions to corporate environmental management, stimulating the development of innovative approaches to environmentally sound product design and manufacturing processes.

While apt at describing the materials and energy flows through industrial ecosystems, however, industrial ecologists do not account for the market system in which these flows are negotiated and executed. Their descriptions, critical to understanding the life-cycle impact of any given product design, for example, do not address what should motivate Firm X to redesign their product, or Firm Y their process, to improve overall environmental performance. Nor do they address the means by which cross-firm financial accommodation might be made for investments to improve overall environmental performance, or how such investments might be governed.

In this paper, we draw attention to these challenges by describing the transactions that take place among the players in a dispersed supply chain as they attempt to jointly produce, use and dispose of products in an environmentally sound manner. In so doing, we hope to extend the work of the industrial ecologists, enriching and increasing the practicality of their models.

1.2   The Computer Industry Supply Chain

We focus our examination on the computer industry supply chain depicted in Figure 2. Raw material converters (e.g., steel mills, virgin plastics producers, chemicals suppliers) and fabricators (e.g., of semiconductors or raw printed circuit boards) provide basic inputs to their downstream counterparts who perform major component assembly (e.g., printed circuit board and disk drive assembly) and final product assembly (e.g., of networking cards and computers). Equipment manufacturers (e.g., of steppers, photolithography equipment) provide the machines used in fabrication and assembly processes, and numerous providers of natural resources (e.g., water, gas, and energy) feed virtually all operations.

Historically, large U.S. computer manufacturers such as DEC, HP and IBM were highly vertically integrated; manufacturing everything from the silicon logs from which silicon wafers were sliced and semiconductor chips produced to the final computer assemblies themselves. Over the past ten years, however, the industry has vertically "disintegrated," as smaller companies such as Apple Computer, Sun Microsystems and Silicon Graphics Inc., have chosen not to vertically integrate from their inception, and the larger companies have shed many of their manufacturing operations, outsourcing them to hundreds of external vendors (Beckman, 1996). As a result, building a computer now entails the work of many different companies and the environmental impact of making computer products is spread both organizationally and geographically across a wide spectrum of organizations.

The locus of environmental impact in terms of process depends upon how much of the supply chain a particular company owns, and where the activities are located along the supply chain. Those companies engaged solely in final assembly and test activities, such as Apple Computer and 3Com Corporation, have insignificant levels of environmental emissions, but primary responsibility for the use and disposal of computer products. In 1995, firms in this category reported no "Toxic Release Inventory" data (TRI, 1996), suggesting no measurable U.S. emissions or emissions below the reportable threshold. Companies engaged in more process-intensive activities further back in the supply chain, however, have much greater environmental impact. U.S. emissions from U.S.-based semiconductor fabrication facilities in 1995, for example, averaged nearly 2 million pounds. The vertically integrated computer manufacturers, such as DEC (now Compaq), Hewlett-Packard, IBM and Motorola vividly display the effects of owning much of the computer supply chain. Companies in this category averaged 1995 TRI emissions of about 6.2 million pounds.[2] (Note that in 1995 all of these companies maintained internal semiconductor fabrication capability.)

The computer industry thus lends itself well to the examination of environmental supply chain management in a vertically disintegrated setting. Assuring the environmental soundness of a product requires a full life-cycle view of its creation, use and disposal (Ehrenfeld, 1994; Ayres 1993). In the dispersed computer industry, we would expect adoption of this life-cycle view to require cooperation across the supply chain.

---

[2] Toxic Release Inventory (TRI) data is only reported for U.S.-based manufacturing facilities. This data is thus biased against those organizations that do most of their manufacturing in the U.S. We don't present these data for rigorous analytical purposes, but rather to illustrate the environmental management challenges in the computer supply chain.

1.3   Our Study

To examine this possibility, we conducted exploratory case interviews with environmental and/or procurement managers at nineteen companies in the industry (listed in Appendix A.) We targeted two "sub-supply chains" in particular: the semiconductor equipment/ semiconductor fabricator/computer assembler supply chain and the disk drive media/disk drive assembler/computer assembler supply chain. Although we talked with a handful of suppliers outside these two sub-chains to flesh out our understanding of the entire chain, focusing on these two sub-chains allowed us to build a more comprehensive "end-to-end" picture. In our interviews, we explored the pressures these companies feel to expand their environmental supply chain management efforts, the programs they have or are putting in place, and their sense of the challenges involved.

We rounded out our understanding by interviewing staff members at Semiconductor Equipment and Materials International (SEMI), a trade organization that has developed environmental standards for the semiconductor equipment industry. And, we also talked with employees of the consulting firm California Environmental Associates who have been active in the Pacific Industry and Business Association (PIBA), another trade organization that has developed industry standards relevant to our study.

2.   INCENTIVES FOR MANAGING THE ENVIRONMENTAL PERFORMANCE OF THE SUPPLY CHAIN

To frame the question of whether or not environmental supply chain management is important in the computer industry, we asked our interviewees what pressures they experienced to pay attention to environmental management issues both within their own organization as well as in those of their suppliers. The answers to these questions allowed us to build a preliminary picture of the extent to which supplier management is an important vehicle for executing environmental improvement programs, and of the paths of propagation throughout the supply chain. Through our interviews, we found two primary incentives for increased attention to environmental management programs in general, and to environmental supply chain management in particular: increasingly stringent governmental regulations, and intensified market pressure experienced largely through the proliferation of environmental standards. Here we briefly explore the nature of these pressures, and describe their effect in the organizations we studied.

2.1 Government Regulations

Traditionally, government regulations have focused on "end-of-pipe" management, requiring that manufacturers take responsibility for reducing pollution emissions and improving the management of solid and hazardous waste materials generated at their own factories. In the U.S., command and control air and water pollution and Superfund regulations enacted in the 1970s have been the most publicly apparent of these efforts, but they have been reinforced by other state, local and federal environmental laws. These regulations have been repeatedly tightened up over time and continue to be made more stringent on a regular basis, forcing manufacturers to take more and more steps to clean up their manufacturing processes. (Gottlieb and Smith, 1995; Gottlieb, Smith, and Roque, 1995; Mazurek, Gottlieb, and Roque, 1995)

The companies most concerned with these more traditional regulations are, as expected, those with the highest levels of emissions. Companies owning processes far back in the supply chain (e.g., semiconductor fabrication, disk drive media production) are acutely aware of the rapidly constricting regulations, and expend considerable effort in reducing waste from their facilities and managing waste that is produced. These efforts extend to the equipment and hazardous waste disposal vendors who support them.

Our interviewees from Intel Corporation made clear the evolving challenge. They, along with other semiconductor manufacturers, are presently developing processes that will go into full production five years from now. Although they can't yet know what environmental regulations will be in place at that time, they can certainly predict that those regulations will be tougher than today's and must specify their new processes accordingly. This, in turn, has significant impact on their equipment suppliers who must meet the challenging specifications. In this case, little of the pressure to engage in environmental management comes from purchasers of semiconductor devices; most of the pressure comes directly from increasingly strict environmental regulations.

Disk drive media manufacturers, similarly, are driven primarily by governmental regulations for the management of hazardous waste. Clear designation of manufacturers as responsible for their hazardous waste, even as it is moved out of their facilities and into the hands of hazardous waste disposal operators, forces them to pay particular attention to the operations of their hazardous waste contractors. Once again, government regulations overwhelm any impact requirements imposed by computer manufacturers might have.

Government regulators throughout the world are augmenting their more traditional emissions regulations to include product content and recyclability. Germany and the Netherlands have led the way in adopting "take-back" policies that give companies cradle-to-grave responsibility for their goods by requiring them to play a role in the ultimate disposal of their products and packaging materials (Kulik, 1994). Many governments, including those of the U.S. and several European countries, now ban or restrict the sale of products that contain hazardous materials, such as mercury-containing switches, nickel-cadmium, lead-acid or mercury-containing batteries, ozone depleting solvents, and polyvinyl chloride packaging materials (Brinkley and Mann, 1997). Though it lags behind northern Europe in this area, the U.S. EPA has established voluntary programs that encourage manufacturers to work more closely with suppliers' design for environment (e.g., Energy Star, which rewards them for improving the energy efficiency of their products).

The effect of these regulatory efforts is clear for computer manufacturers, especially those who target European markets. As one Hewlett-Packard manager noted "It is not an unreasonable assumption that many products in the 'design hopper' today will face mandatory customer- or legislative-driven take-back requirements in major European markets by the time these products reach the end of their useful life" (Bast, 1994: 2). In preparation, companies must ensure that their products can be easily disassembled and recycled which, in many cases, requires engagement of their suppliers in product design. Not surprisingly, we learned that the companies experiencing the most direct pressure from these types of regulations are those delivering products into the end user markets, particularly in Northern European countries. In many cases, these types of governmental regulations complement developing environmental standards and growing market pressure for improved environmental performance.

2.2   Standards Setting and Developing Market Pressure

Much of the growing market demand for environmentally sound products is appearing in the form of "eco-labeling" standards. Several third party organizations, including the German Institute for Quality Assurance and Labeling (RAL) the TUV Rhineland product certification organization, and Tjanstemannens Central Organization (TCO), a Swedish white-collar employee union, have established voluntary design guidelines and certification labels for products in the information industry (Wu, Worhach and Sheng, 1998). Known as eco-labels, these guidelines are receiving increased attention worldwide, and may be adopted as regulatory standards in the European Union (EU).

Other voluntary standards efforts include that of the International Standards Organization (ISO), an organization that gained prominence with ISO 9000, a standard for quality management systems. ISO 14000, the environmental management systems standard, allows firms to be certified for showing that their environmental management systems promote the continuous improvement of their environmental performance and spells out suggested activities in supplier environmental management.

Despite ongoing contention about the structure and application of these standards, it is expected that they will ultimately influence buying behaviors at multiple levels (Environment Watch: Western Europe, 5 April 1996: 1-3). Although consumers have historically shown unwillingness to spend more for improved environmental performance, they are beginning to add environmental performance considerations – facilitated by eco-labels – to their purchasing decisions (Fay, 1992; Miller 1993; Bhate and Lawler, 1997). The U.S. government, a large consumer of electronic products, has developed contracting rules that require Federal contractors to investigate and monitor the environmental performance of subcontractors and vendors (Fenn, 1995). The European Union (EU), in addition to considering adoption of "eco-labeling" standards, is also debating whether to make certification under its currently voluntary EMAS program (an environmental management system certification program) a requirement for firms from which it buys products.

Tangible evidence of the effect of these types of standards is found in the number of environmental management and performance questionnaire computer manufacturers are fielding. Sun Microsystems, for example, reports that in 1993 and 1994 it received requests for information on its environmental practices perhaps once per quarter with most of these requests coming from shareholder activists and environmental organizations. By the end of 1995, however, Sun was fielding such requests on a weekly basis with the majority of the inquiries coming from purchasers or potential purchasers of its products.[3] Other firms report similar experiences, several suggesting that the majority of the inquiries arise from their Northern European customers. (Interestingly, the lack of standardization of these inquiries is causing considerable frustration as companies attempt to fulfill similar, but not quite the same, requests.) One computer manufacturer has adopted a proactive strategy to gain eco-label certification on all of its products.

These market pressures, in combination with new government regulations on recyclability and product take-back, are propagating through the computer supply chain. The first tier suppliers (e.g., disk drive manufacturers, printed circuit board assemblers) are themselves fielding

---

[3] Interview with Environmental Management personnel at Sun Microsystems, 1995.

questionnaires or inquiries from the computer manufacturers regarding their operations. In some cases, these inquiries are directly driven from questions raised by the end customers themselves in their questionnaires.

Figure 3 summarizes our findings on the propagation of environmental management pressures throughout the supply chain. Those companies owning significant internal manufacturing capability – the highly vertically integrated companies and those farthest back in the supply chain – are largely driven by traditional governmental regulations on emissions and waste management. They, in turn, are working closely with critical suppliers (e.g., of equipment or hazardous waste disposal services) to manage their environmental impact of their processes.

Companies with meaningful presence in Northern European markets experience market pressure, particularly from emerging eco-labeling standards, most acutely. They, too, are turning to their suppliers to support activities that allow them to better manage the environmental impact of their products. Their direct suppliers are starting to report that these pressures are trickling down to their level, and suggest that they will pass along requirements for improved performance to their suppliers as well. To date, however, our interviews suggest that those companies in the "middle" of the supply chain feel much less direct pressure than the computer manufacturers, and in some cases none at all.

How are these pressures propagated? In the next section, we describe the range of environmental supply chain management programs undertaken by these organizations.

3.   ORGANIZATIONAL RESPONSES TO ENVIRONMENTAL PRESSURES

Until recently, few manufacturers in any industry considered their suppliers' environmental performance to be relevant to either the selection of trading partners or to the subsequent structuring of the buyer-supplier relationship. Our interviews suggest that a new and different set of behaviors is emerging in response to the regulatory, standards and market pressures described above.

The firms we studied are at different stages in the development of their environmental supplier management programs as characterized in Figures 4a and 4b. Virtually all of the companies include a clause in their corporate procurement contract requiring suppliers to comply with local environmental, health and safety regulations, but the extent to which they engage in more proactive programs varies. Seven of the companies have taken little or no action on the environmental supply chain management front as yet, while the remaining companies have programs of various levels of sophistication, from use of simple environmental performance metrics to highly-involved "design for environment" programs. Several companies have had active environmental supply management programs in place for some time. IBM, for example, has been practicing aspects of proactive supplier environmental management since the late 1970s, while Intel and HP have been doing so since the late 1980s and early 1990s. Others are in the process of developing their programs and plan to implement in the next year or so. Sun Microsystems, for instance, has developed an environmental management module, but not yet incorporated it into their corporate supplier management program. Silicon Valley Group, a supplier of equipment to the semiconductor industry, is in the midst of integrating environmental requirements into existing programs to manage supplier quality. In the following section, we describe the spectrum of environmental supplier management programs in some detail.

3.1 The Role of Relational Contracting

To fully understand advances in environmental supplier management programs, it is important to appreciate the broader context in which they are embedded. Over the past several years, computer manufacturers have, in general, moved away from short-term, arm's length contracts with a great many vendors towards long-term, customized relationships with a smaller number of vendors. These longer-term relationships often entail a more extensive exchange of information (e.g., about product designs, production schedules, cost structures), and are thus conducive to the joint undertaking of environmental management projects as well.

Environmental supplier management, when undertaken, leverages these existing structures, but does not, on its own dictates adoption of new structures. The primary reason for this is administrative overhead and cost minimization. All our informants make clear that environmental supplier management must make sound business sense for their firms to undertake it. Selling this business proposition typically requires that it fit within existing systems and structures.

At Sun Microsystems, for example, the Environmental, Health and Safety (EHS) staff discovered that Sun's Supplier Management team was shrinking its supplier base and developing long term strategic partnerships. This led EHS to "de-emphasize the notion of using environmental criteria for supplier selection, and highlight their use in continuous improvement efforts." To win acceptance of their plan, they modified elements of the Supplier Management Program's continuous improvement tool, the "Supplier ScoreCard," rather than suggest more radical changes. In their presentation to the Vice President of Supplier Management, EHS staff presented a "value added" list which "showed how scoring suppliers on environmental practices could enhance [Sun's] relationship with key suppliers, reduce risks of major revenue interruptions, and drive down suppliers' - and hence Sun's costs -- over time." In short, the EHS staff did everything possible to adapt the new program to the infrastructure of the existing program and to present it as an enhancement. Their effort was successful in winning the support of managers in the supplier management program and in the executive suite.[4]

Relational contracting – as we refer to the longer-term, more intense relationships – is critical to many of the specific programs that are being undertaken in the environmental arena.

3.2   Company-Specific Programs

3.2.1   Contract Clauses and Metrics.

All companies we interviewed maintain basic corporate contracts.[5] A typical structure involves an umbrella contract that includes commonly applied basic compliance requirements. Product-specific requirements are included in an addendum or "product specification" attached to the umbrella contract. All companies we interviewed include a basic compliance clause such as the following in their corporate contract: "Seller represents and warrants that in all respects, the manufacture and sale of the Products comply, and will throughout the term of this Agreement comply, with all applicable environmental laws, regulations and other regulatory requirements."

Several companies expressly avoid specifying expectations for supplier environmental management practice that exceed compliance standards set by federal, state and/or local regulatory agencies fearing this would imply legally responsibility for environmental malfeasance on the part of their suppliers. Most communicate such expectations instead through such non-contractual means as supplier selection, training, and management programs.

Many companies in the computer industry employ a standard set of metrics for measuring supplier performance that include technology, quality, responsiveness, delivery and cost characteristics, known in several organizations as the "TQRDC" metrics. HP, IBM, SGS-Thomson, Silicon Valley Group and 3Com have also integrated an environmental, or "E",

---

[4] Interview with Environmental Management personnel, Sun Microsystems, 1997.

[5] A glaring exception to this finding is BayNetworks. They have been acquiring companies at such a rapid rate in the past year or so that they have been unable to consolidate the various procurement activities to a single contract.

metric creating "TQRDCE" programs. The "E" component typically asks suppliers to develop an environmental improvement policy, create implementation plans for carrying out the policy, and define metrics for measuring their progress. Suppliers generally self-report against these metrics. Poor performance often leads to discussion between purchaser and supplier, but rarely to disqualification of the supplier in the short run. The manufacturers we talked with were clear that there are trade-offs to be made among these metrics, and emphasized that they will work with their suppliers to identify appropriate trade-offs and develop improvement programs to meet overall objectives.

3.2.2   Supplier Qualification and Auditing Programs.

Several companies are moving beyond basic contractual and rough measurement approaches as their concern about their suppliers' environmental performance increases. Companies such as HP, IBM, DEC, Western Digital, Solectron, Intel and SGS-Thomson include environmental performance prominently in their supplier qualification process. As one of our Intel informants asserted, supplier qualification is the process in which one has the most leverage. Often qualification requirements are largely compliance driven as they are at DEC and Western Digital, but other times they are broader. HP, for example, "favors environmentally progressive suppliers," and includes questions in their qualification process to identify such suppliers.

Several companies follow up their qualification process with regular monitoring or audits. Once again, fearing legal repercussions, companies are rather informal in the conduct of the audits. Some issue regular questionnaires (e.g., DEC, IBM), while others have site visit and auditing programs (e.g., SGS-Thomson). Some prioritize their vendors based on perceived environmental risk,' visiting or auditing those of higher risk more often. Waste management vendors, for example, typically fall into the high-risk category for companies like Komag and Silicon Valley Group.

The most advanced implementation of supplier qualification and auditing programs encourages suppliers to build and continuously improve their environmental management systems. Encouraged by standards programs such as ISO 14000 and EMAS certification, buyers are increasingly taking it upon themselves to request that their suppliers invest in development of their own internal environmental management programs. Some, such as Silicon Valley Group, go beyond to offer venues in which suppliers can share information with one another and the buyer regarding program implementation.

In all cases, environmental standards beyond compliance are apparently applied flexibly. HP and IBM, for example, are willing to trade off a vendor's substandard "E" performance for superior performance elsewhere in the "TQRDC" spectrum. Substandard performance across the board, however, is grounds for dismissal. Environmental performance goals are negotiated well in advance of implementation through letters, brochures, questionnaires and personal conversation.

3.2.3   Advanced Programs.

The most proactive companies actively engage in programs that can be characterized as "design for environment" or DfE programs with their suppliers.   These programs may focus on improving the environmental performance of a process or a product.

3.2.3.1   Product Improvement.
Product-specific environmental requirements can be handled at a corporate level through inclusion of "banned materials" in a corporate contract. DEC, HP, IBM, Sun Microsystems, and Quantum all include a specific list of materials that are not to be used in product sold to them in their corporate contracts. Other specifications are applied in more limited fashion to specific components or materials.

At HP, for example, a Corporate Procurement Strategy Board overseeing procurement of plastics developed criteria for the specially formulated plastics resin used to launch the first recycled plastic DeskJet printer in 1995. The criteria required that vendors use post-consumer plastics taken back from HP, develop a resin incorporating 25% recycled content, price the resin at par with or below the price of comparable virgin plastic, improve the printer design for thinner wall applications, and use no PBBs or PBDE in compliance with specific country requirements (Choong, 1996).

Typically, when product specific criteria are developed the vendor is included in the process. Environmental requirements are not included in the product specification until both buyer and supplier agree that they are both technically feasible and economically viable. If the goal cannot be met, buyer and supplier work towards a mutually agreeable solution.

3.2.3.2   Process Improvement.
Process improvement, too, can be handled at a general or more specific level. The Montreal Protocol ban on ozone-depleting substances, for example, led several organizations to include specifications for compliance in their corporate contracts. IBM went further, providing proprietary

techniques, processes and equipment to their vendors, sometimes free of charge.

Other process improvement efforts are considerably more specific. The semiconductor manufacturers, for example, work very closely with their equipment suppliers to manage the environmental performance of their own processes. In concerted efforts that begin five years or more before the processes become operational, the semiconductor companies (often in collaborative efforts described below) establish environmental performance requirements for the equipment that their suppliers will develop. These include performance expectations for electrical power, heat load, total exhaust, process cooling water, city water and ultrapure water. They then work closely with their equipment suppliers to see that these requirements are met.

Early in the development process, engineers from Intel, for example, will take up residence at their equipment suppliers' sites to work with them in early identification and resolution of performance problems, environmental and other. Later, engineers from the suppliers will move with their equipment to Intel's factories to trouble-shoot the implementation of their equipment and ensure achievement of the performance standards. Intel claims that most, if not all, of their environmental standards are met through this highly collaborative, interactive process.

3.3   The Importance of Industry Standards

One of the critical challenges that a number of the firms we studied face is that of keeping the costs of their environmental management programs in line. Because environmental performance is not yet viewed as a significant differentiator in the marketplace, several prominent companies in the computer industry have chosen to collaborate, developing shared industry standards that allow them to minimize the costs of implementing environmental programs.

Working through two trade organizations, the Pacific Industry and Business Association (PIBA) and the Computer Industry Quality Conference (CIQC), HP, IBM and Sun, among others, developed a supplier environmental management tool to standardize the process by which firms roll out and administer programs to improve their vendor's environmental management practices. Promulgated in late 1996, the PIBA/CIQC questionnaire sets forth a standardized set of questions for computer manufacturers to ask their vendors. The questions are designed to give buyers a read on what environmental management policies and practices individual vendors are currently following, as well as a system for analyzing the progress these vendors are making towards implementing better policies, programs, and practices. The questionnaires also provide a standardized mechanism by which purchasing firms can communicate their heightened environmental expectations to vendors and a means to initiate discussions that can enable buyer and seller to collaborate to meet those expectations.

In the semiconductor sector, Intel has driven the Semiconductor Equipment and Materials International (SEMI) task forces to develop industry standards that govern the environmental quality of materials and equipment used in the manufacture of semiconductors. The S-2 standard first promulgated in 1990 and currently undergoing revision, requires equipment manufacturers to document the environmental, health, and safety hazards inherent in their equipment and to develop plans to reduce the hazards. The S-10 standard establishes rules for ranking environmental health and safety hazards, thus providing a basis for buyers and vendors to approach the issue consistently.

Both the S-2 standard and the PIBA/CICQ questionnaire establish a set of questions or procedures that force vendors to identify the environmental deficiencies of their equipment or management systems and develop plans for correcting these deficiencies. They do not, however, dictate how the vendor is to make the corrections. Instead, the standards leave it to the vendor to develop its own plans for improving the environmental quality of its management system or equipment design in collaboration with the purchasing firm.

4.   CONCLUSION

Companies in the computer industry are starting to feel increasing pressure to manage not only their environmental issues, but those of their supply chain as well. Although much of the pressure to engage in environmental supplier management has come from governmental regulations in the past, there is a recent shift to market-based pressures that is beginning to propagate through the supply chain.

There is a broad spectrum of approaches to managing the environmental performance of suppliers ranging from simple clauses in purchasing contracts to highly involved and sophisticated design for environment programs. While the literature on industrial ecology suggests that these relationships must exist to fully manage product life cycles, little is said about what form the relationships should take. We hope that our work will prompt others to examine this issue as well and improve our understanding of how environmental supplier management is best done.

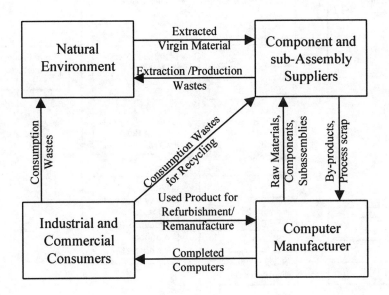

Figure 1: Industrial Ecology View of the computer industry supply chain

Figure 2: Computer Industry Supply Chain

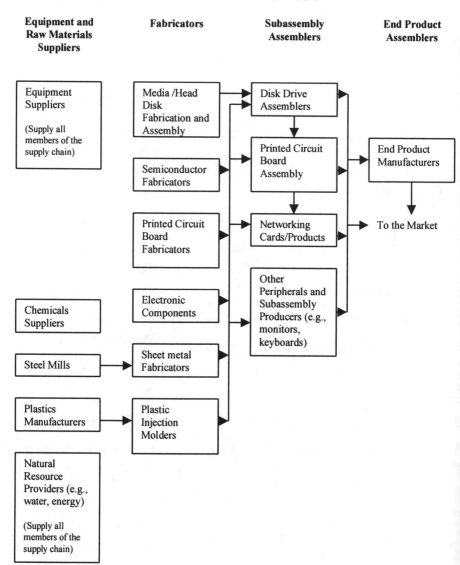

FIGURE 3: Propagation of Environmental Pressures throughout the
Computer Industry Supply Chain

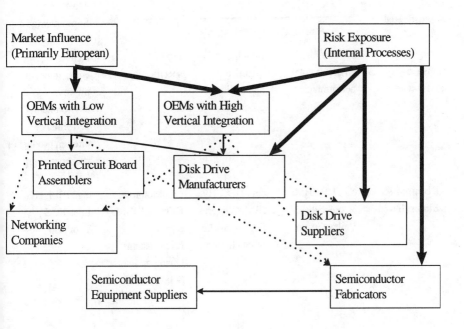

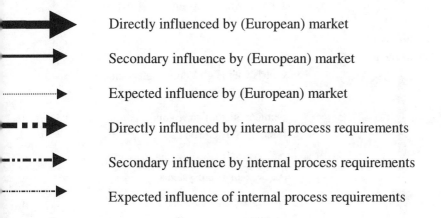

Directly influenced by (European) market

Secondary influence by (European) market

Expected influence by (European) market

Directly influenced by internal process requirements

Secondary influence by internal process requirements

Expected influence of internal process requirements

FIGURE 4a: Spectrum of Environmental Management Programs[6]

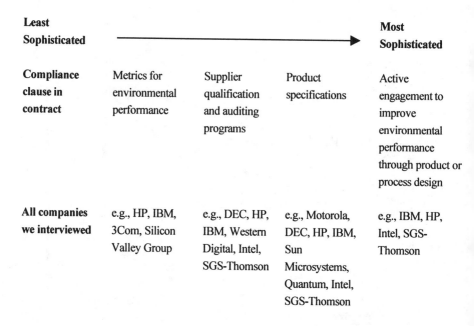

Least Sophisticated				Most Sophisticated
Compliance clause in contract	Metrics for environmental performance	Supplier qualification and auditing programs	Product specifications	Active engagement to improve environmental performance through product or process design
All companies we interviewed	e.g., HP, IBM, 3Com, Silicon Valley Group	e.g., DEC, HP, IBM, Western Digital, Intel, SGS-Thomson	e.g., Motorola, DEC, HP, IBM, Sun Microsystems, Quantum, Intel, SGS-Thomson	e.g., IBM, HP, Intel, SGS-Thomson

Figure 4b: Companies Leading in the Adoption of Environmental Supply Chain Management

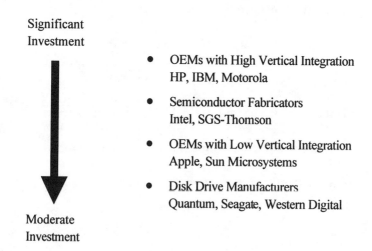

Significant Investment

- OEMs with High Vertical Integration
  HP, IBM, Motorola

- Semiconductor Fabricators
  Intel, SGS-Thomson

- OEMs with Low Vertical Integration
  Apple, Sun Microsystems

- Disk Drive Manufacturers
  Quantum, Seagate, Western Digital

Moderate Investment

[6] Seagate and Apple Computer have asked that we not associate their names with specific environmental actions they have taken, so they are not shown on this chart.

APPENDIX A: INTERVIEWED FIRMS

(Numbers in parentheses show 1997 revenues)

Media/Head Disk Assembly
· Read-Rite Corporation ($1.2B)
· Komag, Inc. ($631M)

Disk Drive Manufacturers
· Seagate Technology* ($8.9B)
· Quantum Corporation ($5.3B)
· Western Digital ($4.2B)

Semiconductor Equipment Suppliers
· Silicon Valley Group, Inc. ($600M)
· Ion Systems ($10M)

Semiconductor Fabricators
· Intel Corporation ($25B)
· SGS-Thomson ($4M)

Computer Manufacturers
· IBM ($78B)
· Hewlett-Packard Company ($43B)
· Motorola ($29.8B)
· DEC (now part of Compaq) ($13B)
· Sun Microsystems, Inc. ($8.6B)
· Apple Computer, Inc.* ($7.0B)
· Silicon Graphics, Inc. ($3.7B)

Printed Circuit Board Assemblers
· Solectron, Corp. ($3.7B)

Networking Products
· 3Com Corporation ($5.6B)
· BayNetworks, Inc. (now part of Nortel) ($2.1B)

*Seagate Technology and Apple Computer asked that their names be listed as participants, but that they not be identified in the text associated with specific activities.

REFERENCES

Ayres, Robert U., "Industrial Metabolism: Theory and Policy," in Braden R. Allenby and Deanna J. Richards, *The Greening of Industrial Ecosystems* (Washington, D.C., 1994), 23-27.

Barney, Jay B. 1991. "Firm Resources and Sustained Competitive Advantage," *Journal of Management*, 17(1): 99-120.

Bast, Cliff, "Hewlett-Packard's Approach to Creating a Life Cycle (Product Stewardship) Program," *Proceedings of the IEEE International Symposium on Electronics and the Environment* (San Francisco, May 1994).

Beckman, Sara L., "Evolution of Management Roles in a Networked Organization: An Insider's View of the Hewlett-Packard Company," *Broken Ladders: Managerial Careers in the New Economy*, ed. Paul Osterman, Oxford University Press, 1996.

Bhate, S. and K. Lawler, "Environmentally friendly products: Factors that influence their adoption," *Technovation* (August 1997), Vol. 17, No. 8, 457-465.

Brinkley, Anne and Tim Mann, "Documenting Product Environmental Attributes," *Proceedings of the IEEE International Symposium on Electronics and the Environment* (San Francisco, May, 1997).

Choong, Hsia, "Procurement of Environmentally Responsible Material," *Proceedings of the IEEE International Symposium on Electronics and the Environment* (Dallas, May 1996).

Ehrenfeld, John R., "Industrial Ecology: A Strategic Framework for Product Policy and Other Sustainable Practices," paper prepared for Green Goods: The Second International Conference and Workshop on Product Oriented Policy (Stockholm, 1994).

Environment Watch: Western Europe, "Draft ISO Standard 14020 on General Principles for All Environmental Labels and Declarations," (5 April 1996): 2.

Fay, W., "The Environment's Second Wave," *Marketing Research: A Magazine of Management and Applications* (December 1992), Vol. 4, No. 4, pp. 44-45.

Fenn, Scott A., "Green Heat," 98 *Technology Review,* (July 1995): 62-63.

Gottlieb, Robert and Maureen Smith, The Pollution Control System: Themes and Frameworks," in Robert Gottlieb (ed.), *Reducing Toxics: A New Approach to Policy and Industrial Decision Making* (Washington, D.C. and Covelo, Ca.: Island Press, 1995).

Gottlieb, Robert, Maureen Smith, and Julie Roque, "By Air, Water, and Land: The Media Specific Approach to Toxics Policies," in Robert Gottlieb (ed.), *Reducing Toxics: A New Approach to Policy and Industrial Decision Making* (Washington, D.C. and Covelo, Ca.: Island Press, 1995).

Kulik, Ann, "Producers to be Held Accountable for Product Waste," *World Wastes* (November 1994) 37(11): 10-14.

Lowe, Ernest, "Industrial Ecology -- An Organizing Framework for Environmental Management," *Total Quality Environmental Management* (Autumn 1993): 73-85.

Janice Mazurek, Robert Gottlieb, and Julie Roque, "Shifting to Prevention: The Limits of Current Policy," in Robert Gottlieb (ed.), *Reducing Toxics: A New Approach to Policy and Industrial Decision Making* (Washington, D.C. and Covelo, Ca.: Island Press, 1995).

Miller, C., "Conflicting Studies Still Have Execs Wondering What Data to Believe," *Marketing News* (June 7, 1993), Vol. 27, No. 12, 1-12.

O'Rourke, Dara, Lloyd Connelly, and Catherine P. Koshland, "Industrial Ecology: A Critical Review," (University of California Berkeley, March 7, 1996).

Richards, Deanna J., Braden R. Allenby and Robert A. Frosch. 1994. "The Greening of Industrial Ecosystems: Overview and Perspectives, " in Braden R. Allenby and Deanna J. Richards, *The Greening of Industrial Ecosystems* (Washington, D.C., 1994).

Tibbs, Hardin, "Industrial Ecology: An Environmental Agenda for Industry" (1991)Arthur D. Little, Inc.

Tibbs, Hardin, "Industrial Ecology: An Agenda for Environmental Management," Pollution Prevention Review (Spring 1992): 167-180.

Wu, Tse-sung, Paul Worhach and Paul Sheng, "The Role of Environmental Specifications in the Electronics Product Development Cycle," Working Paper of the Consortium on Green Design and Manufacturing, Department of Mechanical Engineering, University of California Berkeley, 1998.

AUTHORS

Sara Beckman is on the faculty of the University of California's Haas School of Business where she teaches new product development and manufacturing and operations management. She has also worked for the Hewlett-Packard Company, where one of her positions was managing environmental, health and safety programs throughout the company. Dr. Beckman has B.S., M.S. and Ph.D. degrees from the Department of Industrial Engineering and Engineering Management at Stanford University and a M.S. in Statistics from the same institution.

Christine Rosen is Associate Professor of Business and Public Policy at the Walter A. Haas School of Business at the University of California, Berkeley. She has a Ph.D. in history from Harvard University. She teaches an MBA elective on Corporate Environmental Management and has written on the implications of industrial ecology for the writing of business history. She is currently writing a book on the history of pollution control in the U.S from the industrial revolution to the present.

Janet Bercovitz is an Assistant Professor of Management at Duke University's Fuqua School of Business. She teaches graduate business courses in both entrepreneurship and strategic management. She did her Ph.D. work in Business and Public Policy at the Haas School of Business at the University of California where she previously received her M.B.A. She also holds a B. S. degree from the College of Chemistry at the University of California, Berkeley.

Chapter 17

Life Cycle Assessment

Christian N. Madu
Department of Management & Management Science
Lubin School of Business
Pace University, New York

This chapter discusses the "cradle-to-grave" approach to environmental management of products or processes. It focuses on tracing the environmental burden presented by a product and process and tracks the management of its raw materials, energy and material usage and waste management with a view to developing programs for environmental improvement. Life cycle assessment (LCA) is discussed by looking at its three main stages namely inventory analysis, impact assessment, and improvement assessment.

1.   INTRODUCTION

Life cycle assessment (LCA) is a tool to evaluate the total environment impact of a product or process through every stage of its life. This involves a cradle-to-grave approach of the product or process since its environmental impact is assessed from the point of obtaining the raw materials for product development through production, distribution, and its final disposal or recycling. Environmental impact assessment is a major issue today in many countries. In fact, in most countries, before any major project can be commissioned, environmental impact assessment is required. This great concern for environmental impact assessment can be attributed to the greater environmental challenge that we are faced with today. There is a growing desire for sustainable development, which is a philosophy that is based on the concept that "less is more." Through this, the goal is to use less of raw materials and energy resources in generating products, processes or services and to also create less wastes through recycling, less use of packaging materials which will make less demand on landfills, and less emission of gases to the atmosphere. Thus, this concept adopts a holistic view of the environment by examining environmental quality from the point that the total system operation is environmentally friendly.

LeVan [1995] traced the history of life cycle assessment to 1969 and noted that the first life cycle analysis was conducted on beverage containers. The aim of this analysis was to determine the type of container that had the least impact on natural resources and the environment. This led to the documentation of the energy and material flows although the environmental impact was not determined. Since this initial work, LCA has been broadened to focus on inventorying of energy supply and demand for fossil and renewable alternative fuels. Thus, the focus of LCA is no longer inward with a concentration on the direct influence of the product but also outward to consider the energy and natural resources input during the product's life cycle as well as potential impacts of its usage on the environment. Also, the increasing concern about limited landfill spaces and the health risks associated with pollution have generated the need for a more holistic view of environmental impact assessment.

2. DEFINITION

There are two major definitions of life cycle assessment. These definitions are provided by the Society of Environmental Toxicology and Chemistry (SETAC) and the International Organization for Standards (ISO) who have been active in developing guidelines for LCA. SETAC defines life cycle assessment as:

"An objective process to evaluate the environmental burdens associated with a product, process or activity by identifying and quantifying energy and materials used and wastes released to the environment, to assess the impact of those energy and materials uses and releases on the environment, and to evaluate and implement opportunities to affect environmental improvements. The assessment includes the entire life-cycle of the product, process or activity, encompassing extracting and processing raw materials; manufacturing, transportation, and distribution; use/reuse/maintenance; recycling; and final disposal."

ISO's definition appears in the ISO 14040.2 Draft: Life Cycle Assessment - Principles and Guidelines and is defined as:

"A systematic set of procedures for compiling and examining the inputs and outputs of materials and energy and the associated environmental impacts directly attributable to the functioning of a product or service throughout its life cycle." This goal is accomplished by taking the following steps:

- compiling an inventory of relevant inputs and outputs of a system;
- evaluating the potential environmental impacts associated with those inputs and outputs;
- interpreting the results of the inventory and impact phases in relation to the objectives of the study.

There are three major components of life cycle assessment. These are: Inventory analysis, impact assessment, and improvement assessment. LCA is a way of making the manufacturer to take responsibility for its products. It induces a design discipline that aims at achieving more value for less where the definition of value is expanded to include the potential impacts of the product or service on the environment. The designer focuses on design option that is environmentally sensitive by evaluating the product's demand for limited resources, energy, and disposal requirements at every stage of the product's life. Emphasis is on potential environmental burdens, energy consumption, and environmental releases. The manufacturer also takes a product stewardship or cradle-to-grave approach in evaluating the product, process, or activity. Environmental impacts include the expedition and use of limited natural resources, the pollution of the atmosphere, land, water or air, ecological quality (i.e., noise), ecological health, and human health and safety issues at each stage of the product's life cycle. We shall now, discuss the three stages of LCA.

3. THREE COMPONENTS OF LIFE CYCLE ASSESSMENT

As we mentioned above, there are three major parts to life cycle assessment. These three parts are discussed below:

3.1 Life Cycle Inventory Analysis

The aim of life cycle inventory analysis is to quantify energy and raw material requirements, atmospheric emissions to land, water and air (environmental burdens), generation of solid wastes, and other environmental releases that may result throughout the life cycle of the product, process or activity within the system boundary. These environmental problems affect the quality of the environment and in many ways, the public pays the cost of environmental burdens. The costs of these environmental burdens are often difficult to estimate since they are not all direct costs. Some of the costs may even not be detected until several years after the damage has been done. There are both economic and social costs that are involved. In order to conduct life cycle inventory analysis, we must associate the environmental burdens with functional units. In other words, it should be measurable. For example, we need to be able to use a standardized measurement in quantifying waste or raw material and energy consumption. We can for example, measure the carbon emission to the atmosphere in metric tons or the per unit weight of solid waste from a particular geographical location. The functional unit should provide information on the composition of waste both in terms of material type and relative weight [Kirkpatrick 1999].

Inventory analysis is the thrust of LCA. The normal production process involves actually three main steps: inputs, transformation, and outputs. Each of these steps is a major source of environmental burden and environmental releases. By looking at each of these steps, the process of data collection for inventory analysis can be enhanced:

Input - The input stage involves the acquisition of raw materials and energy resources. Inputs can also come in the form of transfers from other processes. For example, a recycled product can be a source of raw material for producing new product or semi-finished product from a different production source.

Transformation - The transformation process normally deals with the process to convert the input into a desired output. The transformation process also involves energy consumption as well as information flow. Further, wastes could be created through the process as a result of systemic problems with the process itself.

Output - Output may be in the form of finished product, which is shipped out to the consumer, or semi-finished product that becomes input in another process. Also, at the output stage, two types of outcomes can be expected: products that meet the quality guidelines and those that fail the quality requirements. There is therefore, the potential that waste may be generated at this stage both in terms of raw material consumption and energy that is used to generate such wastes.

It is therefore important that these three stages of the production process be evaluated in order to generate an inventory of raw material usage and energy consumption as well as environmental releases.

In the quest for environmentally conscious manufacturing, one of the popular strategies today is to seek for better environmental alternatives. For example, polyethylene and glass, which one is more environmentally friendly? Or, should cloth baby diapers be used in place of paper diapers? One important information that is generated in life cycle inventory analysis is known as the *table of impacts*. This is a table that presents the impacts from the possible production of two materials. For example, we can look at the emissions and solid wastes generated in the production of 1kg of polyethylene and compare it to that for the production of 1kg of glass. However, such evaluation cannot really suggest to us which alternative is better without taking a systemic view of the entire production process. For example, in a study by Johnson [1994], he noted that cloth diapers will require more chemical releases and water usage for cotton while softwood pulp for paper diapers will require more energy requirements. These two options: cloth diaper and paper diaper creates environmental burdens and it is difficult to compare these environmental burdens. So, how does one make a trade-off between these two fibers? There are therefore, a number of problems that make it difficult to conduct life cycle assessment. Some of these problems as they relate to inventory analysis were identified by Product Ecology Consultants [1999] and are discussed below:

3.1.1   Problems with life cycle inventory analysis

* Boundary conditions - It is difficult to define the system's boundary. For example, how far should one go in identifying inputs and outputs that relate to a particular product or process? Based on SETAC guidelines, components that comprise less than 5 percent of the inputs should be excluded. This is however, problematic since it is based on the assumption that the 5 percent component in a product will not have a significant environmental burden. If we go by the ABC rule, there is the potential that a very small fraction of the components may indeed, contribute to the majority of the environmental burden observed. LeVan [1995] presented a good example by noting that the electricity used for particular activity may be a small part of the input. However, if such

electricity is generated from a high sulphur coal plant, its environmental burden could be enormous.

- System boundary condition - There is also the possibility that the links to certain products may be traced infinitum. Kirkpatrick [1999] notes for example, that the production of polyethylene involves the extraction of crude oil, which is transported in a tanker. The tanker is made of steel, and the raw material required for steel is extracted. If we continue, we can see a long product chain that grows larger and larger and becomes more complex to analyze. Thus, a line must be drawn on what constitutes the system's boundary. This will generally not include capital goods.
- Multi-product processes - Some processes are designed to generate multiple products. In such cases, it is not easy to allocate and assign environmental burdens and releases to the different products.
- Avoided impacts - When materials are incinerated, energy is normally generated. Such energy is considered an impact but also, saves impacts, as it will no longer be necessary to produce the energy or the material. These avoided impacts are similar to the impacts that would have occurred in the production of material or energy. They are also, deducted from the impacts caused by other processes.
- Geographical variations - This recognizes the fact that environmentally needs may be geographically dependent. We shall present two examples one from LeVan and the other from Product Ecology Consultants. LeVan [1995] notes that in the time of draught in the U.S. Southwest, single-use disposable diapers would be preferred to home-laundered diapers. While these two create environmental burdens however, the need at the time is a prevailing reason for the choice of single-use disposable diaper. Product Ecology Consultants [1999] on the other hand note that an electrolysis plant in Sweden will create less environmental burden than in Holland because hydroelectric power is in abundance in Sweden.
- Data quality - Environmental impact data are often incomplete or inaccurate. The data can also become obsolete and the use of such data may lead to distortion. There is also a problem that some of the environmental burdens may not be known and there may in fact, exist no data on the environmental impacts.
- Choice of technology - Clearly waste, energy consumption, and material releases to the atmosphere can be linked to the type of technology as well as the maintenance of the technology. Poorly maintained vehicles emit more carbon to the atmosphere and so are poorly maintained

manufacturing processes. Also, the precision of such processes may be questionable leading to more creation of wastes. Further, modern technologies meet the new environmental laws and are able to control emissions and some environmental wastes and pollution.

3.2 Life cycle impact assessment

Life cycle impact assessment is a way of interpreting and aggregating the inventory data so they could be useful for managerial decision-making. We mentioned above that one of the important information generated through life cycle assessment is the *table of impacts*. The table of impacts is however, difficult to interpret without further evaluation of the environmental impacts. It is important to evaluate the impacts by assessing their relative contributions to different environmental concerns. Kirkpatrick [1999] lists some of the impact categories that are often considered in environmental impact assessments. These are:

- Resource depletion
- Greenhouse effect (direct and indirect)
- Ozone layer depletion
- Acidification
- Nutrification/eutrophication
- Photochemical oxidant formation.

Other areas that are less well defined were also identified as

- Landfill volume
- Landscape demolition
- Human toxicity
- Ecotoxicity
- Noise
- Odor
- Occupational health
- Biotic resources
- Congestion

All these areas pose environmental hazard and need to be considered in assessing environmental impacts. There are however problems with assessing environmental impacts. One problem is the fact that data is often non-existent and even when data may be available, it may be difficult to accurately determine the extent of the damage to the environment. Another reason is that there is no standardized method of estimating or measuring environmental damages.

The aim of life cycle assessment is pollution prevention rather than pollution control. By identifying the major sources of pollution and their effects on the environment, efforts could be made through product design or product development to prevent waste and reduce the risks to the environment.

3.2.1   Measuring Environmental Impacts

It is difficult to measure environmental impacts. One of the beginning steps is to first identify a complete chain of cause and effect. But as we noted in the system boundary condition, this could become very cumbersome making it difficult to effectively analyze the problem. We recommend here, the use of Fishbone diagram to graphically identify the potential causes and effects. However, we must point out that there will be series of such diagrams since some of the causes or effects will need to be further analyzed. The use of Fishbone diagram offers great opportunity in breaking down the potential causes in to 4 parts or what is known as 4ms: man, machine, material, and method. The fishbone diagram named because of its shape was introduced by Kaoru Ishikawa a Japanese professor of management and it is often referred to as Ishikawa diagram, is an effective tool in organizing problem solving efforts by identifying several factors that may cause a problem. The 4ms listed in the diagram are potential sources of environmental burden. Let us review each case.

1. Man - This constitute the influence of labor in production and service. Obviously, human error can lead to environmental burden and may be a result of poor supervision, inadequate training, inability, or inefficiency. Waste management and minimization could be effectively achieved, and environmental impact could be reduced if labor is better trained and sensitized on environmental issues. It is also labor that will design for environment and make decisions on alternatives or substitutes that are environmentally friendly.

2. Material - Material selection is often the major source of concern in managing for the environment. At the input stage of any process, if the incoming product is not environmentally sound, the out coming products will also be expected not to comply with environmental requirements.

3. Machine - This deals with the transformation process itself. For example, what type of equipment is used to manufacture or produce the product or service? Does this equipment meet environmental standards? What is the emission level or amount of waste that is generated through this process? All these questions address the issue of environmental burden, which must be addressed to effectively assess environmental impacts.

Equipment maintenance issues and its influence on the environment are also discussed at this stage.

4. Method - This deals with the procedure or technique that is currently in use. Are there alternative procedures that can achieve the same goal at reduced environmental burden? For example, design for environment, design for manufacturability, and design for recyclability are all method-oriented issues. Environmental burden may be lessened if certain materials are reusable. Environmental effects for example, is reduced by recycling paper rather than using virgin pulps. Recycling has contributed in reducing the amount of solid wastes discarded to landfills.

The fishbone (cause-and-effect) diagram is a good way of completely analyzing a problem by asking *who, what, when, why,* and *how* questions about factors that appear to be likely sources of environmental burden.

Figure 1. [Fishbone diagram for Environmental Impact Assessment]

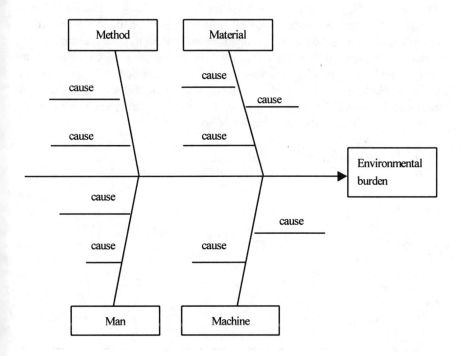

A similar method to the Fishbone diagram is used in Sweden to check cause and effect [Product Ecology Consultants 1999]. This is known as the EPS (Environmental Priority Strategy). The Swiss government on the other hand, uses the Ecopoint method, which measures the distance between the current impact and the target level. This is used to determine the level of seriousness of the impact to the environment.

SETAC presents the procedure for environmental impact assessment that involves three steps:
1. Classification and characterization
2. Normalization
3. Evaluation.

However, only the classification and characterization stage has actually been implemented in practice. We shall discuss them with insight on how they may be improved.

3.2.1.1   Classification and characterization

The fishbone diagram provides a good way of identifying causes and effects. This stage calls for classification of all impacts based on their effects on the environment. Impacts could be grouped based on their contributions to the different environmental burdens such as Resource depletion, Greenhouse effect (direct and indirect), Ozone layer depletion, Acidification, Nutrification/eutrophication, and photochemical oxidant formation. Also, an impact can be classified in more than one area. Thus, an *effect table* may be generated that will be of the form shown below:

Table 1. [Effect Table for Impact Assessment]

Emission	Quantity (kg)	Greenhouse	Ozone layer	Acidifica tion	Nutrificat ion/eutro phication	Photoche mical oxidant formation
Nitrogen dioxide						
Sulphur dioxide						
Carbon dioxide						
Carbon monoxide						
Nitrogen monoxide						
Effect score						

The contribution of each emission to each environmental burden is assessed and its effect score computed. As the intensity of effects may vary among the different chemical compounds, it is important to use a weighting factor. We suggest the use of a systematic weighting scheme such as the Analytic Hierarchy Process (AHP). A pairwise comparison of these chemicals could be conducted to determine their relative contributions to each of the environmental burdens. For example, carbon dioxide and sulfur dioxide, which has the most effect on greenhouse effect? Such weighted factors or priority indices obtained through the AHP could be used to obtain the effect scores. Thus, each quantity will be multiplied by the priority indexes generated through AHP and added up to obtain an environmental burden effect score. Clearly, there are many effect scores that must be generated. Emission for example is one area of environmental burden. The other is resource depletion, landfill usage, etc. The effects could all be compared as well as the relative importance of the different environmental burdens to a particular geographical area. This may form the basis for an informed decision on an environmental policy.

SETAC Workgroups have developed frameworks for life cycle impact assessment (LCIA). They identified the elements for conducting LCIA as classification, characterization, normalization, and valuation. Since SETAC is championing the effort in developing the methodology for life-cycle impact assessment, we shall briefly discuss the work of its two work groups from North America and Europe. The group proposed four major elements for life cycle impact assessment framework. These are classification, characterization, normalization, and valuation.

The classification phase involves creating different categories for inventory results. This will help to distinguish and group impacts for planning purposes.

The characterization phase involves the conversion of inventory results in a category into a category indicator. Equivalent categories are aggregated into a category indicator. The category specific models are constructed using a cause-effect diagram.

The normalization phase involves normalizing the category indicators by dividing them by a reference value. This helps to broaden the scope of the interpretation of the data by comparing the different category indicators.

The valuation phase is based on developing a formal ranking of category indicator results across impact categories. The weights or rank order are subjectively determined.

The use of AHP can also provide an alternative approach to LCIA. We shall therefore, discuss the AHP concept below.

3.2.1.1.1   Analysis of the use of AHP

The Analytic Hierarchy Process (AHP) has three main components: goal, criteria, and alternatives. The goal is what is to be accomplished which in this example, is to select the most environmentally friendly product (i.e., glass or polyethylene). However, this decision depends on several factors denoted as criteria. These factors include the contribution of these products in creating environmental burden such as greenhouse effect, ozone layer depletion, energy consumption, and others. With each of these environmental burdens, there are several other sub-criteria to consider (i.e., emission of gasses that affect the ozone layer or cause the greenhouse effect). All these affect the decision on which of the two products to select. Figure 2 shows the hierarchical network structure of this decision making process.

The use of AHP as a decision tool has been widely published [Saaty 87, Madu & Georgantzas, 1991, Madu 1994, and Madu & Kuei, 1995]. The AHP is defined by its founder Saaty as "a multi-criteria decision method that uses hierarchic or network structures to represent a decision problem and then develops priorities for the alternatives based on the decision makers' judgments throughout the system" (Saaty, 1987, p. 157). The features of AHP that makes it applicable for application in life cycle impact assessment are the following:

- It allows for a systematic consideration of environmental problems by identifying all the major environmental impacts such as the greenhouse effect, depletion of the ozone layer, eutrophication, human or ecological toxicity as well as the factors that may influence them such as the emission of certain types of gases to the atmosphere, use of technology, consumption patterns, etc.
- There are many players when it comes to environmental issues. Environmental policies are not purely technical but also cognizant of the fact that there are many interest groups whose views must be aligned to develop sustainable environmental policies. The AHP makes it easier to consider all these different stakeholders in developing environmental policies.
- Its technique is novel as it deals with issues of consistency in decision-making. Also, priorities generated through AHP can offer a good guide in reaching decisions on the relative importance of the different environmental options.
- Non-technical information can be combined with the more quantitative and scientific information on the environment to reach a decision.
- It helps to breakdown complex problems into levels of complexity that are manageable. As Figure 2 shows, the impact assessment problem can

be broken down into the following parts: goal, criteria, sub-criteria and decision alternatives. This makes it easier to systematically analyze the problem.

- AHP helps to measure the consistency of the decision-maker. Although consistency does not guarantee quality decisions, however, all quality decisions are consistent.

Madu [1999] applied the AHP in analyzing the allocation of carbon emission to inter-dependent industries. The use of AHP as a decision support can help to clarify problems when comparing alternative choices since it will attach relative importance to different environmental impacts.

3.3   Life cycle improvement analysis

Life cycle improvement analysis is akin to the use of continuous improvement strategies. The goal in environmentally conscious manufacturing should be to achieve "zero pollution." This can only be achieved if the entire system processes are continuously improved on. For example, the process of managing the life cycle of a product is a long and an arduous task that starts from raw material acquisition through manufacturing and processing to distribution and transportation, then, recycling, reusing, maintenance, and waste management. Each stage of the product life cycle involves creation of waste, energy consumption, and material usage. A life cycle improvement strategy is needed to identify areas where improvements can be achieved such as a product design that has less demand on material requirement i.e., the reduction in the size of automobiles, replacement of 8 cylinder engines with 4 or 6 cylinder engines; enhanced consumer usage of products by reusing products or creating alternative uses for products, cutting down consumption of fossil fuels, participation in recycling programs, and so on. Life cycle improvement analysis will require the tracking and monitoring of the product through its life cycle to detect areas for continuous improvement. Life cycle improvement analysis has led to a lot of changes in design thus, the design for environment.

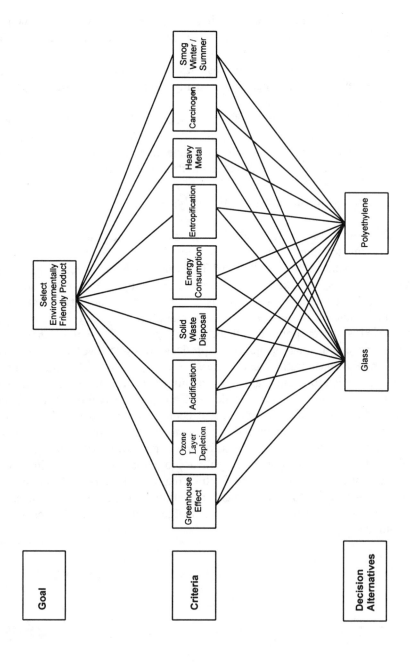

Figure 2. [Selection of Environmentally Friendly Product with AHP]

3.3.1   Design for Environment

Design for environment is a design strategy that ensures the design of environmentally friendly products. This is done by paying attention to the importance of recycling, waste minimization, and reduction in energy consumption. We shall briefly discuss the popular design strategies in designing for environment.

- Design for recyclability - products are now being designed for ease of disassemble so that materials and components can be recovered at the end of the product's life cycle and reused. An example is the Kodak single-usage camera. This process extends the useful life of components in many products and limits the demand for virgin items. This may therefore, lead to the conservation of energy that may be needed to excavate new materials.
- Design for maintainability/durability - increased reliability of products ensures that they will have extended operational life thus reducing the demand for newer generation of products. However, if it is difficult to maintain or repair products, there may be more demand on landfills for disposition of such items. Solid wastes will be created as well as energy wastes.
- Design for pollution prevention - environmentally friendly products are now widely used in the production process to eliminate the amount of toxic and hazardous wastes that are being generated. Further, since products are designed for ease of disassemble and recycling, the amount of wastes generated is significantly reduced.

4.   THE USE OF LIFE CYCLE ASSESSMENT

Life cycle assessment studies should be carefully used. There is the tendency for manufacturers to use such results for marketing purposes. However, any such claims as to which product is better than the other based on life cycle assessment studies may be misleading. There are inherent problems in doing life cycle assessment studies that makes it difficult to use for comparative purposes. Some of the reasons are identified below:
- Life cycle assessment studies cannot definitively state that one product is better than the other if we follow through the product's chain and study all the potential interactions of the product with the environment. It is

difficult to make a decision on which form of pollution is better or preferred to another. For example, is the emission of toxic gases preferred to the use of fossil fuels? No form of pollution should be encouraged. Thus, our target should be to achieve "zero pollution." Until that goal is achieved, it will be incomprehensible to suggest that one form of pollution is better than the other.

- Life cycle assessment will be incomplete with out the consideration of both quantitative and qualitative data. Further, there are cases where either data does not exist or may be incomplete. When we start integrating perceptions and other non-quantitative issues in the decision making process, it becomes difficult to come up with a non-subjective decision.

- Life cycle assessment is limited by both boundary condition and system boundary conditions. Thus, results generated may be geographically dependent rather than universal. So, a claim that one product is better than the other may depend on where the product is being used and may therefore, provide misleading information to consumers.

- Recycling is also a major source of complexity in life cycle assessment studies.

- The systemic structure of life cycle assessment makes it difficult since environmental processes vary spatially (i.e., geographically) and temporally (i.e., in time frame and extent) [SETAC Workgroup].

- There are also assumptions in life cycle assessment studies that there are no thresholds and that there exists a linear response between the system loading and the environment [Fava et al. 1991].

- Other potential problems include the fact that a solution to one form of environmental pollution may actually lead to another form of pollution. Typical example is the attempt to reduce the dependence on landfills for solid wastes. While this may reduce the pollution of air and groundwater supply, however, the burning of such wastes may lead to emission of energy to the atmosphere.

- Like the example we presented earlier, can one conclude that the use of cloth diaper is more environmentally friendly than the disposable paper diaper? Each of these presents an environmental burden.

- Data collection problems can plague the effectiveness of life cycle assessment. Sometimes, it may be difficult to correctly assess and collect information on all the inputs and outputs from a process. Further, the life cycle analysis is greatly affected by the quality of the life cycle inventory analysis. The life cycle inventory analysis is the foundation for life cycle analysis and must be done right.

These drawbacks therefore, make it difficult to use LCA in certain ways. Many have suggested that the adoption of eco-labelling schemes may standardize the comparison of products in terms of their environmentally friendliness. Eco-labelling standards are currently popular in Europe and the Nordic countries as well as Japan but the US does not have a national program for eco-labelling. Eco-labelling is however, not without problems. A major concern is that third-party certification bodies may adopt different and confusing appraisal methods. Eco-labelling is treated in details in Chapter 7 of this book.

LCA however, can be used as a strategic planning tool in the following ways:

- It helps to identify which areas to focus in to achieve the environmental protection strategy. It is a proactive and a systemic way of looking at the company's products and services.
- The cradle-to-grave approach helps to ensure that manufacturers innovate on how to minimize wastes and eliminate environmental pollution. Further, by adopting a cradle-to-grave approach, emphasis is placed on the evaluation of multiple operations and activities throughout a product's life cycle to explore and manage potential sources of environmental pollution.
- The systemic framework adopted enables a functional analysis of the potential impacts of pollution on the environment. Thus, the different effects of energy consumption, material resources utilization, and emissions to the different environmental media such as air, water, and land, and waste disposals can be normalized in order to reach an effective decision on environmental protection.
- Emphasis is in optimization rather than sub-optimization. This is done by looking at a product in its totality by checking the consequences of its interaction and interface with the global environment. A product is therefore seen as interacting with the outer system, receiving and providing feedback and information to it. Thus, a product's quality is not simply measured by its ability to deliver its intended function but its ability to deliver such function at a minimum social cost to the society.
- Life cycle impact assessment can be extended as a means of benchmarking competitors by developing more environmentally friendly substitutes and improving the design for environment.

- Eliminating wastes, cutting down on pollution, and increasing the recyclability of components used in product design can reduce production costs.
- Designing and producing products that meet their environmental needs can improve customer satisfaction and loyalty.

5.   STRATEGIC PLANNING FOR LIFE CYCLE ASSESSMENT

Life cycle assessment can be seen as a strategic tool. It is a tool that will enable the manufacturer to understand the nature of his business and the needs of his customers. Critical questions that are asked in strategic planning include: What business are we in? Who are our customers? And who are our competitors? A well-designed life cycle assessment can enable a manufacturer to address these questions. First, what business are we in? Every business exists for a purpose and one common purpose for manufacturers is to provide goods and services. These goods and services must provide value to customers otherwise, there will be no demand for them. But, what is value? Customer needs are ever changing. Few years back, there was not much concern about the degradation of environment. Once a product meets high "quality" standards, it is expected to do well in the market place. Today's needs are different. Customers are worried about environmental degradation resulting from depletion of limited natural resources, emissions to the different environmental media such as air, water, and land, and the influence of this entire environmental burden on their quality of life. The pressure is on manufacturers to take a cradle-to-grave approach of their products and services. Hence the need for life cycle assessment. As manufacturers grapple with life cycle assessment, they must also come to understand that environmental issues are systemic in nature and are far reaching. A focus on just direct customers will be very narrow. As a result, manufacturers should be concerned with stakeholders rather than customers. Stakeholders are active participants who are affected by the environmental burden and whose action can also affect the role of the manufacturer in contributing to the environmental burden. The issue is who are our competitors and what are they doing? Clearly, manufacturers must strive to be the best or world-class performers in all their operations. This gives them competitive edge. Being the best is a selling point. As we noted earlier, sometimes, LCA is misused by manufacturers who tend to advertise claims that their products may be better than that of competitors based on life cycle assessment. However, there are obvious cases where a manufacturer may be making serious efforts to innovate and reduce its products contribution to the environmental burden and the other sees no need to embark on environmental protection programs. Further, some manufacturers or service providers have achieved remarkable results like in the packaging industry or in digital rather than paper invoicing that they have become companies to be benchmarked. A holistic view of the environmental component of a product or service is not only economical in the long run as it reduces liability costs and other associated costs, but it is also strategic as it positions the manufacturer to compete effectively and expand its market base. While we have discussed the technical aspects of life cycle assessment such as life cycle inventory analysis, life cycle impact assessment, and life cycle improvement analysis, it is also important to look at the non-technical aspects that are strategic in nature.

It is crucial that life cycle assessment is an integral part of any organizational decision making process. The framework presented below has a focus on environmental management which life cycle assessment is a major part of.

5.1 Strategic Framework for Life Cycle Assessment

This framework is broken into three major parts preplanning, evaluation or impact assessment and action implementation/improvement analysis.

5.1.1 Preplanning

The starting point for the framework is the formation of stakeholder team. Although most components of life cycle assessment may be scientific and objective in nature however, there are environmental perceptions and concerns that may not be effectively addressed. People are often concerned about potential impacts. Local conditions such as political and economic issues may influence perceptions. The use of stakeholders in decision-making helps to ensure that the concerns of the stakeholders are considered and perhaps, integrated into the decision-making process. This will make it easier for the stakeholders to accept the final outcome of this process thus making it easier to adopt and implement the final decision. It also helps the manufacturer to expand its scope by considering environmental impacts that may have been neglected internally. For example, stakeholders may take issue with the recycling program or location of landfills.

Participation of the stakeholder team will help the manufacture to hear directly from the "voice of the stakeholder." Thus, the needs and concerns of the stakeholders are better defined and aligned with organizational goals. By working with the stakeholder team, the different environmental media of concern to the stakeholders can be identified and the relationship of the product manufacturing strategies to these media can be better understood through a life cycle inventory analysis of each product strategy. The Analytic Hierarchy Process discussed above can be used to evaluate the different product strategies or scenarios based on the life cycle inventory analysis and the concerns of the stakeholders. This will help to develop a portfolio of product strategies through which an informed decision could be made on potential strategies for adoption. However, there is the issue of cost and feasibility of some strategies. For example, some product alternatives may not be technologically or economically feasible. All these have to be taken into consideration in narrowing down the choices of effective product strategies. The AHP allows establishing priorities for the different product strategies. A systematic consideration of product strategies for implementation can be based on their priority assignments. Once a product strategy is selected, it is matched with design requirements. This phase is accomplished by using the quality function deployment (QFD) as a tool. The QFD if effectively used, will enable the manufacturer to effectively match its capabilities to stakeholders need and also benchmark its design capabilities to that of competitors [Madu 2000].

5.1.2   Evaluation or Impact assessment

This stage involves the evaluation of the product design to ensure that all the significant needs of the stakeholders have been considered in the product design. A prototype or simulated product may be developed in a simulated environment and its potential impacts on the environment and product performance estimated. The simulated impacts will be compared to established standards and targets from the preplanning stage. If the estimated impact from this stage does not conform to expectations, the product design and specifications need to be re-evaluated to identify the source of the problem. Corrective actions are then taken. If however, the simulated impacts conform to expectations, the product could be developed and a random sample of the product may be used for environmental test marketing. However, the problem with this stage is that environmental impacts are often not measured in the short term and may take a long time to show up. However, the use of expert analysis could be helpful in evaluating the potential impacts of the products although there is no guarantee that these potential impacts may be actualized.

5.1.3   Action implementation/improvement analysis

The product is now introduced into the market after it has been certified as meeting the established standards. While the product is in the market, the process of data collection continues. Routine tests are conducted to ensure that the product continues to meet the environmental requirements. Further, the availability of new scientific information may suggest new and more environmentally friendly components that could lead to redesigning of the product to conform to such changes. Also, new legislatures may impose limits on emissions or other environmental burden that may demand a change in product design. Once the product is out in the market place, information gathering must continue to be an ongoing process with the intent of improving the environmental quality of the product.

The Life Cycle Assessment Framework present here is a continuous loop were feedback is frequently being fed into the system. The importance of this feedback look is to ensure that the framework is timely and able to respond quickly to environmental changes. It is a dynamic framework and regards the availability of new information as a necessity in improving the quality of life cycle assessment.

The framework is shown in Figure 3.

6.   LIFE CYCLE COST ASSESSMENT

Life cycle assessment models must address the cost issues. Manufacturers are expected to actively take a cradle-to-grave approach of their products. However, if life cycle assessment is not economical and cannot translate to improvement of the bottom-line, it will be difficult to assure the compliance of manufacturers. What is needed is to be able to speak in the same language as manufacturers. That is, cost and profitability issues must be frequently addressed in conducting life cycle assessment studies.

Life cycle assessment cost is therefore, a method of evaluating costs that are associated with achieving environmental compliance. Such costs may include the conventional costs that may be incurred in implementing warranty programs (i.e., recalls of products that fail to meet environmental standards), social cost to the environment (i.e., costs of clean up and costs of decreased productivity from health and safety factors), liability costs (penalty and legal costs), environmental costs (i.e., environmental pollution costs and health-related costs), costs associated to loss of customer goodwill as well as cost associated to negative campaigns against the manufacturer. These costs can also be grouped into the four types of cost of quality accounting system introduced by Dr. Joseph Juran namely: internal, external, preventive and appraisal costs [Madu 1998].

Prevention costs - involve the cost incurred by ensuring that environmental pollution is prevented. Some examples of this cost include the costs incurred by conducting life cycle assessment as shown in Figure 3, training, designing for environment, product review and preplanning, vendor selection and so on.

External failure costs - these costs are incurred after the product has been manufactured and shipped out to the consumer. Environmental problems found are rectified and the product is returned back to the consumer. Such costs include costs of maintaining warranty, liability, recall, social, loss of customer good will, penalty/fines, complaints, personal and property damages and so on.

Internal failure costs - These costs are incurred internally before the product is shipped out to the end user. Such costs include high levels of emission in the production process, material usage and wastes such as scraps, energy consumption and expenditure, 100% inspection tests and retesting, and so on.

Appraisal costs - This cost is incurred by ensuring that the process and the product meet the target environmental standards or to ensure compliance. Such costs include inspection, equipment calibration, product audit and design qualifications, conformance analysis, and monitoring programs to track wastes and pollution, and environmental quality audits.

When top management understands the cost of poor environmental quality, it will take seriously the effort to improve environmental quality and will therefore, pay attention to life cycle assessment.

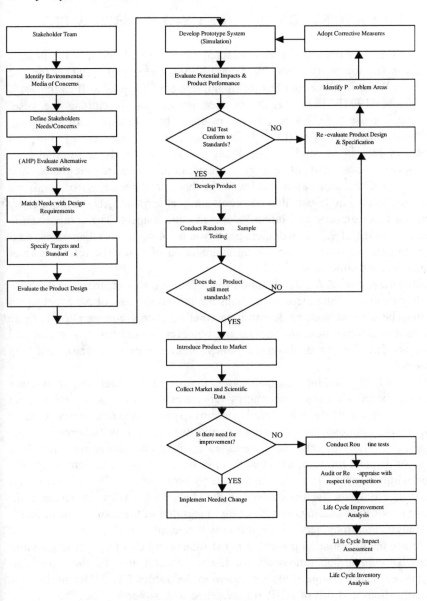

Figure 3. [Strategic framework for life cycle assessment]

7. A CASE STUDY ON LIFE CYCLE ASSESSMENT

The case study presented here is adapted form LeVan [1995]. In her article, she discussed the work done by Franklin Associates Ltd. [1992], for the American Paper Institute and Diaper Manufacturers Group as well as the work by Johnson [1994]. These results provide insight for developing the case. However, we use the AHP to analyze results. The data used here are partly hypothetical and partly estimated from the figures presented by Franklin Associates Ltd. in its studies. The focus of this case study is to conduct a comparative assessment of energy consumption, water requirements, and environmental emissions associated with the three prominent types of children's diaper systems: single-use diapers containing absorbent gels, commercially laundered cloth diapers, and home-laundered cloth diapers. The comparison is based on a usage of 9.7 cloth diapers per day and 5.4 single-use diapers per day. Environmental emissions here are an aggregation of atmospheric, wastewater particulates, and solid waste.

Ultimately, any LCA study should offer a decision support to either decision or policy makers. While there are obvious limitations in some of the conclusions that may be derived such as determining whether energy consumption is more important than environmental emissions, however, a study that is vague and offers no direction to decision or policy makers will only compound the problem.

The work of Franklin Associates Ltd., compared the three diaper systems based on six criteria namely net energy requirements (using LCA method), net energy requirements (using closed thermodynamic balance), water volume requirements, atmospheric emissions, solid waste, and waterborne wastes. Johnson [1994] on the other hand, reported on the input requirements for cloth and paper diapers. Johnson notes that there is more chemical and water usage for cotton while softwood pulp had higher energy requirements. In this case study, we consider only the factors derived in the study by Franklin Associates Ltd., although Johnson's findings can be easily integrated in the framework of AHP. The analytic hierarchy network for this case is presented in figure 4.

We use the information presented in the figures in LeVan's paper to generate a pairwise comparison ratio-scale for the six criteria and the three decision alternatives. These comparisons are shown in the Tables 2 to 7. Discussions on the applications and use of AHP are presented in Chapter 6. Since the data we have in this problem is quantitative, rather than use weight assignments, we used ratios derived from the actual data to derive these tables. The values are then normalized using the method of AHP to obtain the priority indexes for each of the diaper type for any given environmental burden.

Table 2. [Pairwise comparison of the three types of diaper based on solid waste generation]

Type of Diapering system	Single-use diapers	Commercial laundering	Home laundering
Single-use diapers	1	.48	.40
Commercial laundering	2.083	1	.833
Home laundering	2.5	.1.2	1

Table 3. [Pairwise comparison of diaper types based on net energy requirements using LCA methodology]

Type of Diapering system	Single-use diapers	Commercial laundering	Home laundering
Single-use diapers	1	1.21	1.36
Commercial laundering	.826	1	1.12
Home laundering	.735	.892	1

Table 4. [Pairwise comparison of diaper types based on water volume requirements]

Type of diapering system	Single-use diapers	Commercial laundering	Home laundering
Single-use diapers	1	2.25	2.813
Commercial laundering	.444	1	1.25
Home laundering	.355	.8	1

Table 5. [Pairwise comparison of diaper types based on atmospheric emissions]

Types of diapering system	Single-use diapers	Commercial laundering	Home laundering
Single-use diapers	1	1.04	1.8
Commercial laundering	.962	1	1.731
Home laundering	.556	.578	1

Table 6. [Pairwise comparison of diaper types based on wastewater particulates]

Types of diapering system	Single-use diapers	Commercial laundering	Home laundering
Single-use diapers	1	6.571	7.143
Commercial laundering	.152	1	1.087
Home laundering	.140	.920	1

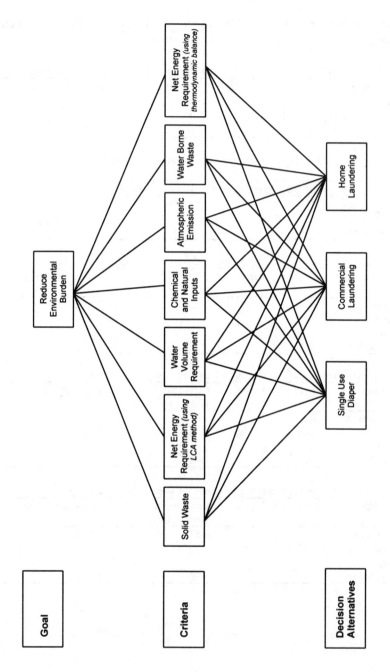

Figure 4. [Analytical hierarchy network for type of diaper selection problem]

Table 7. [Pairwise comparison of diaper types based on net energy requirements using a closed thermodynamic balance]

Types of diapering system	Single-use diapers	Commercial laundering	Home laundering
Single-use diapers	1	.935	1
Commercial laundering	1.07	1	1.07
Home laundering	1	.935	1

Applying the synthesization method of AHP described in Chapter 6, we derive the priority of each diaper type given the six environmental burdens. This result is presented in Table 8.

Table 8. [Priority indexes for the three types of diapering system for each of the environmental burdens]

	Solid waste	Net energy - LCA	Water Vol. Usage	Atmo-spheric Emission	Water-borne waste	Net energy - Thermo.
Single-usage diaper	0.179	0.391	0.556	0.397	0.774	0.326
Comm-ercial laundering	0.373	0.322	0.247	0.382	0.118	0.349
Home laundering	0.448	0.287	0.197	0.221	0.108	0.326

From Table 8, it could be observed that the preferred choice is Home laundering if the only concern is with solid waste disposal since it has the highest priority of 0.448. However, when the concern is with water volume usage, the preferred choice will be single-usage diaper. This is also the preferred choice when our concerns are with atmospheric emission and waterborne wastes.

One problem with the application of the AHP in the Life cycle assessment study is to make a decision on which of the environmental burdens is more preferred to the other. As we noted earlier, such decisions may be affected by system boundary conditions since needs and requirements may vary from different geographical settings. Therefore, results derived may vary depending on where the life cycle assessment is to be applied. This will require us to compare all the six environmental burdens. This is a difficult task since none of the environmental burden should be considered an option. However, given the fact that we must reach a decision that will unavoidably contribute to environmental pollution, it is best to make choices that will lead to minimizing environmental pollution. The type of pairwise comparison conducted for this section of this chapter is described in chapter 6 where weight assignments ranging from 1 to 9 are used. Definitions to these weight assignments are presented in Chapter 6. It is recommended that a group of experts knowledgeable about this problem be used since weights assigned will affect the final recommendation. Table 9 is a presentation of the assignment we have made for the sake of illustration.

Table 9. [Pairwise comparison of environmental burdens]

	Solid waste	Net energy - LCA	Water Volume Usage	Atmospheric Emission	Waterborne waste	Net energy-Thermo.
Solid waste	1	2	4	.25	.2	2
Net energy - LCA	.5	1	2	.167	.125	1
Water Volume Usage	.25	.5	1	.143	.111	.5
Atmospheric Emission	4	6	7	1	.5	3
Waterborne waste	5	8	9	2	1	4
Net energy - Thermo.	.5	1	2	.333	.25	1

Based on the method of synthesization again, the following priorities were derived as shown in Table 10.

Table 10. [Priority indices for types of environmental burden]

Type of environmental burden	Priority indexes
Solid waste	0.114
Net energy requirement - LCA Method	0.061
Water volume requirement	0.037
Atmospheric emission	0.284
Waterborne waste	0.426
Net energy requirement - Thermodynamic	0.078

Type of environmental burden	Priority indexes
balance	

From the results of Table 10, it is clear that the three most important environmental burdens to consider in order of importance are waterborne waste, atmospheric emission, and solid waste. These results will have an effect in determining which of the different types of diaper system will be the preferred choice. We must also note that industrial policies must align with national policies on total environmental quality management. It will of course, be ludicrous if different industrial sectors maintain different priorities on environmental burdens. Priorities generated on environmental burdens should in fact, be in conjunction with policy makers and stakeholders from the different industrial sectors and environmental interest groups.

The final phase of this analysis will be to take the product of Table 8 and Table 10. Since Table 8 is a 3 x 6 matrix and Table 10 is a 6 x 1 matrix, the product of Table 8 x Table 10 will lead to a 3 x 1 matrix that will contain the priorities for the three types of diapering system. This result is presented in Table 11.

Table 11. [Priority indexes for the three types of diapering systems]

Types of diapering systems	Priority indexes
Single-usage diaper	0.533
Commercial laundering	0.257
Home laundering	0.210

Our result suggests that given the six environmental burdens and the comparisons provided by experts on them, the preferred choice should be to implement a single-usage diapering system. The next preferred choice is commercial laundering system.

We must however, point out that this case study is for illustrative purposes only. There may be other critical factors such as the input factors identified by Johnson that were excluded in our analysis. Also, the use of expert judgment in assigning some of the weights introduces subjectivity in the model. The weight assignments may change for different situations and system boundaries. However, this approach provides a guide on how important decisions on LCA studies may be reached. In this chapter, we have excluded some information on the use of AHP such as computing consistency index and how to use AHP for group decision-making. However, these issues need to be explored before applying AHP. Chapter 6 introduced the topic but further reading can be obtained from the following references Madu [1994, 1999], Madu and Georgantzas [1991], Madu and Kuei [1995] and Saaty [1980, 1987].

8.   CONCLUSION

In this chapter, we have introduced the concept of life cycle assessment. We note that this requires product stewardship where the manufacturer takes a cradle to grave approach of its products. There are two main definitions of life cycle assessment that frequently appear in the literature. These two definitions provided by SETAC and ISO have also been discussed. Further, life cycle assessment consists of three main components: life cycle inventory analysis, life cycle impact analysis and life cycle improvement analysis. We have discussed all these and noted that life cycle inventory analysis is the foundation of all life cycle assessment. If that phase is incorrectly done, the entire process will be flawed. We have also identified some of the limitations and benefits of life cycle assessment and we have shown with a case study on types of diapering systems how the analytic hierarchy process (AHP) could be used in the context of life cycle assessment decision-making. We also discussed a strategic framework for LCA.

Another important topic that is discussed in this chapter is life cycle cost assessment. This is an important issue since we depend on the co-operation of businesses that have profit motives to ensure that LCA is successful. As a result, we need to speak in the language business executives understand by exposing them to the cost of poor environmental quality systems.

We also noted that industrial environmental goals must support national policies and in fact, derive from national policies on environmental protection. The guidelines, targets, and priorities on environmental burden should be set together by businesses working together with government agencies. Finally, life cycle assessment can help guide decision-makers to produce environmentally friendly products. It sensitizes them about the needs of their stakeholders and by listening to the voices of the stakeholders, quality decisions on environmental issues that can help improve the bottom-line of the corporation and make it competitive can be derived.

REFERENCES

Fava, J.A., R. Denison, B. Jones, et al., eds., (1991) *A Technical Framework for Life-Cycle Assessments*, SETAC, Washington, D.C. 134 pp.

Franklin Associates, Ltd. (1992*) Energy and environmental profile analysis of children's single use and cloth diapers*, Franklin Associates, Ltd., Prairie Village, Kan. 114 pp.

Johnson, B.W., (1994) "Inventory of Land Management Inputs for Producing Absorbent Fiber for diapers: A comparison of cotton and softwood land management," *Forest Prod. J.* 44(6): 39-45.

Kirkpatrick, N., "Life Cycle Assessment," Retrieved 12/21/99 from http://www.wrfound.org.uk/previous/WB47-LCA.html.

LeVan, S. L., (1995) "Life Cycle Assessment: Measuring Environmental Impact," Presented at the 49th Annual Meeting of the Forest Products Society, Portland, Oregon, June, pp. 7-16.

Madu C. N., (1999) "A Decision Support Framework for Environmental Planning in Developing Countries," *Journal of Environmental Planning and Management*, 42 (3): 287-313.

Madu, C.N. (1996) *Managing Green Technologies for Global Competitiveness*, Quorum Books, Westport, CT.

Madu, C.N., (1994) "A quality confidence procedure for GDSS application in multi-criteria decision analysis," *IIE Transactions*, 26 (3): 31-39.

Madu, C.N., & Georgantzas, N.C., (1991) "Strategic thrust of manufacturing decisions: a conceptual framework," *IIE Transactions*, 23 (2): 138-148.

Madu, C.N., (2000) *House of Quality (QFD) in a Minute*, Fairfield, CT: Chi Publishers, 100 pp.

Madu, C.N. & Kuei, C-H., (1995) "Stability analyses of group decision making," *Computers & Industrial Engineering*, 28(4): 881-892.

Madu, C.N., (1998), "Strategic Total Quality Management," in *Handbook of Total Quality Management*, (ed. C.N. Madu), Boston, MA: Kluwer Academic Publishers, pp 165-212.

Product Ecology Consultants, "Life Cycle Assessment (LCA) Explained," Retrieved from http://www.pre.nl/lca.html on 12/21/99.

Saaty, T.L., (1980) *The Analytic Hierarchy Process*, New York, NY: McGraw-Hill.

Saaty, T.L. (1987) "Rank generation, preservation, and reversal in the analytic hierarchy decision process, *Decision Sciences*, 18: 157-162.

SETAC (North American & Europe) Workgroups, "Evolution and Development of the Conceptual Framework and Methodology of Life-cycle Impact Assessment," SETAC Press, Washington, D.C., 1-14.

Standards Council of Canada (1997) "What will be the ISO 14000 series of international standards - ISO 14000," January.

AUTHOR

Christian N. Madu is Research Scholar, Professor and Management Science Program Chair at Pace University, New York. He is the author of more than 80 research papers, which have been published in leading academic journals. He is also the author of more than six books including Strategic Total Quality

<u>Management</u> (co-author Dr. Kuei), <u>Managing Green Technologies for Global Competitiveness</u>. He recently edited the <u>Handbook of Total Quality Management</u> (London: Chapman & Hall). He is currently working on the <u>Fundamentals of Quality Management Systems</u> (Gordon & Breach Science Publishers) with Dr. Kuei. Dr. Madu is the editor of the *International Journal of Quality Science,* and the North American Editor of the *International Journal of Quality and Reliability Management Journal.* He is also on the editorial board of several journals including *Computers & Operations Research,* and *Engineering Management Journal.* He also serves as a reviewer for many academic journals, publishers and organizations including the National Science Foundation. Dr. Madu also serves as a consultant in the areas of statistics and quality management to several corporations and organizations.

Chapter 18

Multi-Pathway and Cumulative Risk Assessment
Selecting Optimal Pollution Prevention Strategies

Douglas J. Crawford-Brown, Hwong-Wen Ma
*Department of Environmental Sciences and Engineering,
University of North Carolina*

1. INTRODUCTION

Most pollution control strategies can be characterized as a single-medium, single pollutant, approach. Such an approach begins by dividing the environment into three primary compartments: air, water, and land (soil). Pollution problems then are grouped into three categories: air pollution, water pollution, and solid waste pollution problems. In the single-medium approach, these three categories of problems are managed independently without consideration of their interactions. For example, major air pollution sources may be required to install the best available control technologies (BACT) based purely on consideration of the risk from releases to air; industrial wastewater dischargers may be required to meet best available technology (BAT) effluent limitations based solely on the risk from releases to water; and RCRA (Resource Conservation and Recovery Act) and CERCLA (Comprehensive Environmental Response, Compensation, and Liability Act) impose restrictions on the options for land disposal of solid waste (Guruswamy, L, 1990; NRC, 1987; Rao, 1995) based solely on the risk from such routes of disposal. Similarly, the choice of manufacturing processes and pollution control technologies often reflects concern for a single pollutant, while a facility may emit literally dozens of chemicals, metals, etc.

Meeting multiple stringent medium-specific requirements may not necessarily prove best in terms of reducing total risks if there is little consideration of the interactions between environmental media, and of the trade-off needed between releases of different pollutants. A pollution source may release residuals (i.e. the pollution remaining in the waste stream after treatment) to various environmental media. The residuals then are subject to transfer, transport, and transformation both within and between the compartments through natural processes after their release. This movement causes exposures to the pollutant that may not be characterized fully by a risk assessment approach that focuses on a single medium. Focus on a single, primary, medium will produce an underestimate of the risk, since the calculation will not include transfer out of that original medium into the other media through which people may be exposed. At the same time, focusing on the risk from one medium (e.g. air) may cause the selection of a treatment process that minimizes risks through exposures in that medium, but increases risks from other pathways such as the transfer of the pollutant from air to water to the home tap. The result may be selection of a process that fails to minimize the total risk in the environmental system, or simply moves the risk to other populations (Davies, 1986; Irwin, 1992; Travis, 1987).

Because of the complex interrelationship between environmental media, a *systems approach*, which advocates focusing on the entire environmental system as a source of risk, should be used to select risk management strategies (Daellenbach, 1994; Miser and Quade, 1985; van Gigch, 1991). In addition, it is necessary to consider the tradeoff in ri sks from different pollutants that necessarily comes with any re-design of industrial processes. This has been recognized in the United States by a relatively recent requirement for multi-pathway, aggregate and cumulative risk assessment. *Multi-pathway risk assessment* refers to the consideration of all pathways of exposure when calculating risk, removing the restriction of looking only at exposures through the medium into which the pollutant originally was released. A multi-pathway risk assessment provides a richer, and more complete, picture of the risk from a release into a specific medium. It also provides a picture of the total risk from releases into all environmental media, recognizing that a single manufacturing process may release pollutants to the air, water and land simultaneously.

Aggregate risk assessment recognizes that a person may be exposed to a given pollutant (such as chromium) from many sources in addition to the facility being examined in the risk assessment. The goal of environmental and public health protection is to prevent unacceptable risks from all sources of a pollutant. If a person is exposed to chromium from several sources simultaneously (e.g. from two manufacturing facilities), the risk from any one source might be acceptable but the total risk from all sources might be unacceptable. Aggregate risk assessment requires that the risk manager consider the total exposure to a pollutant from all sources, determine if the resulting risk is acceptable and, if not, locate the contributor that is easiest to control. Notice that multi-pathway risk assessments consider a single source and look at all pathways of exposure from that source. Aggregate risk assessments consider multiple sources and, in addition, all pathways from each source. In a sense, multi-pathway risk assessments are used to select control options that are optimal for a single facility, while aggregate risk assessments are used to select control options that are optimal for a suite of facilities, or perhaps even an entire industry.

Finally, a person may be exposed to many pollutants simultaneously. While the risk from each individual pollutant might be acceptable, the total (or cumulative) risk from all pollutants taken as a group may be unacceptable. *Cumulative risk assessment* requires that the risk manager consider exposures to the full range of pollutants released by a facility, and ask whether the total risk from these pollutants is acceptable.

This paper describes the use of a multi-pathway, cumulative, systems approach to selecting waste treatment/disposal and pollution control measures in a sludge management case study in North Carolina. It provides an example of the use of such an assessment in selecting an optimal treatment system in the sense that the total risk to the surrounding population from all pathways, all pollutants and all components of the industrial process is minimized within legal constraints. An important feature of the example is that it is not restricted to justifying a selected treatment system *post hoc* (the dominant use of risk assessment in the past), but rather helps in designing and selecting an optimal treatment system. Since this chapter focuses on risk, the optimization performed here will reflect only risk, and not the more complete picture of risk-cost-benefit that should normally be applied in such cases.

2. A DECISION PROBLEM

The problem to be solved in this case study can be described by identifying the relevant system within which the problem is embedded. Figure 1 depicts the basic relevant system of any similar waste treatment problem by specifying the components of the system. Municipal sludge is a by-product of wastewater treatment, which treats the water that comes from domestic, commercial and industrial (manufacturing) activities. After sludge is produced, the way in which it is managed, as well as the particular pollutants in the sludge and the environmental conditions into which these pollutants are released, will affect the risk to surrounding populations.

To tackle the sludge problem, one needs to recognize the hierarchical nature of a system that provides many levels at which a problem of risk may be tackled. For example, the sludge problem considered here may be divided into four levels of systems as shown at the bottom of Figure 1. Each level might be a candidate for selection of alternative strategies to reduce risk:

- In Level 1, the pollutants already have been released and have traveled through the environmental system. The risk reduction options here refer to moving people so they don't come into contact with the contaminated environmental media. This might be done, for example, by restricting access to land around a facility.

- In Level 2, the sludge has been produced but not yet released into the environment. The risk reduction options here include different methods for treating and/or disposing of the sludge. This might be done, for example, through land application, or through incineration.

- In Level 3, the wastewater has been produced but the method of treatment that ultimately will produce the sludge has not been selected. The risk reduction options include different wastewater treatment methods, each of which produces a different amount of sludge.

- In Level 4, the focus is on the original municipal and industrial processes that produce the wastewater. The risk reduction options are changes in these municipal and industrial processes to lower the concentration of pollutants in this water.

The complexity of a risk assessment used to select optimal risk reduction methods increases as one moves from Level 1 through Level 4, since the number of options increases. While the most creative use of risk assessment includes consideration of all 4 levels, reflecting options as diverse as restricting access to land (Level 1) and changing industrial processes or patterns of consumption (Level 4), the case study presented here would grow too complex to present. As a result, we focus here on examining the waste management system of sludge (system level 2), which includes selecting the method of sludge disposal and the associated air and water pollution control devices. This example also is germane since the waste management system also is the primary focus of most environmental regulations.

Figure 2 shows a typical waste management system for sludge (level 2 of Figure 1). Two primary disposal methods, incineration and land application of the sludge, are shown. The fraction of the sludge treated by each of these two methods may be altered (EPA, 1989). After incineration, ash will be produced, which then is disposed of on land. In addition, an air pollution control device (APCD) may be applied to the waste air from the incinerator, with some of the pollutant escaping to the air, some removed by the filter and applied to land as a solid, and some scrubbed from the air and contained in the wastewater from the scrubber. Notice the wastewater treatment process may also release volatile pollutants into the air or concentrate pollutants into solid form, which then is put onto the land. The wastewater from the incinerator, the pollutant making it past the incinerator filter, and the pollutant applied to land then produce source terms into the environmental system shown at the right in Figure 2.

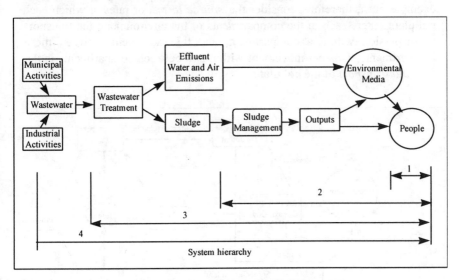

Figure 1. The Relevant System for the Sludge Problem.

A municipal sludge problem can be divided into several components that represent different levels of the system hierarchy. For example, Level 1 only considers human receptors; Level 2 incorporates the environmental media and the sludge management activities; Level 3 further includes the wastewater treatment processes that generate the sludge; and Level 4, includes human activities that release the wastewater.

Once a pollutant has entered either the air, land or water from the treatment facility, it will be moved within and between the various compartments of the environment by the processes that generally move material within and between compartments. These processes are *diffusion, sedimentation, carriage on currents* and *buoyancy* (Crawford-Brown, 1997). If the pollutant remains in a compartment but simply moves around in the compartment (e.g. moving on air currents), the process is one of *transport* or *dispersion*. If the pollutant moves between compartments (e.g. settling from the air onto the soil), the process is one of *transfer*. If the pollutant remains in a compartment but changes form (e.g. through a chemical reaction), the process is one of *transformation*.

The concentration of the pollutant in any one environmental medium (e.g. air) will depend on both the original points at which the pollutant entered the environmental system and the subsequent transfer of the pollutant between the compartments. The risk assessment considered in this example must, therefore, consider the *source terms*, or rates at which each pollutant enters each of the compartments of the environment, the transport or dispersion within a compartment, and the movement between these compartments. This net movement within and between compartments is the *fate and transport* of the pollutant.

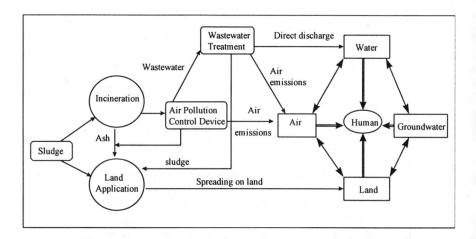

Figure 2. A Sludge Management Decision Influence Diagram.

Beginning from the original sludge, various disposal methods (incineration and land application) and air and water control processes will lead to different release of pollutant into the environmental system of air, water, land and groundwater. The pollutants then may contact the same human (called a receptor) through multiple media and pathways.

Models of fate and transport produce an estimate of the concentration of a pollutant in each of the environmental media (air, water, land and groundwater) and at each location within a compartment. The details of such models are beyond the scope of this chapter, but the model results can be summarized through the concept of *dispersion coefficients*. Consider, for example, a pollutant released to the air from the sludge management facility. It will be transported throughout the atmosphere, producing a concentration in air that is different at each point in space (which we might show by the latitude, longitude and altitude of that point; call these three coordinates x, y and z, respectively) and at each moment in time (t). We will show this concentration here by $C(x,y,z,t)$, which means the concentration of the pollutant at a point in space (x,y,z), and at a time t (e.g. June 3rd at 5 PM). Since the concentration is different in each of the environmental media, we also need a subscript on this concentration. $C_A(x,y,z,t)$ would be the concentration in air at (x,y,z,t); $C_W(x,y,z,t)$ would be the concentration in water at (x,y,z,t); $C_{GW}(x,y,z,t)$ would be the concentration in groundwater at (x,y,z,t); and $C_S(x,y,z,t)$ would be the concentration in soil at (x,y,z,t). The dispersion coefficient at a point (x,y,z,t) in a compartment is the ratio of the concentration produced at that point by the source over the source term or release rate. The units of the dispersion coefficient are concentration per unit source term. For example, if the concentration is in units of grams per liter and the source term is in units of grams per second, the dispersion coefficient will be in units of grams per liter per gram per second. The symbol for the dispersion coefficient is $DC(x,y,z,t)$, which can also be shown with subscripts that refer to the compartment: $DC_A(x,y,z,t)$, $DC_W(x,y,z,t)$, $DC_{GW}(x,y,z,t)$, and $DC_S(x,y,z,t)$, for air, water, groundwater and soil, respectively. The concentration of the pollutant at a point (x,y,z,t) in a compartment then can be calculated by knowing the source term and the dispersion coefficient specific to that point:

(1) $C(x,y,z,t) = ST \times DC(x,y,z,t)$

Once the pollutant is in any of the compartments, humans may be exposed. For example, they may breathe the air, ingest the water or ingest food grown in the soil. It is simple to envision a large number of possible sequences that could lead to these exposures by considering Figure 2. The pollutant might be released to the air and inhaled. It might be released to the air and settle to the soil and be taken up by plants, which then are ingested. Each of these sequences is an exposure pathway, and a multi-pathway assessment considers the total risk from all of these pathways. In the case study presented here, the major exposure pathways are:

For releases to air

Direct inhalation → Inhalation exposure

Deposition from air onto produce → Ingestion exposure

Deposition from air onto plants→ forage by beef and cattle→ Ingestion exposure

Any of the above → breast milk → Ingestion exposure

For releases to soil

Volatilization and re- suspension into air → Inhalation exposure

Direct ingestion of soil→ Ingestion exposure

Root uptake from soil → produce and root vegetables → Ingestion exposure

Root uptake from soil → forage by beef and cattle→ Ingestion exposure

Root uptake → silage/grain→ use to feed beef, dairy, and pork → Ingestion exposure

Ingestion of soil by cattle, pigs and chickens→ Ingestion exposure

Dermal exposure (or absorption through skin in contact with soil)

Any of the above → breast milk → Ingestion exposure

For releases to surface water

Contaminated drinking water supply→ Ingestion exposure

Movement into fish → Ingestion exposure

Bathing in contaminated water → volatilization →   Inhalation exposure

Bathing in contaminated water → Dermal exposure

Any of the above → breast milk → Ingestion exposure

For releases to groundwater

Contaminated drinking water supply→ Ingestion exposure

Bathing in contaminated water → volatilization → Inhalation exposure

Bathing in contaminated water → Dermal exposure

Any of the above → breast milk → Ingestion exposure

Note that there are many exposure pathways, requiring separate models and/or data for each of these pathways.

In addition, there are differences between individuals in the degree to which they are exposed through the different pathways. These differences are summarized by referring to *receptor populations*, or groups of individuals with roughly similar characteristics of exposure. For example, there are *subsistence farmers* highly exposed to the pollutant in food because they grow their own food rather than importing it from other geographic regions that might be far from the source of the pollution (where the dispersion coefficient, and, hence, concentration would be lower). There are *young children* of interest because they play outside more often, or eat more soil, etc. There are groups defined by issues of *environmental justice*, such as the poor and minorities that might be given special consideration for legal, political or moral reasons. Each of these groups makes up a receptor population. The goal of risk management generally is to ensure that the risk to each of these receptor populations is acceptable, taking into account the exposure pathways most relevant for these populations.

Calculating the risk depends on the type of health effect produced by a pollutant: *acute effect* (an effect that occurs rapidly after exposure and then disappears when exposure stops), *chronic effect* (an effect that occurs over a longer period of time and does not necessarily disappear soon after exposure stops), and *cancer*. For all three classes of effect, the first step in calculating risk is to determine the *average daily rate of intake*, or ADRI. This is the rate at which the pollutant is coming into the body per unit body mass. For the example here, exposure pathways resulting in inhalation and ingestion are considered. The value of ADRI for inhalation then can be calculated from the equation (Crawford-Brown, 1997):

(2) $\qquad ADRI = C_A(x,y,z,t) \times IR_{air} \times ED / (AT \times BW)$

where $C_A(x,y,z,t)$ is the concentration of the pollutant in the air at the point in space (x,y,z) and the time (t) at which the receptor population is exposed (e.g. in grams/m^3); IR_{air} is the rate at which an individual in that population inhales air (e.g. in m^3/day); ED is the total period over which the individual is exposed (e.g. in days); AT is the lifetime or averaging time (in days); and BW is the body weight or mass (e.g. in kg). The units of ADRI might then be, for example, grams of pollutant per day per kilogram of body mass. For ingestion of water, a similar equation can be developed:

(3) $\qquad ADRI = C_W(x,y,z,t) \times IR_{water} \times ED / (AT \times BW)$

where everything is the same as in Equation 1 except the concentration in water is used, and IR is the intake rate for water rather than air. Similar equations can be developed for essentially any material inhaled or ingested.

The way in which the estimate of average daily rate of intake is converted to a risk depends on the kind of effect (acute, chronic or cancer). Generally, acute and chronic effects are treated as if there is a value of ADRI below which no effect occurs, and above which each individual develops the effect. This value is the threshold ADRI, which we will show as ADRI $_T$ (the subscript stands for "threshold"). If the ADRI produced by a particular risk management option is below ADRI$_T$, individuals are considered protected against the effect. If the ADRI is above ADRI $_T$, the individuals are not protected. A *hazard index* (HI) for a particular option then can be defined as:

(4)   HI = ADRI/ADRI$_T$

For acute and chronic effects, the risk management goal is to prevent HI from going above 1.0 for most of the population.

For cancer, there is no assumed threshold value of the ADRI. Instead, it generally is assumed that the probability of cancer from the pollutant simply increases in direct proportion to the value of ADRI for that pollutant. The *slope factor* is equal to this proportionality constant, with units of ADRI^{-1}; e.g. (g/d-kg)$^{-1}$. It is found by producing a graph of ADRI (on the x axis) and probability of cancer (on the y axis) and finding the slope (hence, the term slope factor). An example is shown in Figure 3. The probability of cancer then is equal to the product of the ADRI and the slope factor:

(5)   P$_C$ = ADRI x SF

where P$_C$ is the probability of cancer and SF is the slope factor.

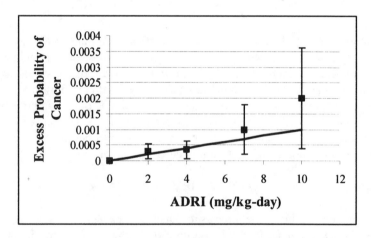

Figure 3. An example of the slope factor for cancer.

This figure shows data (dots with error bars) for the excess probability of cancer, or probability above and beyond the background rate of cancer, as a function of the ADRI. The slope factor in this example is 0.0001 (mg/kg-day)$^{-1}$ as can be seen by noting that the best-fitting curve (solid line) rises to 0.001 when ADRI is 10; the slope then is 0.001/10 or 0.0001.

The calculation of risk for a particular management option then proceeds by (i) estimating the source term into each compartment; (ii) estimating the concentration at each point (x,y,z,t) in each compartment using the dispersion coefficients; (iii) estimating the value of ADRI for each exposure pathway using Equation 2 or 3; (iv) finding the total ADRI from all exposure pathways; and (v) calculating the hazard index (for acute and chronic effects) or probability of cancer for each individual. Two further concepts then complete the assessment: *variability* and *uncertainty analysis*.

Different receptor populations will have different values of the hazard index and/or probability of cancer, since they may be located at different places (which means the concentrations to which they are exposed differ from other populations) and they ingest and/or inhale at different rates (which means the ADRI will be different). Even within a receptor population there will be variation (variability) of the risk. For example, even if one considers only subsistence farmers, different subsistence farmers will have different rates of inhalation, ingestion etc. This variation is *intersubject variability*.

Capturing intersubject variability requires information on the variation of each factor in Equations 1 through 5 within the population (Finley, 1994; Hoffman and Hammonds, 1994). For example, consider the intersubject variability in the rate of ingestion of water in Equation 3. Different individuals ingest water at different rates. This can be summarized by a *probability density function* (usually abbreviated as PDF) showing the fraction of individuals in the population (e.g. the population of subsistence farmers) that ingest water at a rate of 1 liter per day, 2 liters per day, etc. An example PDF is shown in Figure 4.

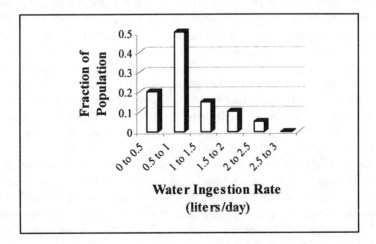

Figure 4. An example of probability density function for the rate of ingestion of water. The height of the bar shows the fraction of the population with a water ingestion rate falling within the upper and lower bounds shown

on the x-axis. The distribution shows the lognormal shape often found in such parameters.

There will be one of these PDFs for every factor that goes into the calculation of the ADRI and the probability of cancer. As a result, individuals in a population who are exposed to precisely the same concentrations in the environmental compartments will have different hazard indices (Equation 4) and probabilities of cancer (Equation 5). This will produce a probability density function for either the hazard index or probability of cancer in the population. The procedure for producing this probability density function, using the example of cancer after exposure to the pollutant in air, is:

- determine the concentration in air, $C_A(x,y,z,t)$ to which the population is exposed;
- select an inhalation rate, IR_A at random from the PDF for inhalation rates;
- select a value of ED, AT, BW and SF from the appropriate PDFs;
- calculate the probability of cancer for that randomly selected individual;
- repeat all 4 steps above for another randomly selected individual.

The result will be a PDF for the estimate of the probability of cancer in the population. This information then usually is summarized as a *cumulative distribution function* (or CDF), which shows the fraction of the population whose probability of cancer is below any particular value. An example is shown in Figure 5. The policy goal is to ensure that a reasonable fraction of the population has a probability of cancer below some limit considered acceptable (e.g. 10^{-4}, or one-in-then-thousand). This fraction varies from organization to organization, but generally is in the range of 95 to 99% of the population, recognizing that there may always some very small fraction of the population that is unusually sensitive (the "bubble babies") and will need to be protected by some means other than controlling releases to the environment. For example, this highly sensitive group may need to be protected by limiting their access to some geographic regions. Notice that in Figure 5, the fraction of the population whose probability of cancer is below 10^{-4} is approximately 0.9, so 90% of the population is protected. A curve similar to Figure 5 could be produced for the hazard index in this population. The policy goal is to ensure that a reasonable fraction of the population has a value of HI lower than 1.0 (since a value above 1.0 indicates the threshold value of ADRI has been exceeded and the effect will occur).

Finally, it is necessary to consider that all of the numerical values used to produce Figure 5 (or the equivalent curve for HI) are uncertain. A full risk assessment produces not only an estimate of the fraction of the population being protected adequately, but a picture of the uncertainty in that estimate. The methods for performing an uncertainty analysis are beyond the level of this chapter, but they produce a confidence interval around the estimate of the fraction of the population protected. For example, consider the prediction from Figure 5 that 90% of the population is protected against a probability of cancer equal to 10^{-4}. The uncertainty analysis could reveal that this fraction lies anywhere between 85% and 94%. The risk manager must consider that the true fraction of the population protected might be as low as 85%, and this might influence the decision as to whether that management strategy (choice of sludge management system) is acceptable.

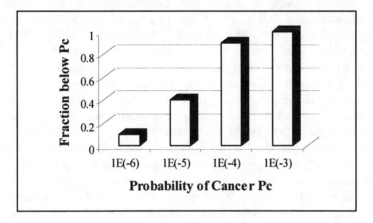

Figure 5. An example of the cumulative distribution function for the probability of cancer in a population. In this example, 100% of the population has a probability below 10^{-3}, but only 90% have a probability below 10^{-4}.

3. THE CASE STUDY METHODOLOGY

The multi-pathway cumulative risk assessment methodology is described below through its four steps: release assessment, exposure assessment, consequence assessment and risk characterization (Covello and Merkhofer, 1993). *Release assessment* characterizes the pollutant source terms associated with each management alternative. Waste-feed rates, destruction rates of incineration, and the removal efficiencies of APCDs and wastewater treatment systems along with the constituents of sludge determine the initial release of pollutants into the air, surface water, and the soil of the land application unit (EPA, 1989). As an example, this study examined four of the most common constituents of sludge (chromium or Cr VI; cadmium or Cd, Bis(2-ethylhexyl)phthalate or BEHP; and trichloroethylene or TCE), and the combustion product of most concern (2,3,7,8- TCDD dioxin) as the risk agents to be analyzed. The concentration of each of these constituents in the original sludge was selected to be representative of those found nationally in typical sludge streams (Lue-Hing, 1992; Ma and Crawford-Brown, 1999). The waste-feed rate, the destruction rates of incineration, and the removal efficiencies of APCDs and wastewater treatment systems determined the initial release of the pollutants into the air, surface water, and soil of the land application unit.

Exposure assessment translated contaminant source terms into estimates of the concentration of each contaminant in the air, surface water, groundwater and soil at the points (x,y,z,t) at which the population is exposed (EPA, 1994). The geographic area surrounding the facility was divided into grid blocks, and the concentration in all 4 environmental compartments estimated at the center of each grid block. In this study, the model for the environmental system was constructed by linking models for the individual compartments so the output files of one model are used as the input files of the next. The model for air dispersion was the Industrial Source Complex-Short Term Version 3 (ISCST3) model. The algorithms for soil and water transport used in two models, the Multi- Pathway Risk Assessment Model (MPRAM) developed at Research Triangle Institute and the Multimedia Environmental Pollutant Assessment System (MEPAS) developed at Battelle, were combined to model multimedia transport and exposure associated with land application and incineration of sludge (Whelan, 1996; Strenge and Chamberlain, 1995).

The outputs of these models were (i) annual average concentrations of air and soil in each grid-block surrounding the modeled sludge management facility over the exposure period; (ii) annual concentrations in the surface waterbodies identified as drinking water and fish sources; and (iii) groundwater concentrations under each grid block (primarily from pollution that moves down through the soil to reach the groundwater). After estimating the pollutant concentrations in each environmental medium in each grid block surrounding the site, exposure scenarios that link an environmental medium and an exposure route were used to estimate values of ADRI for individuals in each grid block. These exposure pathways were described in the previous section and included inhalation, ingestion, and dermal uptake (absorption through the skin). Since biological factors involved in all of these exposure pathways change with age, age-dependent rates of inhalation, ingestion of food and water, and soil contact were used to determine the exposure through each of these pathways and at each separate age. Separate calculations were performed for each grid block surrounding the site, and the population size in each grid block determined from census track data. The inter-subject variability of risk in each separate grid block was generated, then a composite variability distribution generated for the entire exposed population by combining the inter-subject variability distributions for the different grid blocks.

In *consequence assessment*, the exposures identified in exposure assessment are converted to a probability of cancer or a hazard index, using methods described in Section 2 of this chapter. All numerical values of slope factors and threshold values of the ADRI (or, more properly, reference doses in the regulatory arena) were taken directly from the EPA IRIS database (EPA, 1992). The resulting risks then were summarized in the final stage of a risk assessment: *risk characterization*. A Monte Carlo technique (described in Section 2) was used to produce uncertainty and variability distributions for the risk estimates (Hoffman and Hammonds, 1994). The result for each analysis of a mitigation strategy, therefore, was a variability distribution for the risk in the exposed population, and a confidence interval for the prediction of the fraction of the population whose risk was judged acceptable. In this case study, "acceptable" was defined as a probability of cancer below 10^{-6} and a hazard index below 1. In addition, the total number of cancers in the population was determined by multiplying the mean (average) probability of cancer by then size of the exposed population. The total number of acute and chronic effects was determined by calculating the total number of people with a hazard index greater than 1. An individual's probability of effect is the individual risk, and the total number of effects is the population risk. The policy goal generally is to keep the individual risk below an acceptable value and to keep the population risk as low as is feasible.

The decision problem was constructed within an optimization framework in which all necessary management goals are combined in an *objective function*. In this example, the goal was to find a sludge management system that met the following goals while treating the total waste stream for this facility:

- Less than 5% of the population should have a cancer probability above 10^{-6}.
- Less than 5% of the population should have a hazard index above 1.
- The population risk (total number of effects) should be as low as possible.
- There should be approximate risk equity in the sense that the individual risk to minority populations should be within a factor of 2 of the risk to non-minority populations.

The uncertainty was considered by using the upper end of the 90% confidence interval on each estimate. For example, if the 90% confidence interval for the fraction of the population with a hazard index above 1 was [1%, 9%], the value of 9% was selected. This ensures the estimates of risk used were conservative in the sense that they would, if false, lead to an overestimate of the risk.

We then modeled all possible combinations of the fraction of sludge sent to the land disposal unit or incinerated; the choice of APCD; and the treatment of the liquid and/or solid waste from the APCD as shown in Figure 2. Each combination produced a prediction of the four quantities listed above, and the option that satisfied these criteria best was selected (Ma and Crawford-Brown, 1999). Bear in mind that the optimal solutions depend on the decision criteria as to which percentages in variability and uncertainty distributions are selected.

At the top of the figure are the points of release into the environmental system. In this figure, the term "zone" is identical to the term "grid block" in the text. The source data refer to the emission rates into the different environmental compartments.

4.   RESULTS

The models described previously were used to estimate risks from the sludge management facility. Rather than simply calculate risks, we explored the effect of using multi-pathway, cumulative risk assessment in assessing risks and searching for optimal management strategies. This was done by first performing an assessment of risk for each individual pollutant and considering only a single exposure pathway (the *dominant exposure pathway*, or pathway contributing most of the risk). The assessment then was repeated for that single pollutant using the full multi-pathway assessment model. The entire assessment was repeated one more time using all pollutants and all exposure pathways. This analysis helps understand whether the complexity introduced by a multi-pollutant, multi-pathway, cumulative risk assessment really is warranted.

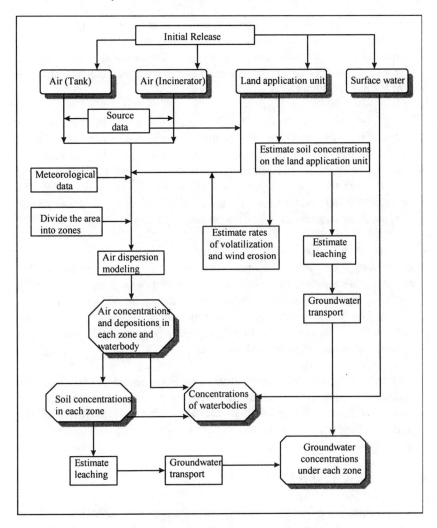

Figure 6. The linked models used in this case study.

Several results emerged from this case study that have a direct bearing on how a risk assessment should be conducted for complex industrial systems emitting multiple pollutants into several compartments simultaneously (Ma and Crawford-Brown, 1999):

- When a single pollutant was considered (e.g. Cr), adding the complexity of a multi-pathway analysis did not significantly affect the choice of an optimal sludge management strategy. Consideration of only the dominant exposure pathway led to the same optimal strategy as that found under a multi-pathway assessment. In addition, if a management strategy satisfied the risk goals for the single pathway analysis, it also satisfied it for the multi-pathway analysis. This indicates that a single exposure pathway dominates the total risk from a management strategy for a single pollutant.

- When all pollutants emitted by the sludge management strategy were considered, strategies that were judged acceptable when each individual pollutant was considered separately became unacceptable. The reason is that the risks posed by each individual pollutant were sufficiently close to 10^{-6} that their cumulative risk rose above 10^{-6}. In these cases, the cumulative risk assessment provided a different optimal strategy than did a risk assessment that focused on single pollutants in isolation.

- When all pollutants were considered, the optimal strategy selected considering only the dominant pathways was not the same as that selected when the multi-pathway assessment was used. In addition, policies that were acceptable in the former case were unacceptable in the latter. The reason is that the underestimation of risk caused by ignoring all but the dominant exposure pathway only became significant when it was compounded over several pollutants.

- When multi-pathway, cumulative risk assessment was used, the population risk produced by the optimal management strategy was almost a factor of 2 lower than the population risk produced by the management strategy found to be optimal by a single pathway approach.

5. DISCUSSION

By combining multi-pathway, cumulative risk analysis and an optimization framework, the selection of waste management strategies can be based on a comprehensive consideration of risk. Such an approach moves beyond the traditional use of risk assessment in justifying decisions made on the basis of other considerations and towards a process in which alternative management strategies are systematically proposed, assessed and ranked.

By considering all pathways of exposure, and all pollutants, simultaneously, it is possible to avoid situations in which risk management strategies simply move pollutants from one point of release in the environmental system to another without significantly affecting risk. In addition, consideration of multiple pathways and multiple pollutants reduces the possibility that a management strategy will be perceived falsely to pose an acceptable risk. Such an approach also ensures that an optimal strategy is found rather than simply an acceptable strategy, leading to possible savings in resources that might be directed towards other sources of risk.

The waste management strategy selected as optimal in this case study depended on the mitigation goals, including the fraction of population protected (a consideration of variability analysis) and the confidence level required in any prediction (a consideration of uncertainty analysis). This leaves open two approaches to making use of results from such an assessment. One possibility is to specify the fraction protected and confidence required before the assessment is produced, and to simple live with whatever management strategy is determined to be optimal. An alternative is to recognize that the fraction protected and confidence required are value judgments and can be traded against other goals (e.g. costs or benefits). In that case, a risk manager is best served by being presented with several alternative management strategies and the fraction protected/confidence levels associated with each. It may prove acceptable to adopt a management strategy that has a slightly lower fraction protected, and/or a slightly lower confidence in predictions, than a second strategy, if this slight decrease is more than offset by a large savings in cost.

This case study focused on a system that included the waste management system, the environmental system, and the exposed population. The amount of sludge to be treated was taken as a given, and the goal was to find a management strategy that optimized risk reduction within the boundaries of this problem. One way to extend the assessment to consider a broader range of risk management strategies would be to enlarge the system boundaries within which the optimal management strategies are searched. For example, the municipal and industrial processes before the generation of sludge could be incorporated to search for the optimum strategy in a broader context. Alternatives such as the modification of production processes and reuse, or changes in consumption patterns, could be compared with the waste management strategies considered here to find the overall optimal solution for an environmental problem.

REFERENCES

Covello, V. T. and M. W. Merkhofer, (1993) Risk Assessment Methods: Approaches for Assessing Health and Environmental Risks, Plenum Press, New York.

Crawford-Brown, D. J. (1997) Theoretical and Mathematical Foundations of Human Health Risk Analysis, Kluwer Academic Publishers, Dordrecht.

Daellenbach, H. G., (1994) Systems and Decision Making, John Wiley & Sons ltd., England.

Davies, J. C., (1986) "A Multimedia Approach to Pollution Control: Cross-Media Problems," in Y. Cohen, ed., Pollutants in a Multimedia Environment, Plenum Press, New York.

EPA (1989) Regulatory Impact Analysis of the Proposed Regulations for Sewage Sludge Use and Disposal, Washington, D.C.

EPA (1992) U.S. EPA's Integrated Risk Information System (IRIS), Office of Research and Development, Cincinnati.

EPA. (1994) Guidance for Performing Screening Level Risk Analyses at Combustion Facilities Burning Hazardous Wastes, Office of Emergency and Remedial Response, Washington, D.C.

Finley, G., (1994) "Recommended Distributions for Exposure Factors Frequently Used in Health Risk Assessment", Risk Analysis, Vol. 14, No. 4, 533-553.

Guruswamy, L., (1990) "Integrated Pollution Control: The Way Forward," Arizona Journal of International and Comparative Law, Vol. 7, No. 2, 173-224.

Hoffman, F. O. and J. S. Hammonds, (1994) "Propagation of Uncertainty in Risk Assessment: The Need to Distinguish Between Uncertainty Due to Lack of Knowledge and Uncertainty Due to Variability," Risk Analysis, Vol. 14, No. 5, 707-712.

Irwin, F. H., (1992) "An Integrated Framework for Preventing Pollution and Protecting the Environment," Environmental Law, Vol. 22, No. 1, 1-76.

Lue-Hing, C. (1992) Municipal Sewage Sludge Management, Technomic Publishing Co., Inc., Lancaster.

Miser, H. J. and E. S. Quade, (1985) Handbook for Systems Analysis: Overview of Uses, Procedures, Applications, and Practice, Elsevier Science Publishing Co.,Inc., New York.

Morgan, M. G. and M. Henrion, (1990) Uncertainty: A Guide to Dealing with Uncertainty in Quantitative Risk and Policy Analysis, Cambridge University Press, New York.

NRC (National Research Council) (1987) Multimedia Approaches to Pollution Control: A Symposium Proceedings, Washington, D.C.

Rao, V. R., (1995) "Risk Management: Time for Innovative Approaches," Environmental Management, Vol. 19, No. 3, 313-320.

Strenge, D. L. and P. J. Chamberlain, (1995) The Multimedia Environmental Pollutant Assessment System (MEPAS): Exposure Pathway and Human Health Impact Assessment Models, prepared for U.S.DOE at Battelle.

Travis, C. C., (1987) "The Extent of Multimedia Partitioning of Organic Chemicals," Chemosphere, Vol. 16, No. 1, 117-125.

van Gigch, J. P., (1991) System Design, Modeling and Metamodeling, Plenum Press, New York.

Whelan, G., 1992) "Overview of the Multimedia Environmental Pollutant Assessment System (MEPAS)," Hazardous Waste & Hazardous Materials, Vol. 9, No. 2, 191-208.

Whelan, G., (1996) The Multimedia Environmental Pollutant Assessment System (MEPAS): Groundwater Pathway Formulations, prepared for U.S.DOE at Battelle.

AUTHORS

Dr. Douglas J. Crawford Brown is a Professor and Associate Director, Educational Programs for the Carolina Environmental Program. He has a BS and MS in Physics and a PhD in Nuclear Science from Georgia Institute of Technology. Dr. Crawford's interests lie at the interface of science, mathematics, philosophy and policy, particularly as these relate to risk-based environmental decisions.

Hwong-Wen Ma (Ph.D., University of North Carolina, 1997) is working in Research Triangle Park at Research Triangle Institute (RTI), a non-profit research institute, as an environmental research scientist.

Chapter 19

Reclamation And Recycling Of Municipal Waste

A sludge dewatering process

Ifeanyi E. Madu

Dr. Ifeanyi Madu is an associate project manager with New York city department of environmental protection and an adjunct assistant professor of management science at Pace University, New York.

> Waste is a material that its producer does not want. Although the product may have value to someone (either in its present or in a converted state), if its producer does not ask for reimbursement for its removal, it is considered to be waste, and at some stage, will enter a waste handling system, either private or public

1. INTRODUCTION

Solid waste, within this context is defined to be the non-gaseous and non-liquid wastes that result from the daily activities of a community's residential, commercial, and industrial sectors [2].

In the United States, more than one hundred and fifty (150) million tons of solid waste is generated every year. New York City, for example generates about twenty-six (26) thousand tons of solid waste daily [3].

Disposal of Municipal waste is an issue of concern not only for Municipal officials but also for the environmental conscious public. In the past, it was a common practice for municipal wastes to be dumped in landfills where it is periodically set on fire to reduce it's organic contents without any regards or concerns to the environment or the health of the public. However, in recent years, due to rapidly dwindling landfill space, uncontrolled dumping of refuse leading to rodent and insect breeding, deterioration of air quality from open burning, deterioration of water quality from drainage through the dump, the general ugliness, rising incineration costs, environmental conscious public awareness, and government regulations, reclamation and recycling has become increasingly more popular in most parts of North

America and in Europe. A decade ago, the Environmental Protection Agency (EPA) established a voluntary guidance program with a goal of twenty five (25) percent recycling of municipal solid waste.

1.1   Reclamation

Reclamation refers to the removal of salvageable items, such as metals and paper, from other wastes to be reused or to serve as raw materials for industry [4]. Industries are already making use of reclaimed waste materials such as metals, paper, and textiles. In the United Kingdom for example, local authorities reclaim about twenty (20) thousand tons of waste paper and about thirty (30) thousand tons of metals annually [5].

1.2   Recycling

Recycling refers to the recovery and reuse of materials from spent products. Recycling has become a major part of environmental policy, largely due to increased costs of solid and hazardous waste disposal, the scarcity of natural resources, and the growing concern over polluted land, water, and air [3].

Recycling operations can either be internal or external. *Internal recycling* is the reuse of materials that are a waste product of the manufacturing process. It is commonly used in the metals industry. *External recycling* on the other hand is the reclaiming of materials from a product that has been worn out or rendered obsolete, for example, the collection of old newspapers and magazines for the manufacture of newsprint or other paper products [3].

Recycling greatly minimizes the land, water, and visual pollution that results from dumping of refuse. However, for recycling programs to succeed, the Federal government must invest greatly on both enhancing the markets for recycled goods and expanding the number of items in the waste stream that can be recycled [6]. The Federal government can ensure markets for recycled products by mandating their use, as appropriate by federal agencies and government contractors. The federal government should establish a national clearinghouse on innovative and successful recycling programs to which all municipalities have access [7].

2.   ENVIRONMENTAL PROTECTION

In the 1990s, Municipalities and many Corporations have come to realize the need to consider environmental impacts of their operations, and take measures to control and reduce waste and pollution rather than wait for government regulations. Quality driven businesses are learning that pollution prevention is often far less costly than regulatory compliance [8]. As in Total Quality Management (TQM) that encompasses the whole organization from the supplier to the customer, companies now see the need to practice Total Environmental Management (TEM) by taking environmental values as an integral part of their corporate cultures and management processes. Although environmental impacts are not always measured in conventional financial terms, they have a special value that companies find increasingly difficult to ignore [9]. Many corporations have realized the need to design and produce recyclable products as a competitive advantage. They have come to realize how huge savings in production costs and energy consumption can be achieved by the recycling of such materials as scrap metals (such as iron, steel, aluminum, copper, lead, and magnesium), waste paper, wood waste, used glass, and plastic scrap [10].

The need to protect the environment is not something new to the general public, businesses/corporations, and municipalities, however, the action and responsibility to protect the environment is something new that have emerged in the recent years. Some of the factors that helped expedite the urgency to protect the environment includes but not limited to the following:

2.1 Design For Environment (DFE):

As in quality control, in which the noted Japanese engineer Yenichi Taguchi advocates robust design, recommending the need to design quality into products by designing products that are insensitive to environmental factors, businesses are now finding it more efficient to design products for disassembly, and recyclability at the outset than to deal with disposal problems at the end of a products life cycle. Many companies now produce reusable products, utilizing waste materials in manufacturing these goods, and consumers on the other hand, reuse the products and collect the materials to recycle out of the waste stream and back to the producers. For example, General Motors designed its Saturn line of cars to be easily disassembled, allowing about ninety-five (95) percent of the automobile to be recycled with less contamination of waste streams. Sonoco, a worldwide packaging products company, introduced a take back policy for many of its paper, plastic, and wood packaging products in the early 1990s that reduced waste disposal problems for its customers [8]. In recent years, Design for Environment has become an integral part of pollution prevention and control in environmental management.

2.2 Government Regulations

In the recent years, many regulations and Acts have been passed at the Federal, State, and Local levels to protect the environment, and many agencies have been charged with the responsibility of monitoring company activities, and enforcing the laws and regulations. Today, in the United States, there are more than one hundred thousand (100,000) federal, state, and local environmental rules and regulations [8]. Responsible companies have no choice but to comply, as the fines can be very substantial and in some cases may include jail time. Most states and municipalities have enacted legislation mandating municipal waste recycling. Some of these legislations are in part due to perceived lack of disposal facilities and the increasing waste generation rates.

2.3   Legal Liabilities

The law empowers citizens to sue both private and industry and government agencies for violating antipollution standards. Many environmental organizations like the National Resources Defense Council, and the Environmental Defense Fund, specialize in bringing lawsuits against offenders.

2.4   Competition

The increased global competition, expansion of the global market, and the proliferation of international trade agreements has being a driving force toward voluntary international standards for environmental quality management [8]. The environmental conscious and informed consumers demand for products that meet or exceed environmental standards have made companies rethink the way they conduct business. Many companies have reverted to producing and packaging products in a way that they can be recycled or reclaimed. This gives a company a competitive advantage over those that ignores the needs of the consumer. Changes in environmental thinking have reshaped businesses and redefined the way businesses are conducted. The need to protect the environment and conserve natural resources is now a value embraced by the most competitive and successful multinational companies [11]. A survey study by the Mckinsey Corporation of four hundred (400) senior executives of companies around the world found that a majority of those surveyed believe that it is their corporation's responsibility to control pollution. About eighty-three (83) percent of those surveyed also agree that corporations have an environmental responsibility for their products even after they leave the plant [8].

2.5   Product Stewardship

This refers to practices that reduce environmental risks or problems resulting from the design, manufacturing, distribution, use, or disposal of products [8]. For example, Germany has take back laws that make companies responsible for reclaiming and recycling their products. Many companies at present are using life-cycle analysis (LCA) to determine ways of reducing or eliminating waste at all stages of their operation. Their waste reduction and elimination efforts range from raw materials acquisition, production, distribution, and customer use to waste reclamation, recycling, reuse, and disposal [12]. In 1976, the United States Congress passed the Resource Conservation and Recovery Act, encouraging States to formulate solid waste recovery plans. Special departments were set up by some States to assist local communities in their recycling efforts [3]. Some States and Cities like New York adopted legislation that gives consumers the option of returning some bottles and cans for recycling in exchange for a small deposit (five cents in N.Y.) paid at the time of purchase.

3.   RECYCLING BENEFITS

In order to achieve the environmental goal of recycling, the public will have to be convinced of the benefits of recycling or their will be some resistance to the need of recycling as evident today. Listed below are some of the benefits of recycling:

(1).   It's potential to reduce fossil fuels consumption, because oil is refined to create plastic resins for such products as soda bottles, food wrappers, and throw-away cameras.

(2).   Aesthetic value- reducing the amount of litter in communities. The Bottles Bill, which requires deposits on recyclable bottles, is one of the most successful litter reduction measures.

(3).   Recycling reduces the amount of household and commercial waste that ends up in already overburdened landfills.

However, the success of any recycling program depends on several factors such as:

- general awareness of the problems caused by solid waste disposal
- an effective, inexpensive method for separating and collecting the recyclable materials
- the economical feasibility for industries to use and market recycled materials.

4.   SLUDGE PROCESSING AND DISPOSAL

Another form of waste that is a threat to the environment if left untreated but can be beneficial if recycled is sludge. Sludge is the solid material, which settles out of the wastewater during treatment. It is the soupy solid left over when Municipal Sewage Treatment Plants finish purifying water used by homes, businesses, and industries [13].

In the treatment of domestic sewage, sludge is produced as a by-product of the treatment process. Sludge is withdrawn from the process at the primary settling tanks, and at the secondary settling tanks as wasted excess activated sludge. This sludge is digested at the treatment plants prior to disposal. This digestion process changes the character of the sludge and produces methane gas as a by-product, which is flared at the plant or used as fuel to generate electricity. Prior to 1988, sludge after digestion was barged to sea and dumped at the 106-mile dumpsite in Atlantic Ocean. On November 18, 1988, following two summers of pollution problems at New Jersey and New York beaches, the public rallied against ocean pollution. The United States Congress heeded the peoples call and passed legislation "The Ocean Dumping Ban Act" which prohibited the ocean dumping of sludge after June 30, 1992, and ordered that by 1994, safe methods be developed to recycle sludge onto farmland. The ban on ocean dumping came at the time Municipalities were faced with diminishing landfills in which to dump the sludge, and as scientists recognized that sludge was polluting the ocean. As a result of the Ocean Dumping Ban Act, alternate sludge disposal methods were therefore, required to dispose of sludge after that date. One method that was proposed and implemented was sludge dewatering - a recycling process. After sludge is recycled into what is known as sludge cake, it has many beneficial applications, and in some cases has been commercialized by some municipalities. Some benefits and applications of recycled sludge other than preventing ocean/land pollution includes:

1.   Recycled sludge is used as fertilizer, and in soil building to restore organic composition of farmlands.
2.   It is used to reclaim soils destroyed by mining.
3.   Recycled sludge is commonly used in landscaping.
4.   It can also be used to clean up superfund sites and inner-city soils contaminated by lead.

Some scientists and scholars however, have argued that recycled sludge has its own problems that must be recognized. Some of these problems as argued are as follows:

1.   Long term poisoning of soils from heavy metals and toxic organic chemicals. Sludge they argue have high levels of heavy metals produced by industries, and from erosion of lead and copper pipes noting evidence that biodegradable materials dumped in landfills do not significantly

decompose even after several decades. However, it was determined that high quality sludge as defined under EPA's regulation Part 503 does not have concentrations of heavy metals high enough to contaminate receiving sites. Part 503 classifies sludge based on levels of pathogens and heavy metals, and limited uses for each level of quality.

2. One major concern of many communities is the fear of disease causing bacteria, since sludge recycling is new and there is not enough data available to ease the fear of the communities.

3. Bad experiences with odorous sludge have led some communities to ban its use.

5. DEWATERING AND RECYCLING OF SLUDGE:
Case Study Of NYC's Dewatering Facility
[Source: NYCDEP]

After Congress passed the Ocean Dumping Ban Act (ODBA) on November 18, 1988, the City of New York on August 10, 1989 entered into a Consent Decree and Enforcement Agreement to develop and implement alternate sludge disposal methods. Under this Decree, the City of New York agreed to cease ocean dumping of twenty (20) percent of the sludge by December 31, 1991, and all ocean-dumping of sludge to cease by June 30, 1992. To meet this decree, the city proposed and implemented the construction of eight (8) dewatering facilities to process sludge: Bowery Bay, Hunts Point, Red Hook, 26th Ward, Jamaica, Oakwood Beach, Tallmans Island, and Wards Island, to process sludge. In this chapter the dewatering process at the 26th Ward facility will be described since it is one of the biggest among the eight plants and of identical design to the other big plants. A schematic diagram (*source: NYCDEP*) of the 26th Ward dewatering facility is shown on figures 1 and 2.

DESCRIPTION OF THE OPERATION
Undewatered Sludge Handling.
After digestion, sludge is stored in sludge storage tanks prior to dewatering. Sludge from 26th Ward as well as three other plants; Rockaway, Coney Island, and Owls Head (see fig. 1) are treated at the 26th Ward dewatering facility. There are provisions for two separate sludge-handling trains because of the different composition of sludge from different parts of the area served. In this manner, sludge can be segregated based on its composition, which allows disposal using alternate methods. Three sludge

storage tanks are provided (fig. 1) to store the sludge prior to dewatering. Tank #1 stores sludge from 26th Ward, Tank #2 stores sludge from Rockaway, and Tank #3 stores sludge from Coney Island. The piping to these tanks is common with isolation valves so that if one tank is down for repair or cleaning, the sludge normally stored in it can be diverted to another tank. The sludge from these tanks is normally pumped to wet wells with three (3) sets of pumps comprising of two (2) pumps per set with one set for each storage tank. Five grinders are installed in each pump set suction lines to macerate any material, rags, etc., that remains in the sludge after digestion. From the wet wells, the sludge flows by gravity to two (2) different piping loops and is pumped from these loops to the centrifuges by two (2) different pumping systems to keep it segregated. A grinder with bypass provisions is installed in the piping from the wet well to each sludge piping loop.

Centrifuges

The sludge dewatering building has thirteen solid bowl centrifuges. These centrifuges rotate and spin the sludge creating elevated "G" forces, which separate the liquid portion of the sludge from the solid portion to form what is called sludge "cake". The liquid portion called centrate flows to the centrate wet well (fig. 1) from where it is then pumped to the aeration tanks for treatment. The solid portion (sludge cake) is conveyed by a combination of belt and screw conveyors to sludge haulage trucks, which transports the sludge cake for disposal.

Polymer Systems

Two (2) polymer storage tanks are provided to store, mix and age the polymer for use in the centrifuges. This polymer comes in two forms, liquid or dry. The selection of which form to use is usually based on actual operation. The selection is made based on the efficiency of dewatering, cost and ease of use. Usually, dry polymer is a less expensive product on a pound for pound basis but it's operation and handling is more difficult. At the 26th Ward dewatering facility liquid polymer is being used. Liquid polymer is a synthetic organic material. It is stored in the storage tanks. It can be either water or oil based and must be mixed in proper proportions and aged prior to use. Usual feed concentrations to the centrifuges are from 0.1 to 0.3 percent by weight or less. Liquid polymer from the storage tanks is mixed with a predetermined amount of water in the polymer mixing and aging tanks. At these tanks, the polymer solution ages prior to use to allow long molecule chains to form, which assists in the agglomeration of the sludge particles. This agglomeration permits sludge dewatering to take place more efficiently (i.e., a dryer sludge cake and clearer centrate formed). To enhance the separation of the solids from the liquid portion of the sludge, the liquid polymer is mixed with the incoming sludge at the centrifuge. Approximately,

95 percent of the sludge solids are separated from the liquid sludge in the centrifuge, and the remaining 5 percent are retained in the centrate.

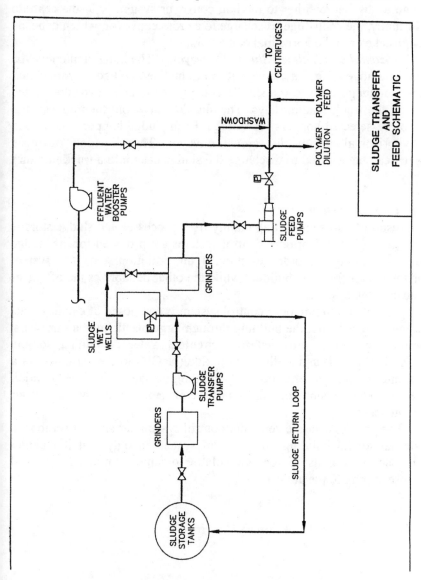

Figure 1: Sludge Transfer and Feed Schematic

Dewatered Sludge Conveying System
To convey the dewatered sludge (sludge cake) to hoppers in the truck loading bays, a dual system of belt conveyors are provided.

. *Cross Conveyors:* Each centrifuge discharges into a hopper located under it. The hoppers are equipped with a diverter, which can divert the centrifuge to discharge either to the belt conveyor system or to the centrate line. Normally, the centrifuges discharge to a cross conveyor, which deposits the sludge cake onto the horizontal conveyors.

. *Horizontal and Inclined Sludge Conveyors:* The horizontal conveyor collects the sludge cake and conveys it to the inclined belt conveyors which elevates the sludge cake to the shuttle conveyors mounted over the sludge hoppers in the truck loading bays. The shuttle conveyors then convey the sludge to pre-selected hoppers for storage. Each sludge hopper is equipped with a horizontal, reversible screw conveyor. The direction of these conveyors can be selected to discharge the sludge cake into a truck in either truck bay.

Odor Control Systems.

As a result of odors emanating mainly from processes and sludge storage functions, two (2) types of odor control systems are provided for the sludge dewatering building, in order to prevent air pollution. One (1) system deodorizes the air from the building while the other deodorizes the off-gases from the centrate system.

. The main building odor control system is composed of exhaust fans, which pump the air from the building through large plastic media scrubbers. A pumping system is provided to recirculate water containing sodium hypochlorite (NaOCl) and sodium hydroxide (NaOH) to the scrubbers in a path counter-current to the flow of building exhaust air. The solution oxidizes the odor causing components of the air before it discharges the air to the atmosphere.

. The process centrate vent odor control system has smaller scrubbers. It is similar to the building odor control system in only that it includes carbon adsorption units to remove volatile organic compounds (VOCs) before discharging to the atmosphere.

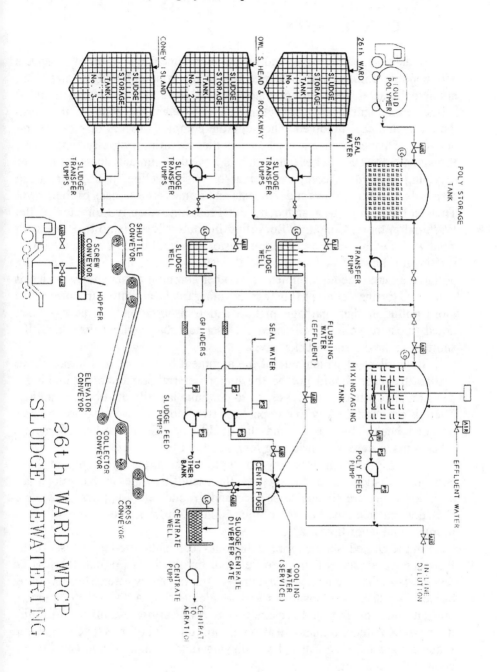

Figure 2: Sludge Transfer and Feed Schematic

6.   CONCLUSION

In as much as recycling and reclamation have shown a lot of prospects and benefits for the need to protect the environment and in managing municipal wastes, recycling opponents/critics have in recent years mounted serious campaigns against recycling asking whether the benefits are worth the effort and cost involved. They cite the complex process of pick up and delivery of paper products, cans and bottles as a drain on city funds [14]. Proponents however believe that this is a weak argument since the cost of recycling is hard to compare to that of not recycling, noting that trash disposals in landfills have hidden costs and immeasurable risks of their own. Their contention is best stated by Pat Franklin; executive director of the Washington based Container Recycling Institute "There is no free ride with what we do with our post-consumer waste, there will be cost associated with it. But you have to add in the social costs – what is the cost of spewing polluted air into the environment". Another common argument by critics as noted by Horrigan et al. [14] is that recycling trucks and infrastructure cause more pollution than garbage pickup and transportation, but as has been noted, most recycling facilities are closer to urban centers than landfills, shortening the distance trucks travel.

Regardless of which side of the argument one is on, it is important to realize that cost should not be the major determining factor in making a decision as to whether to recycle or to continue landfill and ocean dumping. There is no possible way to compare the two. How can we measure the cost of good health, happiness, and the quality of life of the environment's inhabitants? There is no price tag for a healthy life.

The EPA which is entrusted with policing the environment have enough statistical data to suggest that landfills release hazardous air emissions, and are threats to surface and groundwater supplies. There are EPA documented cases of acute injury and death from fires and explosions of municipal landfill gas [23].

In the United States, for recycling and reclamation to be successful, the federal government will have to be more involved. At present, there is no federal law that mandates recycling of municipal waste. The recycling burden is placed on consumers and municipalities. A federal government mandate on recycling and reclamation is necessary to establish uniformity nationwide. A federal mandate also will not run afoul of interstate commerce protections and will create a level playing field so that no one is put at an economic disadvantage.

REFERENCES

N. Morse, E.W. Roth, *Systems Analysis of Regional Solid Waste Handling,* U.S. Department of Health Service, Environmental Health Service, Bureau of Solid Waste Management, 1970.

Kenneth C. Clayton, John M. Huie, *Solid Wastes Management – The Regional Approach,* Ballinger Publishing Company, Cambridge Massachusetts.

Paul J. Allen, *Recycling,* Compton's Encyclopedia Online v3.0, The Learning Company, Inc., 1998.

Gabor Karadi, *Garbage and Refuse Disposal,* Compton's Encyclopedia Online v3.0, The Learning Company, Inc., 1998.

Embassy of the United Kingdom, Washington DC, *United Kingdom: Nature Conservation.,*Countries of the World, Infonautics Corporation, 1998.

U.S. Govt., *Municipal Solid Waste Disposal Crisis,* U.S. Gov't Printing Office, Washington, 1989.

Burt Stallwood, *Hearing before the subcommittee on Transportation and Hazardous material,* June 22, 1989.

Berry, Michael A., Rondinelli, Dennis A., *Proactive Corporate Environment Management: A new Industrial Revolution.* Vol. 12, The Academy of Management Executive, 05-01-1998, pp. 38(13).

T.F.P. Sullivan, *The Greening of American Business – Making Bottom Line Sense of Environmental Responsibility,* (ed.), Rockville, MD: Government Institutes, Inc., 1992.

P.F. Collier, *Industrial Waste Disposal,* A Division of Newfield Publication – Infonautics Corporation.

S. Hart, *Beyond Greening: Strategies for a Sustainable World,* Harvard Business Review, (January-February): 68-77, 1997.

P.S. Dillon, M.S. Baram, *Forces Shaping the Development and Use of Product Stewardship in the Private Sector.* In K. Fischer and J. Schot (Eds), Environment Strategies for Industry: International Perspective on Research Needs and Policy Implications, Washington, DC: Island Press: 329-341, 1993.

Jane Ellen Stevens, *Scientists exchange verbal blows over risk of recycled sewage,* The Dallas Morning News, 03-23-1998, pp. 6D.

Horrigan, Alice; Motavalli, Jim, Talking trash. (Recycling) (Includes related article on history of recycling) (Waste Not, part 1), Vol. 8, E Magazine, 03-13-1997, pp. 28(8).

AUTHOR

Dr. Ifeanyi Madu is an associate project manager with New York City department of environmental protection and an adjunct assistant professor of management science at Pace University, New York.

Chapter 20

Challenging the Future
Ways Towards Sustainable Green Electronics

Jutta Müller, Otmar Deubzer, Hansjörg Griese, Harald Pötter & Herbert Reichl
Fraunhofer Institut Zuverlässigkeit und Mikrointegration
(Franhofer Institute for Reliability and Micro-Ibntegration)

1. TAKING UP THE CHALLENGE: GREEN ELECTRONICS

Microelectronics, on the surface, appears as a clean high technology industry, which creates benefits for mankind. It is estimated that the worldwide sales of electronic products in 1998 was over 1 trillion U.S. dollars. The electronics industry is therefore perceived as a very successful industry. This success, however, takes the electronics industry to the environmental limits: the increased electronic devices require increased resources and the growing amounts of emissions and waste from electronic devices burden the environment.

Natural resources are limited and so is the capacity of the environment to absorb and to assimilate the material consequences of environmental activities such as emissions and waste. There is therefore a need to conserve resources and keep emissions and waste within acceptable limits or to cut them back if they are already beyond these limits. This is what the world community agreed upon at the Earth Summit in Rio de Janeiro in 1992 under the term "sustainable development".

The framework of sustainable development challenges the electronics industry as the touchstone for economic success in the next century.

1.1   Environmental assessment

Where are the environmental hot spots in the electronics industry? To identify them, an ecological assessment needs to be conducted and its results interpreted. Life cycle analysis (LCA) is the most sophisticated tool to identify and compare the environmental impacts caused by products and processes. LCA reviews a product by evaluating its material and energy flows throughout its life. Detailed process data for all stages of the life cycle have to be provided. This makes the application of LCA during product design very complicated since some of the needed information may be unknown. This is common in the electronics industry, where there are many materials and processes involved in the production process. Nevertheless, LCA can be a suitable tool for selected process steps (see section 2.2). Other simplified tools have been developed. They are also evaluated using checklists format. This is discussed in section 3.

1.2   Environmental hot spots in the life chains of electronic products

The Fraunhofer IZM has identified hot spots in the life cycle of a selected electronic product (Figure 1). From this assessment it is clear that the main environmental impact of electronic products occur at the manufacturing phase.

Thus, reducing the environmental impacts of electronics in manufacturing offers the best environmental and very often also economical benefit. In section 2, examples for creating more environmental friendly processes and products are shown.

The next stage of environmental improvement in the electronics industry is to apply life cycle thinking to processes and products. This holistic approach towards green electronics has resulted in the principles of Design for Environment (DfE) (see section 3). Two examples are used to show, how DfE facilitates linking the design phase with the end of life phase and the manufacturing of products (see sections 3.3 and 3.4). Closed loop application and green marketing help to save resources and reduce the amounts of waste in every stage of the product life cycle. So successful enterprises show already today how the electronics industry can meet tomorrow's urgent demands.

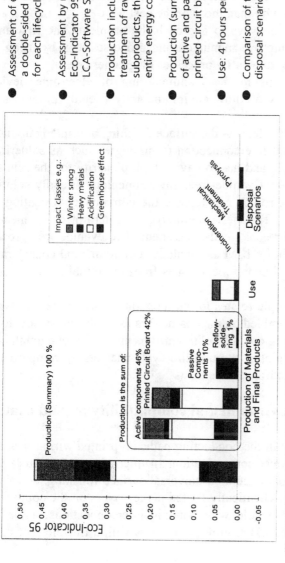

Figure 1: Life Cycle Evaluation of an HDTV-Converter DSP-Unit (use: 4 hours/day in 3 years)
(Source: Ast.)

2.   CREATING GREEN PROCESSES IN THE ELECTRONICS INDUSTRY

New electronic products and systems are characterized by high performance, functionality, high quality and reliability and low costs. Higher integrated and miniaturized devices with new assembly and interconnection technologies are required in order to realize these attributes. This technical progress offers good opportunities to implement environmental improvements to assure that these new technologies are more environmentally friendly than those of the past were.

There are different ways to approach the problem. The first and most obvious one is to improve manufacturing processes, e.g. by using greener materials, or to substitute entire processes by greener ones.

Two relevant steps in the manufacturing of electronic devices are selected in order to show how environmentally friendly processes can be developed.

The first example relates to the surface finishing of printed circuit boards. This involves several interconnection technologies such as soldering, wire and adhesive bonding and has to have a high planarity for the miniaturized printed circuit boards. Cyanide-free, environmentally friendly solutions for gold coating in electronics can replace the common gold coating process with toxic cyanide. The best alternative is determined by means of a simplified model for ecological assessment. The cyanide-free process was developed at Fraunhofer IZM and includes environmental considerations as well as technical and economical issues from the initial planning stages of product development.

The second example relates to the interconnection process in which the components are assembled on the board. This example shows how toxic lead used in manufacturing could be eliminated as more environmentally friendly alternatives are developed through life cycle analysis by comparing energy consumption and landfill requirements.

2.1   More environmental compatibility in gold coating

The copper used in the conducting paths on printed wiring boards needs to be protected against contamination. In microelectronics Au layers gain more and more importance as a universal surface finishing. An electroless chemical process on chemically coated nickel layers on the copper deposits these layers. Reliable gold wire bonding interconnections in the chip on board technology (COB) require gold layers with a thickness of 0,3 µm or more.

First a sudgold layer with a thickness of less than 0,1 µm is generated in a charge exchange process. This thin gold layer has to be reinforced by a chemical reduction process. The key components of such a coating bath are a complexing and a reducing agent. While the complexer keeps the cat-ions of gold in solution, they are reduced only at the active surface by the reducing agent. Thus coating layers of any thickness can be deposited.

State of the art for this kind of gold coating is the application of baths containing cyanide as complexing agent and as free cyanide. But the highly toxic cyanide is a potential danger for occupational safety and for the environment, if it is accidentally released into the atmosphere. Its use therefore, requires high expenses for safety and environmental protection.

The administrative regulations based on the German Water Balance Act (Wasserhaushaltsgesetz) as well as the professional association of the chemical industry in Germany demand that cyanide be replaced with a less harmful substance whenever possible.

Additionally, old bath solutions have to be detoxified before disposal. This is often expensive to achieve. There are different possibilities, e.g. chemical oxidation using sodium hypochlorite (liquid bleach), electrolysis, oxidation using ozone or by hydrogen peroxide with additional ultraviolet radiation. All methods are either costly or increase the salt load of the wastewater. Oxidation by sodium hypochlorite is widely used, however, this method also implies additional risks. In the bath toxic chloro-organic compounds may be formed with organic substances such as tensides and then the AOX-threshold value has to be kept.

Further weakpoints of the cyanide gold bath are its difficult handling, its low thermal stability and its nickel ion sensivity. Finally, because of their alkalinity, these baths are not compatible with the masking process.

There are therefore, various reasons for developing new baths and coating technology. The development objectives for a new Au coating bath are a technically, economically and ecologically optimized bath will have the following characteristics:
* No toxic cyanide as complexing agent
* Can be disposed of in an environmentally friendly manner and at acceptable costs
* a high bath stability against high temperatures and contamination in the bath (e. g. Ni)
* a pH value of 7 or less
* maintains an operating temperature as low as possible
* has coating rates higher than 2 µm/h.

In the following sections, the efforts for an environmentally friendly bath are particularly described in more detail. Finally the results of the ecological, technical/technological and economic optimization are presented.

2.1.1   Selection of more environmentally compatible substances for a gold coating bath

As a first step towards a substitute for cyanide, the literature on complex formations was surveyed to find compounds that form sufficiently stable complexes with Au (bI)-cat-ions. Other important properties that they should have are solubility in water and stability at temperatures up to 80° C.

The pre-selection led to a number of organic sulphur, phosphorous and nitrogen compounds. At this early stage of development, the environmental impact of these compounds was assessed and compared to cyanide. For this the IZM model of ecological assessment "Toxic Potential Indicator" was applied.

After the selection of a green complexer the most environmentally friendly disposal of the bath solutions (after their use) with their high organic load was investigated.

2.1.1.1   Model for the ecological assessment

The ecological assessment was carried out using the Toxic Potential Indicator developed in the Fraunhofer Institute for Reliability and Microintegration in Berlin (IZM-TPI). The IZM-TPI is a simplified tool for the environmental evaluation of materials and products. It is less detailed than a full-scale LCA (Life Cycle Analysis) and does not include the whole life cycle of a material or product. But it can be calculated very quickly and is based on data, which are reliable and easily accessible. The IZM-TPI is based on the following legal threshold values and classifications for substances, the Maximal Workplace Concentration (MAK-value), the risk classification from the German Hazardous Substances Declaration ("R-Werte der Gefahrstoffverordnung") and the Water Pollution Class (Wassergefährdungsklasse). Most of the data can be found in the material safety data sheets provided by the material producer.

All the data is standardized, aggregated and converted into a single value on a scale that ranges from zero (no considerable impact) to 100 (worst rating) per mg. (Figure 2)

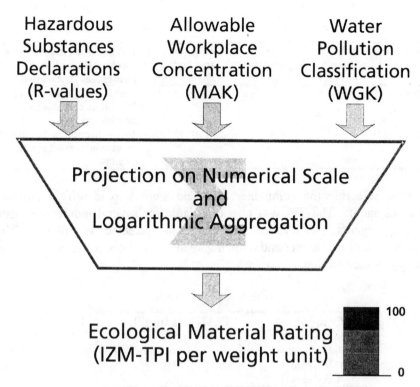

Figure 2: Assessment model for the IZM-TPI

2.1.1.2 Results of the ecological assessment

Table 1 shows some of the complexing agents and their IZM-TPI-values compared to cyanide.

Table 1: Complexing agents and their environmental assessments

Compound	IZM-TPI [mg^{-1}]
Potassium Cyanide	67,5
Trimethylphosphine (TMP)	35,6
Mercaptosuccinic acid (MA)	33,5
2,3-Dimercaptopropane-1-sulfonic acid	1,6
Complexing agent B	0,0

According to the results of the ecological assessment, 2,3-Dimercaptopropane-1-sulfonic acid and the complexing agent B are the preferred substances, but 2,3-Dimercaptopropane-1-sulfonic acid was rejected due to its high cost. For reasons of patent law, the compound B cannot be revealed yet. In **Error! Reference source not found.**, the detailed threshold values and classifications of cyanide and the complexer B are listed.

Table 2: Threshold values and classifications of cyanide and the complexer B

Complexer	MAK-value [ml m^{-3}]	Water pollution class	LD$_{50}$ (or. rat) [mg kg^{-1}]	R-description according to German Hazardous Substances Declaration
Cyanide	5	3	5	Highly toxic in the case of inhalation, contamination, swallowing
Complexer B	-	0	> 15000	Maybe harmful in the case of inhalation, contamination, swallowing

After selecting the complexer, the next step is to identify a suitable reducing agent. The TPI was also calculated for several industrially used reducing agents (see Table 3) given the selected complexer. The combination of complexer and reducing agent must show sufficient chemical stability in the solution.

Table 3: Reducing agents

Compound	IZM-TPI [mg^{-1}]
Hydrazine	77,9
Formaldehyde	20,4
Thiosulfate	0,0
Ascorbic acid	0,0

Table 3 shows that thiosulfate and ascorbic acid have good environmental compatibility and should be preferred. For the combination with the cyanide-free complexer B, thiosulfate was selected.

2.1.2   Disposal of cyanide-free baths

A further aspect of an environmentally friendly bath is in its disposal with little environmental impact.

A closed loop approach for internal recycling of the bath solution is limited to a few cycles, after which, the bath solution has to be disposed of. For the observance of the various regulations the sewage has to be pretreated.

The first step is to remove the gold in the bath through electrochemical process. Before disposing of the gold-free solution, the organic load in the solution must be reduced even if it is non-toxic. Otherwise, the self-cleaning capacity of natural waters could be exceeded. The limit is 600 mg·l^{-1} COD (chemical oxygen demand) for the plating industry in Germany.

In general, biological degradation is preferable to chemical treatment of wastewater with high concentrations of organic compounds. The constituents then can be removed without additional chemical agents. Microorganisms use the pollutants as a source of carbon and/or energy. They are metabolized to Carbon dioxide, water and biomass and thus eliminated from the water.

Figure 3 shows the biodegradation of a cyanide-free plating bath solution as decrease of COD over time. Further investigations showed that the residual COD resulted from the reducing agent in the solution. The complexing agent was totally biodegradable.

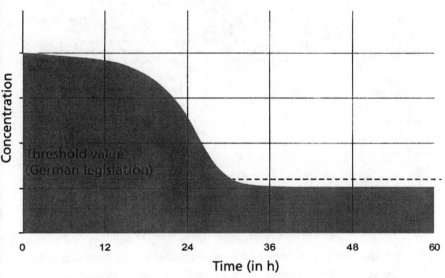

Figure 3: Biodegradation of a worn out bath with the cyanide-free complexer B

2.1.3 Results: Surface and bath properties

The new cyanide-free bath generates high quality gold surfaces for various interconnection technologies. The gold layers are well solderable and bondable with aluminum and gold wire. Figure 4 shows the results of a thermosonic bondability check. All pull forces are higher than 12 cN.

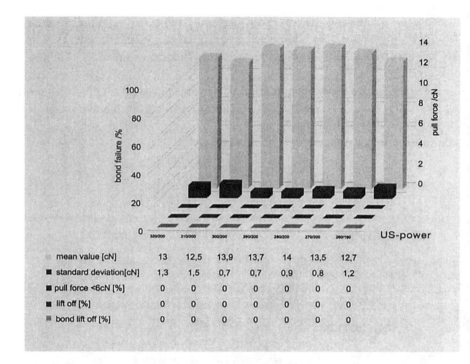

mean value [cN]	13	12,5	13,9	13,7	14	13,5	12,7
■ standard deviation[cN]	1,3	1,5	0,7	0,7	0,9	0,8	1,2
■ pull force <6cN [%]	0	0	0	0	0	0	0
■ lift off [%]	0	0	0	0	0	0	0
■ bond lift off [%]	0	0	0	0	0	0	0

Figure 4: Bonding properties of layers derived from the new cyanide-free plating system: Thermosonic bondability check (33 μm Au wire, 120 °C)

Table 4 shows the properties of a conventional cyanide-plating bath and of the generated layers in comparison to the cyanide-free system.

Table 4: Conventional and new plating solution in comparison

	Conventional plating bath (cyanide containing)	Cyanide-free plating bath
Bath properties		
Environmental compatibility	Strongly toxic and alkaline	Non-toxic and biodegradable
Process temperature [°C]	65...75	30...50
PH-range	13...14	4...6,5
Plating rate [μm/h]	1...2,5	2...5
Gold content [g/l]	2,5	2...6
Bath heating	< 2W/cm	Any
Plating base	Au, Pd or Cu	Au, Ni or Pd
Bath stability	Very sensitive (Ni ions!)	Good
Layer properties		
Roughness [nm]: R_{D-V}/R_{ms}	415/66	415/61
Plastic hardness [N/mm^2]	2160	4080
Au wire bondability	good	Good
Solderability	good	Good

Figure 5 shows a testing board plated in the cyanide-free bath.

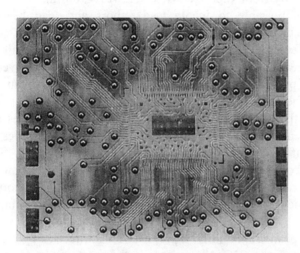

Figure 5: Test board Au-plated with the cyanide-free bath

Considering ecological as well as technical and economical aspects at an early stage of development resulted in an optimized cyanide-free bath for gold coating. It combines technical feasibility and environmental compatibility by the use of non-toxic, partly biodegradable constituents. Furthermore, it is a high-tech surface finishing for lead-free electronic devices.

2.2 Reduced environmental impacts by lead-free electronic assemblies

Up to now mostly tin lead solder has been used to connect electronic devices and components to circuit boards. This is the main reason why electronic equipment contains the heavy metal lead. Further sources are the surface finishing of the printed circuit board, the package leads or the component terminals.

Although the use of lead by the electronics industry is altogether minimal (only appr. $14*10^6$ kg annually worldwide, which are less than 1 %) the potential for lead exposure cannot be ignored. The heavy metal lead accumulates in the human body under chronic exposition and harms the blood picture, as well as the nervous and reproduction system especially of children. Lead in electronic devices endangers humans and the environment in particular by the contamination of groundwater from disposed electronic waste and by wastewater from flux cleaning with aqueous solutions. In the production process at usual soldering temperatures the vapor pressure of lead is so low that no health risks have to be expected from the inhalation of vapors (but possibly from vaporized flux). However, during wave soldering operations, it is possible to inhale lead vapors and lead bearing dust generated by the dross of oxidized solder.

The toxic effects of lead are well known and legislation has initiated and implemented measures for a controlled material flow and even partial prohibition to protect the biosphere. Examples are the ban of lead compounds in fuels and in dyes, the replacement of lead pipes in drinking water supplies, the controlled use of lead bullets for sport rifles and lead electrodes in accumulators. Up to the early nineties the electronics industry had not been affected by administrative regulations on lead. Since the quantity of electronic products has strongly increased, the electronic industry was included into the discussion on a general lead ban. Simultaneously there was a growing demand for an environmentally acceptable handling of used devices.

For the European Community, a proposal for a directive by the General Direction Environment (DG XI) exists on the waste from electrical and electronic equipment demanding a ban of lead in new products from 2004 on.

The Japanese Ministry of International Trade and Industry, MITI, has presented a similar law on take back and recycling of electrical and electronic equipment. It contains no explicit ban of lead, but demands its recovery from the disposed products. Because of the high effort for collection and dismantling especially of small devices, many product responsible Japanese companies have started to develop lead-free interconnection technologies (see Table 5).

In the USA the "Lead Exposure Reduction Law" is effective, which up to now excludes the electronics industry, and the "Lead Tax Law" is still under consideration.

2.2.1   Industrial activities on the way to lead-free electronics

The pressures mounted by legislators and "green" publicity as well as the competition between the companies themselves led to activities in electronics industry on the way to lead-free electronics.

In January 1998, the Japan Electronic Industry Development Association (JEIDA) and the Japan Institute of Electronic Packaging (JIEP) published a roadmap for soldering interconnection technology. Japanese electronic manufacturers also used this for their orientation (see Table 5). In addition, since April 1998 electronic manufacturers, producers of components and solders as well as universities in Japan work on lead-free solders in the NEDO project (2 years, 350 Mio Yen).

Table 5:   Status quo and aims of some manufacturers in the field of lead-free solders

Enterprise	Status quo (Product Example)	Aims
Nortel (IEC Electronics Corp.)	Pb-free telephone with SnCu0,7 solder (Nortel Award of Merit in the category "Innovation" in 1996, Environment Award 1997 of The Financial Post)	
Matsushita (Panasonic)	Pb-free Mini disk player with SnAg3,5Bi4,8 solder from 1. 10. 1998 on Japanese market, from summer 1999 on European market; Label: Produced for the Environment	No Pb containing solders from 2001 on, more Pb-free products (car radios, tvs) from begin of 1999 in Japan, from mid 1999 in Europe
Hitachi		Pb solder reduction to the half from 1997–1999, from 2001 on no more Pb containing solder
NEC		Pb solder reduction to the half until 2002
Toshiba	Pb-free high density cellular Phone (development phase)	Pb-free mobile phones from 2000 on
Sony		Development of Pb-free solders, Pb-free interconnection technology from 2001 (exception: high density electronic packaging)
Ricoh		Use of SnAg3.5 in automotive applications

In the USA today, it is mainly the automobile industry that is interested in lead-free solders. Since 1997 the National Center for Manufacturing Sciences (NCMS) in the USA directs an extensive project on lead-free solders in which 11 industry enterprises (among them Lucent Technologies, Texas Instruments, Rockwell International and Ford Motor Company), institutes and laboratories take part ($19M). The American pcb society (IPC) conducted a conference "IPCWorks '99" in October, 1999 in Minneapolis on the subject of lead-free solders.

European examples concerning lead-free solders are the Brite Euram Projekt IDEALS from 1996 – 1999 carried out by different enterprises like Multicore, GEC, Philips, Siemens, Witmetaal and NMRC - and the recently published latest study of the International Tin Research Institute (ITRI).

Activities for the reduction of lead use are also known from Scandinavian countries, where the ministers of environment have signed a statement on a ban on lead.

The most suitable systems, according to the mentioned studies, contain tin as base metal. Table 6 shows some results.

Table 6: Selected lead-free solders in comparison with SnPb37

Alloy	Use	Properties	
		Melting temperature [°C]	Remarks
SnPb 37	Present standard	183	
SnBi 58	Consumer electronics (ce), tele-communi-cations (tc)	139	Reduced reliability if Pb contaminated because of low melting phases (SnPbBi), only for products with low operating temperature
SnZn 9		199	Tending to oxidation, reduced wetting caused by Zn, corrodible
SnAg 3.5Bi 4.8	ce, tc,astronautics, automotive Matsushita MD-Player	210	Problems with wave soldering, Reduced reliability if Pb contaminated because of low melting phases (SnPbBi)
SnAg 3.5 (also with Cu, e. g. SnAg 3,8Cu 0,7)	ce, tc, astronautics, automotive	221	
SnCu 0.7	ce, tc Nortel Telephone	227	Inert gas atmosphere necessary for wave soldering

The technical feasibility of these lead-free solders has been examined and the costs for them have been estimated. Unfortunately there is no information about their environmental compatibility and health risks.

2.2.2 Studies on the environmental compatibility of lead-free solders

In the following sections, selected alternatives for the common tin/lead soldering connection process are studied regarding their environmental impact. First different Pb-free solders are compared to SnPb37 by means of a Life Cycle Analysis (LCA). Later, the material and energy flow in a SMT reflow soldering process using SnPb37 is compared to conductive adhesive joining using silver filled polymers. Finally the disposal of soldered and adhesive joined pcb is simulated by leaching tests and discussed.

2.2.2.1 Comparison of different lead-free solders to SnPb

Environmental burdens caused by different lead-free solders along their life cycle from raw materials to recycling is studied by a LCA approach and compared to that of SnPb37. Both higher and lower melting solders as well as solders with similar melting temperatures to SnPb37 were included. Tin solders containing high shares of Bismuth (Bi) (e. g. SnBi58; T_m=139 °C) were chosen as a substitute for products with low operating temperatures. Bismuth tin alloys with contents of 5-10% Bi have melting points in the range of SnPb37 (e. g. SnBi9,5 Cu0,5, T_m=198°C). Higher melting solders are important for products with higher operating temperatures. Tin copper (SnCu0,7; T_m=227 °C) and tin silver solders (e. g. SnAg3,5; T_m=221 °C) were chosen for the studies.

In the following paragraphs, a comparative rating of the solder systems SnPb37, SnBi9,5Cu9,5, SnBi58, SnCu0,7 and SnAg3,5 is made. The LCA was carried out according to the "Eco-Indicator 95" method. Figure 6 shows how the Eco-Indicator is formed.

The calculations refer to a real insertion process. 1740 pcb with altogether 1,392,000 components are manufactured daily. Only 0805 SMD chip capacitors were assumed to be on the pcb to facilitate the comparisons. Assumptions and simplifications:

The same flux is assumed for all solder systems. Producers do not publish detailed flux compositions and expensive chemical analyses were not carried out.

The mass share of the metallic components is a total of 90% totally in all solder pastes examined. The volume of a SnPb37 soldered joint was calculated to be 0,14 mm³ per component and assumed to be constant for all joints. Due to their different specific gravities this assumption results in different masses of the respective solders, which affects the eco-indicator results (Table 7).

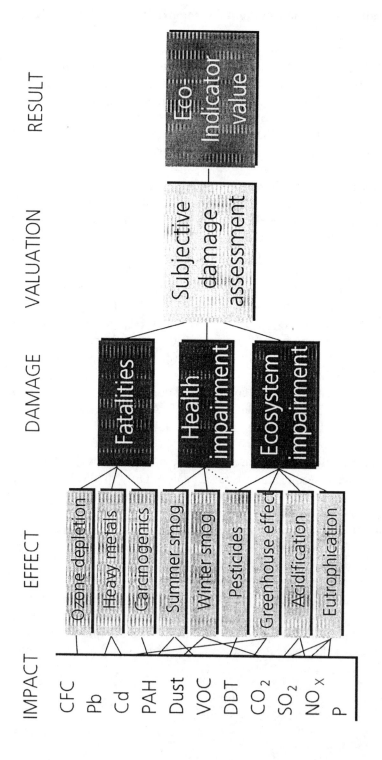

Figure 6: Ecological assessment with Eco-Indicator 95

Table 7:   Specific gravity of different solders

Solder	SnPb 37	SnBi 9,5Cu 0,5	SnBi 58	SnCu 0,7	SnAg 3,5	SnZn 9
Spec. Gravity [g cm^{-3}]	8,800	7,550	8,750	7,321	7,422	7,28

The same processes of solder paste production and product assembly and disposal were assumed for the regarded solder systems. They only differ in their inputs and outputs. Different energy consumptions for different soldering temperatures are not considered due to insufficient data!

The eco-indicator evaluation of the solder metals (Figure 7) shows that lead has the highest environmental burden, whereas tin and silver have little impact only. Merely 10 % of the material and energy flows in the bismuth refining processes were allocated to Bi because of the mass share of 10 % Bi (main product) and 90 % Pb (by-product) in the process outputs.

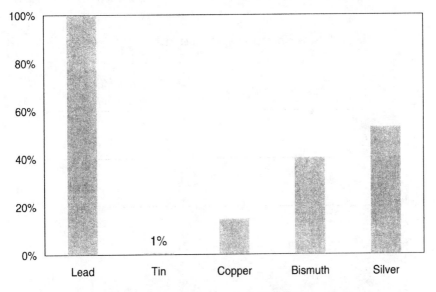

Figure 7: Eco-Indicator per 1 kg of the metal

The results for the solder materials are shown in Figure 8. SnCu0,7 and SnAg3,5 are the environmentally , the most friendly alloys. Taking into account the lower specific gravity of SnCu0,7 and SnAg3,5 compared to SnPb37, the result is even more in favor of this lead-free solder (Figure 9).

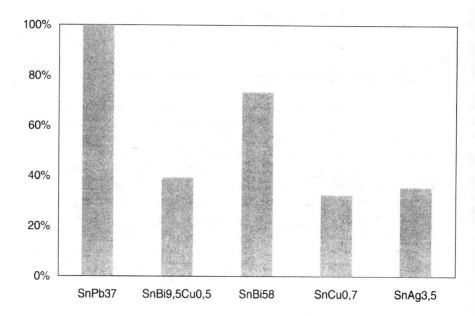

Figure 8: Eco-Indicator of solder materials per 1 kg of alloy

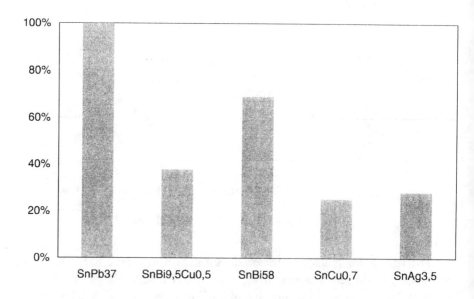

Figure 9: Eco-Indicator of solder materials per solder joint

2.2.2.2 Comparison of material and energy balances of SnPb37 soldering and lead-free conductive adhesive joining

In this section conductive adhesives are examined as another substitute for SnPb soldering in SMT. Material and energy balances are set up for reflow soldering with SnPb37 and adhesive joining with silver filled epoxy. The daily production of a model pcb populated only with 0805 thick film capacitors was chosen as the reference quantity. All inputs and outputs that are different for the two processes are considered, from printing (stencil and screen printing) to final testing including cleaning processes and repair. Also included in the investigation are the terminal areas of the components and surface finish of the board. Components for soldering need SnPb surfaces on nickel and silver/palladium for their contact areas. As silver conductive adhesive joining on tin surfaces still is not reliable, the components for adhesive joining carry silver/palladium with a slightly higher thickness. The pcb for soldering has a SnPb finish. For adhesive joining, it is chemically nickel/gold plated. Nitrogen serves as a protective atmosphere for soldering. The solder is exposed to temperatures of up to 230 °C for a short time whereas the screen-printed adhesive hardens within three minutes at 150 °C. Figures 10 and 11 summarize material and energy balances.

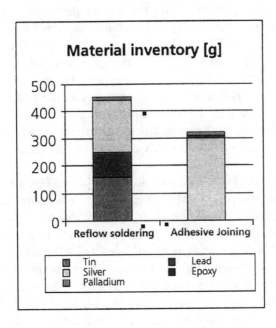

Figure 10: Materials in reflow soldered and adhesive joined interconnections on pcb's

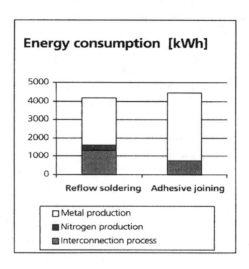

Energy consumption [kWh]

Figure 11: Energy balance for reflow soldering and adhesive joining

The different environmental impacts result from different material flows of lead, tin and silver. Energy consumption in conductive adhesive joining is lower than in soldering due to lower process temperatures, but if the energy consumption during raw material extraction is taken into account, the difference is negligible. The study therefore shows no ecological advantage for either of the processes as far as energy consumption is concerned. Regarding the material flow, both the lower material throughput (because of a lower specific gravity of the adhesive compared to the soldering paste) and the use of silver instead of the toxic lead favors the use of adhesive joining. This general statement has nevertheless to be verified in each individual case with regard to technological suitability and cost effectiveness.

2.2.2.3   Disposal of soldered and adhesive-joined circuit boards

To gain more information about the different environmental impacts of soldered and adhesive joined pcb, their behavior during waste disposal was examined. Especially the effect of the leachate on the heavy metal load was examined. For this, leaching tests of soldered and adhesive joined pcb were carried out under different conditions as close to the reality as possible. The leaching tests were carried out according to the standardized German DEV method (Deutsches Einheitsverfahren, DIN 38414), the NEN 7349 method (Niederländische Einheitsnorm) from the Netherlands and the American TCLP method (Toxicity Characteristic Leaching Procedure). Information about the methods and the necessary modifications are summarized in Table 8.

Table 8: Comparison of leaching test methods

Leaching test	DEV	NEN	TCLP
Coverage	Solid and pasty material, sludge	Incineration residues, waste	Waste, inorganic and organic material
Grain size	Coarse grain	< 3 mm	Coarse grain
Agitation	Shaking	Shaking	Shaking
Solvent	Distilled water	Deion. water acidified with nitric acid	Buffered acetic acid
Solvent/material relation	10:1	20:1	20:1
pH-value	Not defined	pH 4	pH 4,93 or 2,88
Number of passes	1	5	1
Duration	24 h	23 h	18 h
Filtration	0,45 μm membrane	0,45 μm membrane	0,6-0,8 μm glass fiber filter
Modifications of the standardized test	Acidification, miniaturization (pH 4 and 7), grain size 1 cm and < 3 mm, shaking and stirring	One-pass-test, grain size 1 cm, shaking and resting	Multiple passes, grain size < 3 mm, shaking and resting

The leachate was analyzed by ICP-OES-Spectrometry (Inductively Coupled Plasma – Optical Emission Spectrometry).

The quantities of the relevant metals tin, lead and silver leached in the DEV test are below 0,01 ppm near the determination limit. But these results are not very close to reality because distilled water is used as the solvent. At a lower pH like in landfills, the results change dramatically (Figure 12).

The results from the other methods (which by standard use acidic solvents, see table 8) show a high lead burden compared to the other metals (up to 100 times higher) and thus a high risk for the groundwater.

The results also make it probable that SnAg solders, like the studied adhesives, are less susceptible to leaching and therefore are more environmentally compatible than solders containing lead.

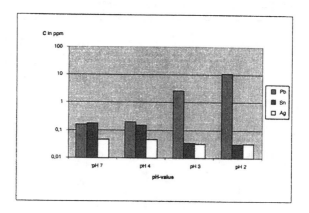

Figure 12: pH-dependence of lead, tin and silver leaching

Altogether, the results show that from an ecological point of view the substitution of lead containing solders by alloys of tin, copper, silver and also bismuth and zinc should be preferred. Neither the toxic cadmium and antimony nor the scarce indium are suitable alternatives.

SnAg solders with and without bismuth or copper and SnCu solders are materials that are now used in certain lead-free mass products and are technologically suitable for many purposes. Moreover, these materials are more environmentally friendly. The industry has to decide whether these slightly more expensive but more efficient products can achieve recognition in the market.

3.   CONSIDERING THE LIFE CYCLE – DESIGN FOR ENVIRONMENT AND PRODUCT LOOPS

3.1   Why Design for Environment?

While improving the processes is one way to reduce the environmental impacts in manufacturing of electronics, Design for Environment (DfE) is the complementary equivalent on the product side. Design for Environment avoids waste in manufacturing by appropriate construction and choice of materials.

Furthermore DfE reduces the environmental impacts of electronic devices in the end of life phase. Parts and components can be recovered, materials recycled, waste minimized and costs reduced due to life cycle thinking behind DfE. DfE links the end of life phase to the manufacturing phase. Reclaimed components and recycled materials substitute new ones in manufacturing, which must be manufactured with high expenses and high environmental impacts. DfE also helps to comply with take back regulations and to minimize the product life cycle costs.

3.2   What is DfE?

DfE is based on the idea to take into consideration and to minimize environmental impacts throughout the life cycle of a product and to transform these environmental issues to the product design. The principles of DfE are adopted from the very beginning, the first step towards a new product.

DfE represents a bunch of different measures. These general principles must be specified for specific products based on the results of environmental and also economical assessments and evaluations. Thus, the most appropriate materials for the product have to be identified as well as the most appropriate recycling and re-use concepts and product specific marketing strategies. Interactions between these different aspects also have to be taken into consideration. For example, the materials must not only be environmentally friendly, but also appropriate for the identified recycling and marketing strategies and vice versa.

3.2.1   How DfE Affects Product Features – General Principles

Avoiding environmental impacts, facilitating re-use and recycling and reducing the life cycle costs are the overall targets of DfE. The following product design features can achieve these targets:

1. Choosing materials that can easily be recycled
2. Choosing non-toxic materials
3. Labeling of plastics
4. Choosing materials with low energy intensity
5. Reducing material intensity
6. Reducing number of materials
7. Specifying the product for the most appropriate way of recycling and re-use
8. Interconnections that can easily and quickly be disassembled, preferably snap fits
9. Unify and reduce types of interconnections

10. Minimizing the energy consumption of manufacturing processes and products
11. Facilitating repair and upgrading by modularization

To take full advantage of DfE it has to be complemented by marketing strategies corresponding to the assessed DfE priorities verified in a product. Two examples will show how the general principles of DfE were applied on concrete products and how this has influenced business.

3.3 Design for Environment and Closed Loops –Two Examples

Siemens Nixdorf, a big German manufacturer of information systems, and Wuseltronik, a small company producing inverters for solar modules in Berlin, re-designed products applying DfE principles. Both German manufacturers additionally created a marketing strategy that allows them to take full advantage of the DfE product features. With this proceeding the two companies defused environmental and also economical hot spots in the life cycle of their products.

3.3.1 Computers in Loops

The average lifetime of a Personal Computer is about 3 to 5 years. After this time, although it is still functioning, the technical progress makes it contribute to the ever-growing amount of electric and electronic scrap. This scrap, however, on one hand contains numerous valuable electronic components and materials as well as scarce and expensive materials. On the other hand it burdens the environment when disposed of at landfill sites or incinerated.

Siemens Nixdorf (now Siemens) did the obvious thing and placed the first eco-PC on the market in the beginning of the nineties. As the first PC the product met the criteria for eco-computers and was awarded the Blue Angel, the German label for environmentally preferable products. All Siemens computers from then on have been designed and manufactured following these principles.

Siemens also took up the product responsibility and takes back computer products after use. Siemens uses the company's own recycling center to ensure the re-use, recycling and disposal of these products.

Figure 13: Blue Angel for Eco-Computers

3.3.2   DfE – the Cradle of Environmentally Friendly Products

Siemens re-designed the computer and the keyboard. The energy consumption decreased to less than 50 % in the use phase. It complies with the U.S. Energy Star conditions.

The computer is also designed for upgrading thus lengthening the lifetime of the product and reducing environmental impacts in manufacturing. About 90 % of the computer can be recycled. This was assured with different measures:

- Special interconnection technology facilitating re-use of electronic components and recycling
- No halogenated flame retardants in the cases disturbing recycling of plastics and endangering the environment
- Reduction of number of plastics; casing consists of one plastic type only
- Labeling of plastics for easy identification for proper recycling
- Weight reduction from 16 kg to 6 kg saving resources and reducing waste
- Replacing all interconnects by snap fits mainly for easy repair and disassembly
- Reduction of number of parts from 87 to 29 for easy assembly and disassembly.

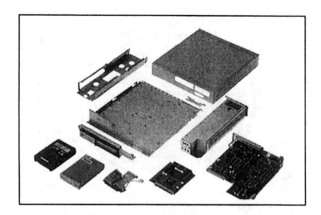

Figure 14: Disassembled Siemens Eco-PC (Source: Siemens AG)

These measures result in quick (dis)-assembly. The assembly time could be reduced from 33 to 7 minutes, the disassembly time from 18 to 4 minutes. So costs are reduced in manufacturing as well as in the end of life phase enabling higher shares of re-use and recycling.

3.3.3   3 Steps Towards More Sustainability

Customers can hand over their used computers to Siemens Nixdorf. This is free of charge when this computer is a Siemens Blue Angel eco-computer. Siemens adopts a three-step strategy to recovery and reuse of these computers.

First Siemens tries to find a customer for the computer after upgrading it. These computers can be sold for about 10 to 33 % of the initial price. The customers for these products are price-conscious users who want to extend an existing installation without investing in new software and training of staff or users who do not need the latest computer models for their demands.

In the second step, computers that cannot be sold are disassembled in the Siemens recycling center in Paderborn (Germany). Their parts and components are reclaimed for re-use like power supply units, entire printed wiring boards and valuable electronic components. These parts are tested according to the quality demands for new products. Appropriate parts are used as spare parts for upgrading and repair of other computers. They are then sold worldwide and other manufacturers install them in other electronic products.

In the last step, the rest of the computers that can neither be re-used nor recycled are brought to material recycling. The materials then partially go back into the Siemens manufacturing plants for computers.

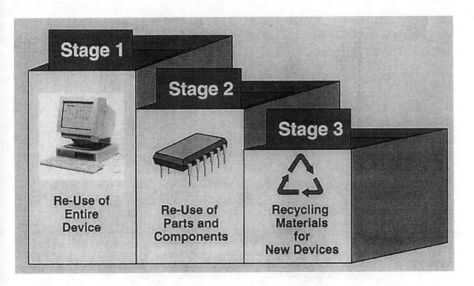

Figure 15: Siemens Recycling Concept (Source: Siemens AG)

In total more than 90 % of the Siemens eco-computers can be re-used and recycled due to consequent DfE and marketing strategies.

Concerning the eco-computers, the recycling center covers its expenses through the sale of computers, components and recycled materials. Additionally, Siemens gains experience on how to improve the DfE of its computers to reduce environmental impacts and life cycle costs. This experience may become crucial in the market when free-of-charge take back regulations are introduced.

3.4 Green Products – Success not only for the Global Players

Wuseltronik GmbH, a small enterprise in Berlin, developed a green inverter for solar energy. The device was consequently designed according to the principles of DfE. Non-recyclable materials as well as toxic substances were avoided. Flame-retardants were eliminated as much as possible. Wuseltronik also successfully applied natural materials in this inverter. Finally, the device can easily and quickly be disassembled.

For this inverter Wuseltronik was awarded the Environmental Prize of Berlin in 1999.

Figure 16: Wuseltronik's Inverter NEG 1200
(Source: Wuseltronik Forschung und Entwicklung)

Wuseltronik now is planning and implementing an innovative and cost neutral marketing and recycling concept for this inverter.

3.4.1   Shared Profits - Innovative Marketing for Innovative Products

Wuseltronik takes back used inverters free of charge. The inverters are disassembled. Components pass through rigorous quality controls. Appropriate components are re-used in new inverters.

Expensive components like the case, the annular core transformator and the accumulator choke coil are very cost-effective in re-use. Detailed failure studies proved that these durable components are not responsible for failures. Their lifetime by far exceeds the average usage time of the inverter. So these components can be re-used saving costs and avoiding waste and environmental impacts in the manufacturing of the inverters.

Other components are used as spare parts for the repair of inverters. The rest of the inverters goes into material recycling or has to be disposed of.

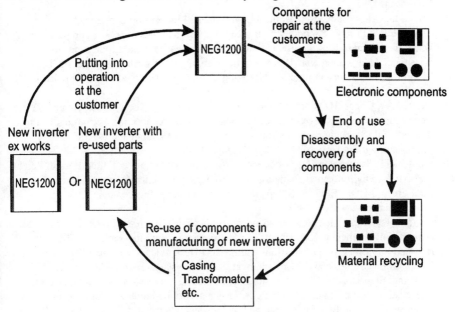

Figure 17: Wuseltronik Marketing Concept (Source: Wuseltronik Forschung und Entwicklung)

Both re-use and the design of the inverter make the manufacturing costs decrease for about 10 %. The customers choose between the cheaper inverters with re-used components and more expensive ones without. The guarantee period is always 3 years. For an extra charge the customer also may decide to buy an inverter with 5 years of guarantee irrespectively whether the inverter is completely new or has re-used parts.

Wuseltronik's inverter is a green product for a green technology setting ecological as well as economical standards.

REFERENCES

With IZM participation:

Ast, Stefan: Vergleich der Umweltwirkungen spezifischer Entsorgungsverfahren für bestückte Leiterplatten am Beispiel der DSP-Unit eines HDTV-Konverters; Fraunhofer IZM, Berlin 1997

Deubzer, Otmar; Middendorf, Andreas: Environmentally Friendly Materials and Processes for SMT-Reflow Soldering and Conductive Adhesive Joining in Comparison; Proc. Int. Conf. On Electronic Assembly: Materials and Process Challenges, Atlanta/Georgia, June 16 - 18, 1998

Griese, Hansjörg; Müller, Jutta; Sietmann, Richard: Kreislaufwirtschaft in der
Elektronikindustrie – Konzepte, Strategien, Umweltökonimie; VDE-Verlag, Berlin,
Offenbach, 1997

Müller, Jutta; Griese, Hansjörg: Reduced Environmental Impacts by Lead Free Electronic
Assemblies ?; Proc. ICPWorks'99, Minneapolis, October 26-28, 1999, in preparation

Müller, Jutta; Schmidt, Ralf; Griese, Hansjörg; Hannemann, Monika; Lindemann, Jörg;
Wiesmann, Udo; Zuber, Karl-Heinz: Environmentally Acceptable Gold Coating; Proc.
12th European Microelectronics & Packaging Conference, Harrogate/Yorkshire/ England,
June 7 – 9, 1999

Nissen, Nils F.; Griese, Hansjörg; Middendorf, Andreas; Müller, Jutta; Pötter, Harald; Reichl,
Herbert: Environmental Screening of Packaging and Interconnection Technologies; Proc.
International Symposium on Environmentally Conscious Design and Inverse
Manufacturing: EcoDesign '99, Tokyo/Japan, February 1 – 3, 1999

Nissen, Nils F.; Griese, Hansjörg; Middendorf, Andreas; Müller, Jutta; Pötter, Harald; Reichl,
Herbert: An Environmental Comparison of Packaging and Interconnection Technologies;
Conf. Record 1998 IEEE Int. Symp. on Electronics and the Environment, Oak
Brook/Illinois, May 4 - 6, 1998

Nissen, Nils F.; Griese, Hansjörg; Middendorf, Andreas; Müller, Jutta; Pötter, Harald; Reichl,
Herbert: Environmental Assessments of Electronics: A New Model to Bridge the Gap
Between Full Life Cycle Evaluations and Product Design; Conf. Record 1997 IEEE Int.
Symp. on Electronics and the Environment, San Francisco/California, May 5 - 7, 1997

Ram, Benno; Stevels, Ab; Griese, Hansjörg; Middendorf, Andreas; Müller, Jutta; Nissen, Nils
F.; Reichl, Herbert: Environmental Performance of Mobile Products; Conf. Record 1999
IEEE Int. Symp. on Electronics and the Environment, Danvers/Ms., May 11 - 13, 1999

Further references:

Charles, Harry K.: Microelectronics. Rising to the Environmental Challenge?; Proc. 12th
European Microelectronics & Packaging Conference, Harrogate/Yorkshire/ England, June
7 – 9, 1999

Fiksel, J.: Design for Environment; McGraw-Hill, New York, ect., 1996 (ISBN 0-07-020972-
3)

Goedkoop, M. et al.: The Eco Indicator '95 Weighting Method; ISBN 90-72130-77-4

Hauschild, M.: Environmental Assessment of Products; Kluwer-Verlag, 1998

Murphy, C. F.; Sandborn, P. A.; Schuldt, G. H.; Lott, J. W.: An Environmental and
Economical Comparison of Additive and Subtractive Approaches for Printed Wiring
Board Fabrication; Conf. Record 1997 IEEE Int. Symp. on Electronics and the
Environment, San Francisco/California, May 5 - 7, 1997

Nagel, M. H.: Environmental Quality Related to Printed Board Suppliers of an Original
Equipment Manufacturer (OEM); Proc. CARE Innovation '98, Vienna/Austria, November
16 - 19, 1998

Quella, Ferdinand (Hrsg.): Umweltverträgliche Produktgestaltung; Publicis-MCD-Verlag;
Erlangen, München 1998: S. 134

Steinhilper, R.: Remanufacturing - The Ultimate Form of Recycling; Stuttgart, Fraunhofer-
IRB-Verlag, 1998

Van Nes, C. N.; Stevels, Ab: Selecting green design strategies on the basis of eco-efficiency
calculations; 4th Int. Seminar on Life Cycle Engineering, Life Cycle Networks, Berlin,
June 26/27, 1997

Index